Introduction to Stochastic Calculus Applied to Finance

Second Edition

CHAPMAN & HALL/CRC
Financial Mathematics Series

Aims and scope:
The field of financial mathematics forms an ever-expanding slice of the financial sector. This series aims to capture new developments and summarize what is known over the whole spectrum of this field. It will include a broad range of textbooks, reference works and handbooks that are meant to appeal to both academics and practitioners. The inclusion of numerical code and concrete real-world examples is highly encouraged.

Series Editors

M.A.H. Dempster
Centre for Financial Research
Judge Business School
University of Cambridge

Dilip B. Madan
Robert H. Smith School
of Business
University of Maryland

Rama Cont
Center for Financial Engineering
Columbia University
New York

Published Titles

American-Style Derivatives; Valuation and Computation, *Jerome Detemple*

Engineering BGM, *Alan Brace*

Financial Modelling with Jump Processes, *Rama Cont and Peter Tankov*

An Introduction to Credit Risk Modeling, *Christian Bluhm, Ludger Overbeck, and Christoph Wagner*

Introduction to Stochastic Calculus Applied to Finance, Second Edition, *Damien Lamberton and Bernard Lapeyre*

Numerical Methods for Finance, *John A. D. Appleby, David C. Edelman, and John J. H. Miller*

Portfolio Optimization and Performance Analysis, *Jean-Luc Prigent*

Robust Libor Modelling and Pricing of Derivative Products, *John Schoenmakers*

Structured Credit Portfolio Analysis, Baskets & CDOs, *Christian Bluhm and Ludger Overbeck*

Understanding Risk: The Theory and Practice of Financial Risk Management, *David Murphy*

Proposals for the series should be submitted to one of the series editors above or directly to:
CRC Press, Taylor and Francis Group

CHAPMAN & HALL/CRC FINANCIAL MATHEMATICS SERIES

Introduction to Stochastic Calculus Applied to Finance

Second Edition

Damien Lamberton
Bernard Lapeyre

CRC Press
Taylor & Francis Group
Boca Raton London New York

CRC Press is an imprint of the
Taylor & Francis Group, an **informa** business
A CHAPMAN & HALL BOOK

Chapman & Hall/CRC
Taylor & Francis Group
6000 Broken Sound Parkway NW, Suite 300
Boca Raton, FL 33487-2742

First issued in paperback 2022

ISBN 13: 978-1-03-247781-7 (pbk)
ISBN 13: 978-1-58488-626-6 (hbk)

DOI: 10.1201/9781420009941

Library of Congress Cataloging-in-Publication Data

Lamberton, Damien.
    [Introduction au calcul stochastique appliqué à la finance. English]
    Introduction to stochastic calculus applied to finance / Damien Lamberton and Bernard Lapeyre. -- 2nd ed.
      p. cm. -- (Chapman & Hall/CRC financial mathematics series)
    Includes bibliographical references and index.
    ISBN 978-1-58488-626-6 (alk. paper)
    1. Investments--Mathematics. 2. Stochastic analysis. 3. Options (Finance)--Mathematical models. I. Lapeyre, Bernard. II. Title. III. Series.

HG4515.3.L3613 2008
332.64'530151922--dc22
                                             2007031483

Visit the Taylor & Francis Web site at
http://www.taylorandfrancis.com

and the CRC Press Web site at
http://www.crcpress.com

Preface to the second edition

The topic of mathematical finance has been growing rapidly since the first edition of this book. For this new edition, we have not tried to be exhaustive on all new developments but to select some techniques or concepts that could be incorporated at reasonable cost in terms of length and mathematical sophistication. This was partly done by adding new exercises. The main addition concern:

- complements on discrete models (Rogers' approach to the Fundamental Theorem of Asset Pricing, super-replication in incomplete markets, see chapter 1 exercises 1 and 2),

- local volatility and Dupire's formula (see Chapter 4),

- change of numéraire techniques and forward measures (see Chapter 1 and Chapter 6),

- the forward libor model (BGM model, see Chapter 6),

- a new chapter on credit risk modelling,

- an extension of the chapter dealing with simulation with numerical experiments illustrating variance reduction techniques, hedging strategies and so on.

We are indebted, in addition to those cited in the introduction, to a number of colleagues whose suggestions have been helpful for this new edition. In particular we are grateful to Marie-Claire Quenez, Benjamin Jourdain, Philip Protter and, for the chapter on credit risk, to Monique Jeanblanc and Rama Cont (whose lectures introduced us to this new area) and to Aurélien Alfonsi.

Contents

Introduction

The objective of this book is to give an introduction to the probabilistic techniques required to understand the most widely used financial models. In the last few years, financial quantitative analysts have used more sophisticated mathematical concepts, such as martingales or stochastic integration, in order to describe the behavior of markets or to derive computing methods.

In fact, the appearance of probability theory in financial modeling is not recent. At the beginning of this century, Bachelier (1900), in trying to build up a "Theory of Speculation", discovered what is now called Brownian motion. From 1973, the publications by Black and Scholes (1973) and Merton (1973) on option pricing and hedging gave a new dimension to the use of probability theory in finance. Since then, as the option markets have evolved, Black-Scholes and Merton results have developed to become clearer, more general and mathematically more rigorous. The theory seems to be advanced enough to attempt to make it accessible to students.

Options

Our presentation concentrates on options, because they have been the main motivation in the construction of the theory and still are the most spectacular example of the relevance of applying stochastic calculus to finance. An option gives its holder the right, but *not the obligation*, to buy or sell a certain amount of a financial asset, by a certain date, for a certain strike price.

The writer of the option needs to specify:

1. the type of option: the option to buy is called a *call* while the option to sell is a *put*;

2. the underlying asset: typically, it can be a stock, a bond, a currency and so on;

3. the amount of an underlying asset to be purchased or sold;

4. the expiration date; if the option can be exercised at any time before maturity, it is called an *American* option but, if it can only be exercised at maturity, it is called a *European* option;

5. the exercise price which is the price at which the transaction is done if the option is exercised.

The price of the option is the *premium*. When the option is traded on an organised market, the premium is quoted by the market. Otherwise, the problem is to price the option. Also, even if the option is traded on an organized market, it can be interesting to detect some possible abnormalities in the market.

Let us examine the case of a European call option on a stock, whose price at time t is denoted by S_t. Let us call T the expiration date and K the exercise price. Obviously, if K is greater than S_T, the holder of the option has no interest whatsoever in exercising the option. But, if $S_T > K$, the holder makes a profit of $S_T - K$ by exercising the option, i.e., buying the stock for K and selling it back on the market at S_T. Therefore, the value of the call at maturity is given by

$$(S_T - K)_+ = \max(S_T - K, 0).$$

If the option is exercised, the writer must be able to deliver a stock at price K. It means that he or she must generate an amount $(S_T - K)_+$ at maturity. At the time of writing the option, which will be considered as the origin of time, S_T is unknown and therefore two questions have to be asked:

1. How much should the buyer pay for the option? In other words, how should we price at time $t = 0$ an asset worth $(S_T - K)_+$ at time T? That is the problem of *pricing* the option.

2. How should the writer, who earns the premium initially, generate an amount $(S_T - K_+$ at time T? That is the problem of *hedging* the option.

Arbitrage and put/call parity

Answers the above questions require some modelling. The basic one, which is commonly accepted in every model, is the absence of arbitrage opportunity in liquid financial markets, i.e. there is no riskless profit available in the market. We will translate that into mathematical terms in the first chapter. At this point, we will only show how we can derive formulae relating European put and call prices from the no arbitrage assumption. Consider a put and a call with the same maturity T and exercise price K, on the same underlying asset which is worth S_t at time t. We shall assume that it is possible to borrow or invest money at a constant rate r.

Let us denote by C_t and P_t respectively the prices of the call and the put at time t. Because of the absence of arbitrage opportunity, the following equation called *put/call parity* is true for all $t < T$

$$C_t - P_t = S_t - Ke^{-r(T-t)}.$$

To understand the notion of arbitrage, let us show how we could make a riskless profit if, for instance,

$$C_t - P_t > S_t - Ke^{-r(T-t)}.$$

At time t, we purchase a share of stock and a put, and sell a call. The net value of the operation is

$$C_t - P_t - S_t.$$

If this amount is positive, we invest it at rate r until time T, whereas if it is negative we borrow it at the same rate. At time T, two outcomes are possible:

- $S_T > K$: the call is exercised, we deliver the stock, receive the amount K and clear the cash account to end up with a wealth $K + e^r(T - t)(C_t - P_t - S_t) > 0$.

- $S_T \leq K$: we exercise the put and clear our bank account as before to finish with wealth $K + e^{T-t}(C_t - P_t - S_t > 0$.

In both cases, we locked in a positive profit without making any initial endowment: this is an example of an arbitrage strategy.

There are many similar examples in the book by Cox and Rubinstein (1985). We will not review all these formulae, but we shall characterize mathematically the notion of a *financial market without arbitrage opportunity*.

Black-Scholes model and its extensions

Even though no-arbitrage arguments lead to many interesting equations, they are not sufficient in themselves for deriving pricing formulae. To achieve this, we need to model stock prices more precisely. Black and Scholes were the first to suggest a model whereby we can derive an explicit price for a European call on a stock that pays no dividend. According to their model, the writer of the option can hedge himself perfectly, and actually the call premium is the amount of money needed at time 0 to replicate exactly the payoff $(S_T - K)_+$ by following their dynamic hedging strategy until maturity. Moreover, the formula depends on only one non-directly observable parameter, the so-called *volatility*.

It is by expressing the profit and loss resulting from a certain trading strategy as a stochastic integral that we can use stochastic calculus and, particularly, Itô formula, to obtain closed form results. In the last few years, many extensions of the Black-Scholes approach has been considered. From a thorough study of the Black-Scholes model, we will attempt to give to the reader the means to understand those extensions.

Contents of the book

The first two chapters are devoted to the study of discrete time models. The link between the mathematical concept of martingale and the economic notion of arbitrage is brought to light. Also, the definition of complete markets and the pricing of options in these markets are given. We have decided to adopt the formalism of Harrison and Pliska (1981) and most of their results are stated in the first chapter, taking the Cox, Ross and Rubinstein model as an example. The second chapter deals with American options. Thanks to the theory of optimal stopping in a discrete time set-up, which uses quite elementary methods, we introduce the reader to all the ideas that can be developed in continuous time.

Chapter 3 is an introduction to the main results in stochastic calculus that we will use in Chapter 4 to study the Black-Scholes model. As far as European options are concerned, this model leads to explicit formulae. But, in order to analyze American options or to perform computations within more sophisticated models, we need numerical methods based on the connection between option pricing and partial differential equations. These questions are addressed in Chapter 5.

Chapter 6 is a relatively quick introduction to the main interest rate models and Chapter 7 looks at the problems of option pricing and hedging when the price of the underlying asset follows a simple jump process.

In these latter cases, perfect hedging no longer possible and we must define a criterion to achieve optimal hedging. These models are rather less optimistic than the Black-Scholes model and seem to be closer to reality. However, their mathematical treatment is still a matter of research, in the framework of so-called *incomplete markets.*

Finally, in order to help the student to gain a practical understanding, we have included a chapter dealing with the simulation of financial models and the use of computers in the pricing and hedging of options. Also, a few exercises and longer questions are listed at the end of each chapter.

This book is only an introduction to a field that has already benefited from considerable research. Bibliographical notes are given in some chapters to help the reader to find complementary information. We would also like to warn the reader that some important questions in financial mathematics are not tackled. Amongst them are the problems of optimization and the questions of equilibrium for which the reader might like to consult the book by Duffie (1988).

A good level in probability theory is assumed to read this book. The reader is referred to Dudley (2002)) and Williams (1991) for prerequisites. However, some basic results are also proved in the Appendix.

Acknowledgments

This book is based on the lecture notes of a course taught at *l'Ecole des Ponts* since 1988. The organisation of this lecture series would not have been possible without the encouragement of N. Bouleau. Thanks to his dynamism, CERMA (Applied Mathematics Institute of ENPC) started working on financial modeling as early as 1987, sponsored by *Banque Indosuez* and subsequently by *Banque Internationale de Placement.*

Since then, we have benefited from many stimulating discussions with G. Pagès and other academics at CERMA, particularly O. Chateau and G.

Caplain. A few people kindly read the earlier draft of our book and helped us with their remarks. Amongst them are S. Cohen, O. Faure, C. Philoche, M. Jeanblanc and X. Zhang. Finally, we thank our colleagues at the university and at INRIA for their advice and their motivating comment: N. El Karoui, T. Jeulin, J.F. Le Gall and D. Talay.

Chapter 1

Discrete-time models

The objective of this chapter is to present the main ideas related to option theory within the very simple mathematical framework of discrete-time models. Essentially, we are following the first part of the paper by Harrison and Pliska (1981). Cox, Ross and Rubinstein's model is detailed at the end of the chapter in the form of a problem with its solution.

1.1 Discrete-time formalism

1.1.1 Assets

A discrete-time financial model is built on a finite probability space $(\Omega, \mathscr{F}, \mathbb{P})$ equipped with a filtration, i.e. an increasing sequence of σ-algebras included in $\mathscr{F}$: $\mathscr{F}_0, \mathscr{F}_1, \ldots, \mathscr{F}_N$. The σ-algebra $\mathscr{F}_n$ can be seen as the information available at time n and is sometimes called the σ-algebra of events up to time n. The horizon N will often correspond to the maturity of the options. From now on, we will assume that $\mathscr{F}_0 = \{\emptyset, \Omega\}$, $\mathscr{F}_N = \mathscr{F} = \mathscr{P}(\Omega)$, where $\mathscr{P}(\Omega)$ denotes the collection of all subsets of the finite sample space Ω, and we also assume that $\mathbb{P}(\{\omega\}) > 0$, for $\omega \in \Omega$. Working with a finite probability space avoids some technicalities: for instance, all real-valued random variables are integrable.

The market consists of $(d+1)$ financial assets, whose prices at time n are given by the positive random variables $S_n^0, S_n^1, \ldots, S_n^d$, which are measurable with respect to $\mathscr{F}_n$ (investors know past and present prices but obviously not the future ones). The vector $S_n = (S_n^0, S_n^1, \ldots, S_n^d)$ is the vector of prices at time n. The asset indexed by 0 is the *riskless asset* and we set $S_0^0 = 1$. If the return of the riskless asset over one period is constant and equal to r, we will obtain $S_n^0 = (1+r)^n$. The coefficient $\beta_n = 1/S_n^0$ is interpreted as the discount factor (from time n to time 0): if an amount β_n is invested in the riskless asset at time 0, then one dollar will be available at time n. The assets indexed by $i = 1 \ldots d$ are called *risky assets*.

1.1.2   Strategies

A *trading strategy* is defined as a stochastic process (i.e. a sequence in the discrete case)

$$\phi = ((\phi_n^0, \phi_n^1, \ldots, \phi_n^d))_{0 \leq n \leq N}$$

in $\mathbb{R}^{d+1}$, where ϕ_n^i denotes the number of shares of asset i held in the portfolio at time n. The sequence ϕ is assumed to be *predictable*, i.e.

$$\forall i \in \{0, 1, \ldots, d\} \begin{cases} \phi_0^i \text{ is } \mathscr{F}_0\text{-measurable} \\ \text{and, for } n \geq 1, \quad \phi_n^i \text{ is } \mathscr{F}_{n-1}\text{-measurable.} \end{cases}$$

This assumption means that the positions in the portfolio at time n, namely ϕ_n^0, ϕ_n^1,..., ϕ_n^d, are decided with respect to the information available at time $(n-1)$ and kept until time n, when new quotations are available.

The *value of the portfolio* at time n is the scalar product

$$V_n(\phi) = \phi_n.S_n = \sum_{i=0}^{d} \phi_n^i S_n^i.$$

Its *discounted value* is

$$\tilde{V}_n(\phi) = \beta_n(\phi_n.S_n) = \phi_n.\tilde{S}_n,$$

where $\beta_n = 1/S_n^0$ and $\tilde{S}_n = (1, \beta_n S_n^1, \ldots, \beta_n S_n^d)$ is the vector of *discounted prices*. By considering discounted prices, we take the price of the non-risky asset as a monetary unit or *numéraire* (see Exercise 3 for an introduction to change of numéraire techniques).

A strategy is called *self-financing* if the following equation is satisfied for all $n \in \{0, 1, \ldots, N-1\}$:

$$\phi_n.S_n = \phi_{n+1}.S_n.$$

The interpretation is the following: at time n, once the new prices $S_n^0, \cdots, S_n^d$ are quoted, the investor readjusts his positions from ϕ_n to ϕ_{n+1} without bringing or consuming any wealth.

Remark 1.1.1. The equality $\phi_n.S_n = \phi_{n+1}.S_n$ is obviously equivalent to

$$\phi_{n+1}.(S_{n+1} - S_n) = \phi_{n+1}.S_{n+1} - \phi_n.S_n,$$

or to

$$V_{n+1}(\phi) - V_n(\phi) = \phi_{n+1}.(S_{n+1} - S_n).$$

At time $n+1$, the portfolio is worth $\phi_{n+1}.S_{n+1}$ and $\phi_{n+1}.S_{n+1} - \phi_{n+1}.S_n$ is the net gain caused by the price changes between times n and $n+1$. Hence, the profit or loss realized by following a self-financing strategy is only due to the price moves.

The following proposition makes this clear in terms of discounted prices.

Proposition 1.1.2. *The following are equivalent:*

(i) *The strategy ϕ is self-financing.*

(ii) *For any $n \in \{1, \ldots, N\}$,*

$$V_n(\phi) = V_0(\phi) + \sum_{j=1}^{n} \phi_j \cdot \Delta S_j,$$

where ΔS_j is the vector $S_j - S_{j-1}$.

(iii) *For any $n \in \{1, \ldots, N\}$,*

$$\tilde{V}_n(\phi) = V_0(\phi) + \sum_{j=1}^{n} \phi_j \cdot \Delta \tilde{S}_j,$$

where $\Delta \tilde{S}_j$ is the vector $\tilde{S}_j - \tilde{S}_{j-1} = \beta_j S_j - \beta_{j-1} S_{j-1}$.

Proof. The equivalence between (i) and (ii) results from Remark 1.1.1. The equivalence between (i) and (iii) follows from the fact that $\phi_n.S_n = \phi_{n+1}.S_n$ if and only if $\phi_n.\tilde{S}_n = \phi_{n+1}.\tilde{S}_n$. $\square$

This proposition shows that, if an investor follows a self-financing strategy, the discounted value of his portfolio, and hence its value, are completely defined by the initial wealth and the strategy $(\phi_n^1, \ldots, \phi_n^d)_{0 \leq n \leq N}$ (this is only justified because $\Delta \tilde{S}_j^0 = 0$). More precisely, we can prove the following proposition.

Proposition 1.1.3. *For any predictable process $((\phi_n^1, \ldots, \phi_n^d))_{0 \leq n \leq N}$ and for any $\mathscr{F}_0$-measurable variable V_0, there exists a unique predictable process $(\phi_n^0)_{0 \leq n \leq N}$ such that the strategy $\phi = (\phi^0, \phi^1, \ldots, \phi^d)$ is self-financing and its initial value is V_0.*

Proof. The self-financing condition implies

$$\begin{aligned}
\tilde{V}_n(\phi) &= \phi_n^0 + \phi_n^1 \tilde{S}_n^1 + \cdots + \phi_n^d \tilde{S}_n^d \\
&= V_0 + \sum_{j=1}^{n} \left(\phi_j^1 \Delta \tilde{S}_j^1 + \cdots + \phi_j^d \Delta \tilde{S}_j^d \right),
\end{aligned}$$

which defines ϕ_n^0. We just have to check that ϕ^0 is predictable, but this is obvious if we consider the equation

$$\begin{aligned}
\phi_n^0 = V_0 + \sum_{j=1}^{n-1} &\left(\phi_j^1 \Delta \tilde{S}_j^1 + \cdots + \phi_j^d \Delta \tilde{S}_j^d \right) \\
&+ \left(\phi_n^1 \left(-\tilde{S}_{n-1}^1 \right) + \cdots + \phi_n^d \left(-\tilde{S}_{n-1}^d \right) \right).
\end{aligned}$$

$\square$

1.1.3   Admissible strategies and arbitrage

We did not make any assumption on the sign of the quantities ϕ_n^i. If $\phi_n^0 < 0$, we have borrowed the amount $|\phi_n^0|$ in the riskless asset. If $\phi_n^i < 0$ for $i \geq 1$, we say that we are *short* a number ϕ_n^i of asset i. In this model, short-selling and borrowing are allowed, but, by the following *admissibility* condition, the value of the portfolio must remain non-negative at all times.

Definition 1.1.4. A strategy ϕ is admissible if it is self-financing and if $V_n(\phi) \geq 0$ for any $n \in \{0, 1, \ldots, N\}$.

The investor must be able to pay back his debts (in the riskless or the risky assets) at any time. The notion of *arbitrage* (possibility of a riskless profit) can be formalised as follows:

Definition 1.1.5. An arbitrage strategy is an admissible strategy with zero initial value and non-zero final value.

In other words, an arbitrage starts with a zero initial value and achieves a nonnegative value at all times, with strictly positive probability of the final value being positive. Most models exclude any arbitrage opportunity, and the objective of the next section is to characterize these models with the notion of martingale.

1.2   Martingales and arbitrage opportunities

In order to analyze the connections between martingales and arbitrage, we must first define a *martingale* on a finite probability space. The conditional expectation plays a central role in this definition, and the reader can refer to the appendix for a quick review of its properties.

1.2.1   Martingales and martingale transforms

In this section, we consider a finite probability space $(\Omega, \mathscr{F}, \mathbb{P})$, with $\mathscr{F} = \mathscr{P}(\Omega)$ and $\forall \omega \in \Omega, \mathbb{P}(\{\omega\}) > 0$, equipped with a filtration $(\mathscr{F}_n)_{0 \leq n \leq N}$ (without necessarily assuming that $\mathscr{F}_N = \mathscr{F}$, nor $\mathscr{F}_0 = \{\phi, \Omega\}$). A sequence $(X_n)_{0 \leq n \leq N}$ of random variables is adapted to the filtration if, for any n, X_n is $\mathscr{F}_n$-measurable.

Definition 1.2.1. An adapted sequence $(M_n)_{0 \leq n \leq N}$ of real-valued random variables is

- a martingale if $\mathbb{E}(M_{n+1}|\mathscr{F}_n) = M_n$ for all $n \leq N - 1$;

- a supermartingale if $\mathbb{E}(M_{n+1}|\mathscr{F}_n) \leq M_n$ for all $n \leq N - 1$;

- a submartingale if $\mathbb{E}(M_{n+1}|\mathscr{F}_n) \geq M_n$ for all $n \leq N - 1$.

These definitions can be extended to the multidimensional case: for instance, a sequence $(M_n)_{0 \leq n \leq N}$ of $\mathbb{R}^d$-valued random variables is a martingale if each component is a real-valued martingale.

In a financial context, saying that the price $(S_n^i)_{0 \leq n \leq N}$ of the asset i is a martingale implies that, at each time n, the best estimate (in the least-square sense) of S_{n+1}^i is given by S_n^i.

The following properties are easily derived from the previous definition and stand as a good exercise to get used to the concept of conditional expectation:

1. $(M_n)_{0 \leq n \leq N}$ is a martingale if and only if
$$\mathbb{E}(M_{n+j}|\mathcal{F}_n) = M_n \quad \forall j \geq 0.$$

2. If $(M_n)_{n \geq 0}$ is a martingale, then for any n: $\mathbb{E}(M_n) = \mathbb{E}(M_0)$.

3. The sum of two martingales is a martingale.

4. Obviously, similar properties can be shown for supermartingales and submartingales.

Definition 1.2.2. An adapted sequence $(H_n)_{0 \leq n \leq N}$ of random variables is predictable if, for all $n \geq 1$, H_n is $\mathcal{F}_{n-1}$-measurable.

Proposition 1.2.3. *Let $(M_n)_{0 \leq n \leq N}$ be a martingale and $(H_n)_{0 \leq n \leq N}$ a predictable sequence with respect to the filtration $(\mathcal{F}_n)_{0 \leq n \leq N}$. Denote $\Delta M_n = M_n - M_{n-1}$. The sequence $(X_n)_{0 \leq n \leq N}$ defined by*
$$X_0 = H_0 M_0$$
$$X_n = H_0 M_0 + H_1 \Delta M_1 + \cdots + H_n \Delta M_n \quad \text{for } n \geq 1$$

is a martingale with respect to $(\mathcal{F}_n)_{0 \leq n \leq N}$.

(X_n) is sometimes called the *martingale transform* of (M_n) by (H_n). A consequence of this proposition and Proposition 1.1.2 is that if the discounted prices of the assets are martingales, the *expected value* of the wealth generated by following a self-financing strategy is equal to the initial wealth.

Proof. Clearly, (X_n) is an adapted sequence. Moreover, for $n \geq 0$,
$$\mathbb{E}(X_{n+1} - X_n | \mathcal{F}_n)$$
$$= \mathbb{E}(H_{n+1}(M_{n+1} - M_n) | \mathcal{F}_n)$$
$$= H_{n+1}\mathbb{E}(M_{n+1} - M_n | \mathcal{F}_n) \text{ since } H_{n+1} \text{ is } \mathcal{F}_n\text{-measurable}$$
$$= 0.$$

Hence
$$\mathbb{E}(X_{n+1}|\mathcal{F}_n) = \mathbb{E}(X_n|\mathcal{F}_n) = X_n,$$

which shows that (X_n) is a martingale. $\qquad\square$

The following proposition is a very useful characterization of martingales.

Proposition 1.2.4. *An adapted sequence of real-valued random variables* (M_n) *is a martingale if and only if for any predictable sequence* (H_n), *we have*

$$\mathbb{E}\left(\sum_{n=1}^{N} H_n \Delta M_n\right) = 0.$$

Proof. If (M_n) is a martingale, the sequence (X_n) defined by $X_0 = 0$ and, for $n \geq 1$, $X_n = \sum_{j=1}^{n} H_j \Delta M_j$ for any predictable process (H_n) is also a martingale, by Proposition 1.2.3. Hence, $\mathbb{E}(X_N) = \mathbb{E}(X_0) = 0$. Conversely, we notice that if $j \in \{1, \ldots, N\}$, we can associate the sequence (H_n) defined by $H_n = 0$ for $n \neq j+1$ and $H_{j+1} = 1_A$, for any $\mathscr{F}_j$-measurable A. Clearly, (H_n) is predictable and $\mathbb{E}\left(\sum_{n=1}^{N} H_n \Delta M_n\right) = 0$ becomes

$$\mathbb{E}(1_A(M_{j+1} - M_j)) = 0.$$

Therefore $\mathbb{E}(M_{j+1}|\mathscr{F}_j) = M_j$. $\qquad\square$

1.2.2 Viable financial markets

Let us get back to the discrete-time models introduced in the first section.

Definition 1.2.5. A market is viable if there is no arbitrage opportunity.

The following result is sometimes referred to as the *Fundamental Theorem of Asset Pricing.*

Theorem 1.2.6. *The market is viable if and only if there exists a probability measure* $\mathbb{P}^*$ *equivalent*[1] *to* $\mathbb{P}$ *such that the discounted prices of assets are* $\mathbb{P}^*$-*martingales.*

Proof. (a) Let us assume that there exists a probability $\mathbb{P}^*$ equivalent to $\mathbb{P}$ under which discounted prices are martingales. Then, for any self-financing strategy (ϕ_n), Proposition 1.1.2 implies

$$\tilde{V}_n(\phi) = V_0(\phi) + \sum_{j=1}^{n} \phi_j . \Delta \tilde{S}_j.$$

Thus, by Proposition 1.2.3, $(\tilde{V}_n(\phi))$ is a $\mathbb{P}^*$-martingale. Therefore, $\tilde{V}_N(\phi)$ and $V_0(\phi)$ have the same expectation under $\mathbb{P}^*$:

$$\mathbb{E}^*(\tilde{V}_N(\phi)) = \mathbb{E}^*(\tilde{V}_0(\phi)).$$

---

[1]Recall that two probability measures $\mathbb{P}_1$ and $\mathbb{P}_2$ are equivalent if and only if for any event A, $\mathbb{P}_1(A) = 0 \Leftrightarrow \mathbb{P}_2(A) = 0$. Here, $\mathbb{P}^*$ equivalent to $\mathbb{P}$ means that, for any $\omega \in \Omega$, $\mathbb{P}^*(\{\omega\}) > 0$.

If the strategy is admissible and its initial value is zero, then $\mathbb{E}^*(\tilde{V}_N(\phi)) = 0$, with $\tilde{V}_N(\phi) \geq 0$. Hence $\tilde{V}_N(\phi) = 0$ since $\mathbb{P}^*(\{\omega\}) > 0$, for all $\omega \in \Omega$.

(b) The proof of the converse implication is more tricky. Denote by Γ the set of all non-negative random variables X such that $\mathbb{P}(X > 0) > 0$. Clearly, Γ is a convex cone in the vector space of real-valued random variables. The market is viable if and only if for any admissible strategy ϕ, $V_0(\phi) = 0 \Rightarrow \tilde{V}_N(\phi) \notin \Gamma$.

(b1) To any admissible process $(\phi_n^1, \ldots, \phi_n^d)$ we associate the process defined by

$$\tilde{G}_n(\phi) = \sum_{j=1}^{n} \left(\phi_j^1 \Delta \tilde{S}_j^1 + \cdots + \phi_j^d \Delta \tilde{S}_j^d \right),$$

which is the cumulative discounted gain realised by following the self-financing strategy $\phi_n^1, \ldots, \phi_n^d$. According to Proposition 1.1.3, there exists a (unique) process (ϕ_n^0) such that the strategy $((\phi_n^0, \phi_n^1, \ldots, \phi_n^d))$ is self-financing with zero initial value. Note that $\tilde{G}_n(\phi)$ is the discounted value of this strategy at time n, and, because the market is viable, the fact that this value is non-negative at any time, i.e $\tilde{G}_n(\phi) \geq 0$ for $n = 1, \ldots, N$, implies that $\tilde{G}_N(\phi) = 0$. The following lemma shows that even if we do not assume that all the $\tilde{G}_n(\phi)$'s are non-negative, we still have $\tilde{G}_N(\phi) \notin \Gamma$.

Lemma 1.2.7. *If a market is viable, any predictable process* $(\phi^1, \ldots, \phi^d)$ *satisfies*

$$\tilde{G}_N(\phi) \notin \Gamma.$$

Proof. Let us assume that $\tilde{G}_N(\phi) \in \Gamma$. First, if $\tilde{G}_n(\phi) \geq 0$ for all $n \in \{0, \ldots, N\}$, the market is obviously not viable. Second, if the $\tilde{G}_n(\phi)$'s are not all non-negative, we define $n = \sup\{k | \mathbb{P}(\tilde{G}_k(\phi) < 0) > 0\}$. It follows from the definition of n that

$$n \leq N - 1, \quad \mathbb{P}(\tilde{G}_n(\phi) < 0) > 0 \text{ and } \forall m > n, \quad \tilde{G}_m(\phi) \geq 0.$$

We can now introduce a new process ψ:

$$\psi_j(\omega) = \begin{cases} 0 & \text{if } j \leq n \\ \mathbf{1}_A(\omega)\phi_j(\omega) & \text{if } j > n, \end{cases}$$

where A is the event $\{\tilde{G}_n(\phi) < 0\}$. Because ϕ is predictable and A is $\mathscr{F}_n$-measurable, ψ is also predictable. Moreover,

$$\tilde{G}_j(\psi) = \begin{cases} 0 & \text{if } j \leq n \\ \mathbf{1}_A(\tilde{G}_j(\phi) - \tilde{G}_n(\phi)) & \text{if } j > n; \end{cases}$$

thus, $\tilde{G}_j(\psi) \geq 0$ for all $j \in \{0, \ldots, N\}$ and $\tilde{G}_N(\psi) > 0$ on A. That contradicts the assumption of market viability and completes the proof of the lemma. $\square$

(b2) The set $\mathscr{V}$ of random variables $\tilde{G}_N(\phi)$, with ϕ a predictable process in $\mathbb{R}^d$, is clearly a vector subspace of $\mathbb{R}^\Omega$ (where $\mathbb{R}^\Omega$ is the set of real-valued random variables defined on Ω). According to Lemma 1.2.7, the subspace $\mathscr{V}$

does not intersect Γ. Therefore, it does not intersect the convex compact set $K = \{X \in \Gamma \mid \sum_\omega X(\omega) = 1\}$, which is included in Γ. As a result of the convex sets separation theorem (see the Appendix), there exists $(\lambda(\omega))_{\omega \in \Omega}$ such that:

1. $\forall X \in K, \quad \sum_\omega \lambda(\omega) X(\omega) > 0.$

2. For any predictable ϕ,

$$\sum_\omega \lambda(\omega) \tilde{G}_N(\phi)(\omega) = 0.$$

>From Property 1, we deduce that $\lambda(\omega) > 0$ for all $\omega \in \Omega$, so that the probability $\mathbb{P}^*$ defined by

$$\mathbb{P}^*(\{\omega\}) = \frac{\lambda(\omega)}{\sum_{\omega' \in \Omega} \lambda(\omega')}$$

is equivalent to $\mathbb{P}$.

Moreover, if we denote by $\mathbb{E}^*$ the expectation under measure $\mathbb{P}^*$, Property 2 means that, for any predictable process (ϕ_n) in $\mathbb{R}^d$,

$$\mathbb{E}^* \left(\sum_{j=1}^N \phi_j \Delta \tilde{S}_j \right) = 0.$$

It follows that for all $i \in \{1, \ldots, d\}$ and any predictable sequence (ϕ_n^i) in $\mathbb{R}$, we have

$$\mathbb{E}^* \left(\sum_{j=1}^n \phi_j^i \Delta \tilde{S}_j^i \right) = 0.$$

Therefore, according to Proposition 1.2.4, we conclude that the discounted prices $(\tilde{S}_n^1), \ldots, (\tilde{S}_n^d)$ are $\mathbb{P}^*$-martingales.

1.3   Complete markets and option pricing

1.3.1   Complete markets

A *European option*[2] with maturity N can be characterized by its payoff h, which is a non-negative $\mathscr{F}_N$-measurable random variable. For instance, a *call* on the underlying S^1 with strike price K will be defined by setting $h = \left(S_N^1 - K \right)_+$. A *put* on the same underlying asset with the same strike price K will be defined by $h = \left(K - S_N^1 \right)_+$. In these two examples, which are actually the two most important in practice, h is a function of S_N only. There are some options dependent on the whole path of the underlying asset, i.e. h is a function of $S_0, S_1, \ldots, S_N$. That is the case of the so-called *Asian options*,

---

[2]Or, more generally, a contingent claim.

where the strike price is equal to the average of the stock prices observed during a certain period of time before maturity.

Definition 1.3.1. The contingent claim defined by h is attainable if there exists an admissible strategy worth h at time N.

Remark 1.3.2. In a viable financial market, we just need to find a *self-financing* strategy worth h at maturity to say that h is attainable. Indeed, if ϕ is a self-financing strategy and if $\mathbb{P}^*$ is a probability measure equivalent to $\mathbb{P}$ under which discounted prices are martingales, then $(\tilde{V}_n(\phi))$ is also a $\mathbb{P}^*$-martingale, being a martingale transform. Hence, for $n \in \{0, \dots, N\}$, $\tilde{V}_n(\phi) = \mathbb{E}^*(\tilde{V}_N(\phi)|\mathscr{F}_n)$. Clearly, if $\tilde{V}_N(\phi) \geq 0$ (in particular if $V_N(\phi) = h \geq 0$), the strategy ϕ is admissible.

Definition 1.3.3. The market is complete if every contingent claim is attainable.

To assume that a financial market is complete is a rather restrictive assumption that does not have such a clear economic justification as the no-arbitrage assumption. The interest of complete markets is that it allows us to derive a simple theory of contingent claim pricing and hedging. The Cox-Ross-Rubinstein model, which we shall study in the next section, is a very simple example of a complete market model. The following theorem gives a precise characterization of complete, viable financial markets.

Theorem 1.3.4. *A viable market is complete if and only if there exists a unique probability measure $\mathbb{P}^*$ equivalent to $\mathbb{P}$, under which discounted prices are martingales.*

The probability $\mathbb{P}^*$ will appear to be the *computing tool* whereby we can derive closed-form pricing formulae and hedging strategies.

Proof. (a) Let us assume that the market is viable and complete. Then, any non-negative, $\mathscr{F}_N$-measurable random variable h can be written as $h = V_N(\phi)$, where ϕ is an admissible strategy that replicates the contingent claim h. Since ϕ is self-financing, we know that

$$\frac{h}{S_N^0} = \tilde{V}_N(\phi) = V_0(\phi) + \sum_{j=1}^{N} \phi_j . \Delta \tilde{S}_j.$$

Thus, if $\mathbb{P}_1$ and $\mathbb{P}_2$ are two probability measures under which discounted prices are martingales, $(\tilde{V}_n(\phi))_{0 \leq n \leq N}$ is a martingale under both $\mathbb{P}_1$ and $\mathbb{P}_2$. It follows that, for $i = 1, 2$,

$$\mathbb{E}_i(\tilde{V}_N(\phi)) = \mathbb{E}_i(V_0(\phi)) = V_0(\phi),$$

the last equality coming from the fact that $\mathscr{F}_0 = \{\emptyset, \Omega\}$. Therefore,

$$\mathbb{E}_1\left(\frac{h}{S_N^0}\right) = \mathbb{E}_2\left(\frac{h}{S_N^0}\right)$$

and, since h is arbitrary, $\mathbb{P}_1 = \mathbb{P}_2$ on the whole σ-algebra $\mathscr{F}_N$, which is assumed to be equal to $\mathscr{F}$.

(b) Let us assume that the market is viable and incomplete. Then, there exists a random variable $h \geq 0$ that is not attainable. We call $\check{\mathscr{V}}$ the set of random variables of the form

$$U_0 + \sum_{n=1}^{N} \phi_n.\Delta\tilde{S}_n, \tag{1.1}$$

where U_0 is $\mathscr{F}_0$-measurable and $\left((\phi_n^1, \ldots, \phi_n^d)\right)_{0 \leq n \leq N}$ is an $\mathbb{R}^d$-valued predictable process.

It follows from Proposition 1.1.3 and Remark 1.3.2 that the variable h/S_N^0 does not belong to $\check{\mathscr{V}}$. Hence, $\check{\mathscr{V}}$ is a strict subset of the set of all random variables on $(\Omega, \mathscr{F})$. Therefore, if $\mathbb{P}^*$ is a probability equivalent to $\mathbb{P}$ under which discounted prices are martingales, and if we define the following scalar product on the set of random variables $(X, Y) \mapsto \mathbb{E}^*(XY)$, we notice that there exists a non-zero random variable X orthogonal to $\check{\mathscr{V}}$. We now define

$$\mathbb{P}^{**}(\{\omega\}) = \left(1 + \frac{X(\omega)}{2\|X\|_\infty}\right)\mathbb{P}^*(\{\omega\})$$

with $\|X\|_\infty = \sup_{\omega \in \Omega}|X(\omega)|$. Because $\mathbb{E}^*(X) = 0$, that defines a new probability measure equivalent to $\mathbb{P}$, and different from $\mathbb{P}^*$. Moreover,

$$\mathbb{E}^{**}\left(\sum_{n=1}^{N} \phi_n.\Delta\tilde{S}_n\right) = 0$$

for any predictable process $\left((\phi_n^1, \ldots, \phi_n^d)\right)_{0 \leq n \leq N}$. It follows from Proposition 1.2.4 that $(\tilde{S}_n)_{0 \leq n \leq N}$ is a $\mathbb{P}^{**}$-martingale. $\square$

1.3.2 Pricing and hedging contingent claims in complete markets

The market is assumed to be viable and complete and we denote by $\mathbb{P}^*$ the unique probability measure under which the discounted prices of financial assets are martingales. Let h be an $\mathscr{F}_N$-measurable, non-negative random variable and ϕ be an admissible strategy replicating the contingent claim hence defined, i.e.

$$V_N(\phi) = h.$$

The sequence $(\tilde{V}_n)_{0 \leq n \leq N}$ is a $\mathbb{P}^*$-martingale, and consequently

$$V_0(\phi) = \mathbb{E}^*(\tilde{V}_N(\phi)),$$

that is $V_0(\phi) = \mathbb{E}^* \left(h/S_N^0 \right)$ and, more generally,

$$V_n(\phi) = S_n^0 \mathbb{E}^* \left(\frac{h}{S_N^0} \Big| \mathscr{F}_n \right), \qquad n = 0, 1, \ldots, N.$$

At any time, the value of an admissible strategy replicating h is completely determined by h. It seems quite natural to call $V_n(\phi)$ the *value of the option* at time n: that is the wealth needed at time n to replicate h at time N by following the strategy ϕ. If, at time 0, an investor sells the option for

$$\mathbb{E}^* \left(\frac{h}{S_N^0} \right),$$

he can follow a replicating strategy ϕ in order to generate an amount h at time N. In other words, the investor is *perfectly hedged*. The number $\mathbb{E}^*(h/S_N^0)$ is called the *fair price* of the option at time 0.

Remark 1.3.5. It is important to notice that the computation of the option price only requires the knowledge of $\mathbb{P}^*$ and not $\mathbb{P}$. We could have just considered a *measurable* space $(\Omega, \mathscr{F})$ equipped with the filtration $(\mathscr{F}_n)$. In other words, we would only define the set of all possible states and the evolution of the information over time. As long as the probability space and the filtration are specified, we do not need to find the *true* probability of the possible events (say, by statistical means) in order to price the option. The analysis of the Cox-Ross-Rubinstein model will show how we can compute the option price and the hedging strategy in practice.

1.3.3 Introduction to American options

Since an American option can be exercised at any time between 0 and N, we shall define it as a non-negative sequence (Z_n) adapted to $(\mathscr{F}_n)$, where Z_n is the immediate profit made by exercising the option at time n. In the case of an American call on the stock S^1, with strike price K, $Z_n = (S_n^1 - K)_+$; in the case of the put, $Z_n = (K - S_n^1)_+$. In order to define the price of the option associated with $(Z_n)_{0 \leq n \leq N}$, we shall think in terms of a backward induction starting at time N. Indeed, the value of the option at maturity is obviously equal to $U_N = Z_N$. At what price should we sell the option at time $N - 1$? If the holder exercises straight away, he will earn Z_{N-1}, or he might exercise at time N, in which case the writer must be ready to pay the amount Z_N. Therefore, at time $N - 1$, the writer has to earn the maximum between Z_{N-1} and the amount necessary at time $N - 1$ to generate Z_N at time N. In other words, the writer wants the maximum between Z_{N-1} and the value at time $N - 1$ of an admissible strategy paying off Z_N at time N, i.e. $S_{N-1}^0 \mathbb{E}^*(\tilde{Z}_N | \mathscr{F}_{N-1})$, with $\tilde{Z}_N = Z_N/S_N^0$. As we see, it makes sense to price the option at time $N - 1$ as

$$U_{N-1} = \max(Z_{N-1}, S_{N-1}^0 \mathbb{E}^*(\tilde{Z}_N | \mathscr{F}_{N-1})).$$

By induction, we define the American option price for $n = 1, \ldots, N$ by

$$U_{n-1} = \max\left(Z_{n-1}, S_{n-1}^0 \mathbb{E}^*\left(\frac{U_n}{S_n^0}\bigg|\mathscr{F}_{n-1}\right)\right).$$

If we assume that the interest rate over one period is constant and equal to r,

$$S_n^0 = (1 + r)^n$$

and

$$U_{n-1} = \max\left(Z_{n-1}, \frac{1}{1+r}\mathbb{E}^*(U_n|\mathscr{F}_{n-1})\right).$$

Let $\tilde{U}_n = U_n/S_n^0$ be the discounted price of the American option.

Proposition 1.3.6. *The sequence $(\tilde{U}_n)_{0 \leq n \leq N}$ is a $\mathbb{P}^*$-supermartingale. It is the smallest $\mathbb{P}^*$-supermartingale that dominates the sequence $(\tilde{Z}_n)_{0 \leq n \leq N}$.*

Note that, in contrast to the European case, the discounted price of the American option is generally not a martingale under $\mathbb{P}^*$.

Proof. ▷From the equality

$$\tilde{U}_{n-1} = \max(\tilde{Z}_{n-1}, \mathbb{E}^*(\tilde{U}_n|\mathscr{F}_{n-1})),$$

it follows that $(\tilde{U}_n)_{0 \leq n \leq N}$ is a supermartingale dominating $(\tilde{Z}_n)_{0 \leq n \leq N}$. Let us now consider a supermartingale $(\tilde{T}_n)_{0 \leq n \leq N}$ that dominates $(\tilde{Z}_n)_{0 \leq n \leq N}$, and prove by backward induction that it dominates $(\tilde{U}_n)$. Clearly, $\tilde{T}_N \geq \tilde{U}_N$, and, if $\tilde{T}_n \geq \tilde{U}_n$, we have

$$\tilde{T}_{n-1} \geq \mathbb{E}^*(\tilde{T}_n|\mathscr{F}_{n-1}) \geq \mathbb{E}^*(\tilde{U}_n \mid \mathscr{F}_{n-1})$$

whence

$$\tilde{T}_{n-1} \geq \max(\tilde{Z}_{n-1}, \mathbb{E}^*(\tilde{U}_n|\mathscr{F}_{n-1})) = \tilde{U}_{n-1}.$$

$\square$

1.4   Problem: Cox, Ross and Rubinstein model

The Cox-Ross-Rubinstein model is a discrete-time version of the Black-Scholes model. It considers only one risky asset whose price is S_n at time n, $0 \leq n \leq N$, and a riskless asset whose return is r over one period of time. To be consistent with the previous sections, we denote $S_n^0 = (1 + r)^n$.

The risky asset is modelled as follows: between two consecutive periods the price changes by a factor that is either d or u, with $0 < d < u$:

$$S_{n+1} = \begin{cases} S_n u \\ S_n d. \end{cases}$$

Although the notation refers to moves upwards or downwards through the letters u and d, we do not assume $u > 1$ or $d < 1$. The initial stock price S_0 is given. The set of possible states is then $\Omega = \{d, u\}^N$. Each N-tuple represents the successive values of the ratio $S_{n+1}/S_n, n = 0, 1, \ldots, N - 1$. We also assume that $\mathscr{F}_0 = \{\emptyset, \Omega\}$ and $\mathscr{F} = \mathscr{P}(\Omega)$. For $n = 1, \ldots, N$, the σ-algebra $\mathscr{F}_n$ is generated by the random variables $S_1, \ldots, S_n$: $\mathscr{F}_n = \sigma(S_1, \ldots, S_n)$. The assumption that each singleton in Ω has a strictly positive probability implies that $\mathbb{P}$ is defined uniquely up to equivalence. We now introduce the variables $T_n = S_n/S_{n-1}$, for $n = 1, \ldots, N$. If $(x_1, \ldots, x_N)$ is one element of $\Omega, \mathbb{P}\{(x_1, \ldots, x_N)\} = \mathbb{P}(T_1 = x_1, \ldots, T_N = x_N)$. As a result, knowing $\mathbb{P}$ is equivalent to knowing the joint distribution of the N-tuple $(T_1, T_2, \ldots, T_N)$. We also remark that for $n \geq 1$, $\mathscr{F}_n = \sigma(T_1, \ldots, T_n)$.

1. Show that the discounted price $(\tilde{S}_n)$ is a martingale under $\mathbb{P}$ if and only if $\mathbb{E}(T_{n+1}|\mathscr{F}_n) = 1 + r, \forall n \in \{0, 1, \ldots, N - 1\}$.

   The equality $\mathbb{E}(\tilde{S}_{n+1}|\mathscr{F}_n) = \tilde{S}_n$ is equivalent to $\mathbb{E}(\tilde{S}_{n+1}/\tilde{S}_n|\mathscr{F}_n) = 1$, since $\tilde{S}_n$ is $\mathscr{F}_n$-measurable and this last equality is actually equivalent to $\mathbb{E}(T_{n+1}|\mathscr{F}_n) = 1 + r$.

2. Deduce that r must belong to $(d - 1, u - 1)$ for the market to be arbitrage-free.

   If the market is viable, there exists a probability $\mathbb{P}_*$ equivalent to $\mathbb{P}$, under which $(\tilde{S}_n)$ is a martingale. Thus, according to Question 1,

   $$\mathbb{E}^*(T_{n+1}|\mathscr{F}_n) = 1 + r,$$

   and therefore $\mathbb{E}^*(T_{n+1}) = 1 + r$. Since T_{n+1} is equal to either of d and u with non-zero probability, we necessarily have $(1 + r) \in (d, u)$.

3. Give examples of arbitrage strategies if the no-arbitrage condition derived in Question 2 is not satisfied.

   Assume for instance that $r \leq u - 1$. By borrowing an amount S_0 at time 0, we can purchase one share of the risky asset. At time N, we pay the loan back and sell the risky asset. We realised a profit equal to $S_N - S_0(1 + r)^N$ which is always positive, since $S_N \geq S_0 d^N$. Moreover, it is strictly positive with non-zero probability. This is an arbitrage opportunity. If $r \geq u - 1$, we can make a riskless profit by short-selling the risky asset.

4. From now on, we assume that $r \in (d - 1, u - 1)$ and we write $p = (u - 1 - r)/(u - d)$. Show that $(\tilde{S}_n)$ is a $\mathbb{P}$-martingale if and only if the random variables $T_1, T_2, \ldots, T_N$ are independent, identically distributed (IID) *and* their distribution is given by $\mathbb{P}(T_1 = d) = p = 1 - \mathbb{P}(T_1 = u)$. Conclude that the market is arbitrage-free and complete.

   If the T_i's are independent and satisfy $\mathbb{P}(T_i = d) = p = 1 - \mathbb{P}(T_i = u)$, we have

   $$\mathbb{E}(T_{n+1}|\mathscr{F}_n) = \mathbb{E}(T_{n+1}) = pd + (1 - p)u = 1 + r,$$

   and thus $(\tilde{S}_n)$ is a $\mathbb{P}$-martingale, according to Question 1.

Conversely, if for $n = 0, 1, \ldots, N - 1$, $\mathbb{E}(T_{n+1}|\mathscr{F}_n) = 1 + r$, we can write

$$d\mathbb{E}(\mathbf{1}_{\{T_{n+1}=d\}}|\mathscr{F}_n) + u\mathbb{E}(\mathbf{1}_{\{T_{n+1}=u\}}|\mathscr{F}_n) = 1 + r.$$

Then, the equality

$$\mathbb{E}(\mathbf{1}_{\{T_{n+1}=d\}}|\mathscr{F}_n) + \mathbb{E}(\mathbf{1}_{\{T_{n+1}=u\}}|\mathscr{F}_n) = 1,$$

implies that $\mathbb{E}(\mathbf{1}_{\{T_{n+1}=d\}}|\mathscr{F}_n) = p$ and $\mathbb{E}(\mathbf{1}_{\{T_{n+1}=u\}}|\mathscr{F}_n) = 1 - p$. By induction, we prove that for any $x_i \in \{d, u\}$,

$$\mathbb{P}(T_1 = x_1, \ldots, T_n = x_n) = \prod_{i=1}^{n} p_i,$$

where $p_i = p$ if $x_i = d$ and $p_i = 1 - p$ if $x_i = u$. This shows that the variables T_i are IID under measure $\mathbb{P}$ and that $\mathbb{P}(T_i = d) = p$.

We have shown that the very fact that $(\tilde{S}_n)$ is a $\mathbb{P}$-martingale uniquely determines the distribution of the N-tuple $(T_1, \ldots, T_N)$ under $\mathbb{P}$, hence the measure $\mathbb{P}$ itself. Therefore, the market is arbitrage-free and complete.

5. We denote by C_n (resp. P_n) the value at time n of a European call (resp. put) on a share of stock, with strike price K and maturity N.

   (a) Derive the put/call parity equation

   $$C_n - P_n = S_n - K(1 + r)^{-(N-n)}$$

   from the put/call prices given in their conditional expectation form. Denote by $\mathbb{E}^*$ the expectation with respect to the probability measure $\mathbb{P}^*$ under which $(\tilde{S}_n)$ is a martingale. We have

   $$\begin{aligned} C_n - P_n &= (1 + r)^{-(N-n)}\mathbb{E}^*((S_N - K)_+ - (K - S_N)_+|\mathscr{F}_n) \\ &= (1 + r)^{-(N-n)}\mathbb{E}^*(S_N - K|\mathscr{F}_n) \\ &= S_n - K(1 + r)^{-(N-n)}, \end{aligned}$$

   where the last equality comes from the fact that $(\tilde{S}_n)$ is a $\mathbb{P}^*$-martingale.

   (b) Show that we can write $C_n = c(n, S_n)$, where c is a function that depends on K, d, u, r and p.
   When we write $S_N = S_n \prod_{i=n+1}^{N} T_i$, we get

   $$C_n = (1 + r)^{-(N-n)}\mathbb{E}^*\left(\left(S_n \prod_{i=n+1}^{N} T_i - K\right)_+ \middle| \mathscr{F}_n\right).$$

   Since under the probability $\mathbb{P}^*$, the random variable $\prod_{i=n+1}^{N} T_i$ is independent of $\mathscr{F}_n$, and since S_n is $\mathscr{F}_n$-measurable, Proposition A.2.5 in the Appendix allows us to write $C_n = c(n, S_n)$, where c is the function defined by

   $$\frac{c(n, x)}{(1 + r)^{-(N-n)}} = \mathbb{E}^*\left(x \prod_{i=n+1}^{N} T_i - K\right)_+ \qquad (1.4.1)$$

   $$= \sum_{j=0}^{N-n} \frac{(N-n)!}{(N-n-j)!j!}p^j(1-p)^{N-n-j}(xd^j u^{N-n-j} - K)_+.$$

(c) Prove that the function c satisfies the recursive equations

$$c(n, x) = \frac{(pc(n+1, xd) + (1-p)c(n+1, xu))}{1+r}, \quad n = 0, \ldots, N-1.$$

By conditioning with respect to $(T_{n+2}, \ldots, T_N)$ in (1.4.1), we have

$$\frac{c(n, x)}{(1+r)^{-(N-n)}} = p\mathbb{E}^* \left(xd \prod_{i=n+2}^{N} T_i - K \right)_+$$

$$+ (1-p)\mathbb{E}^* \left(xu \prod_{i=n+2}^{N} T_i - K \right)_+.$$

6. Show that the replicating strategy of a call is characterized by a quantity $H_n = \Delta(n, S_{n-1})$ at time n, where Δ will be expressed in terms of the function c.

We denote by H_n^0 the number of riskless assets in the replicating portfolio. We have

$$H_n^0(1+r)^n + H_n S_n = c(n, S_n).$$

Since H_n^0 and H_n are $\mathscr{F}_{n-1}$-measurable, they are functions of $S_1, \ldots, S_{n-1}$ only and, since S_n is equal to $S_{n-1}d$ or $S_{n-1}u$, the previous equality implies

$$H_n^0(1+r)^n + H_n S_{n-1}d = c(n, S_{n-1}d)$$

and

$$H_n^0(1+r)^n + H_n S_{n-1}u = c(n, S_{n-1}u).$$

Subtracting one from the other, it turns out that

$$\Delta(n, x) = \frac{c(n, xu) - c(n, xd)}{x(u-d)}.$$

7. We can now use the model to price a call or a put with maturity T on a single stock. In order to do that, we divide the interval $[0, T]$ into N subintervals, so that we can refer to the N-period model discussed above and we study the asymptotic case when N goes to infinity, $r = RT/N$, $\log(d/(1+r)) = -\sigma\sqrt{T/N}$ and $\log(u/(1+r)) = \sigma\sqrt{T/N}$. The real number R is to be interpreted as the *instantaneous interest rate* at all times between 0 and T. Indeed, we have $e^{RT} = \lim_{N\to\infty}(1+r)^N$. The number $\sigma^2 T$ can be seen as the limit variance, under the measure $\mathbb{P}^*$, of the random variable $\log(S_N)$, when N goes to infinity. The number σ^2 is the limit variance of the increase in the log-price over a time interval with unit length.

(a) Let $(Y_N)_{N\geq 1}$ be a sequence of random variables equal to

$$Y_N = X_1^N + X_2^N + \cdots + X_N^N,$$

where, for each N, the random variables X_i^N are IID, belong to the set

$$\{-\sigma\sqrt{T/N}, \sigma\sqrt{T/N}\},$$

and their mean is equal to μ_N, with $\lim_{N\to\infty}(N\mu_N)=\mu$. Show that the sequence (Y_N) converges in law towards a Gaussian variable with mean μ and variance $\sigma^2 T$.

We just need to study the convergence of the characteristic function ϕ_{Y_N} of Y_N. We have

$$\phi_{Y_N}(\xi) = \mathbb{E}(\exp(i\xi Y_N)) = \prod_{j=1}^{N}\mathbb{E}\left(\exp\left(i\xi X_j^N\right)\right)$$

$$= \left(\mathbb{E}\left(\exp\left(i\xi X_1^N\right)\right)\right)^N$$

$$= (1 + i\xi\mu_N - \sigma^2 T\xi^2/2N + o(1/N))^N.$$

Hence, $\lim_{N\to\infty}\phi_{Y_N}(\xi)=\exp(i\xi\mu - \sigma^2 T\xi^2/2)$, which proves the convergence in law.

(b) Give explicitly the asymptotic prices of the put and the call at time 0.

For a fixed N, the put price at time 0 is given by

$$P_0^{(N)} = (1 + RT/N)^{-N}\mathbb{E}^*\left(K - S_0\prod_{n=1}^{N}T_n\right)_+$$

$$= \mathbb{E}^*\left((1 + RT/N)^{-N}K - S_0\exp(Y_N)\right)_+$$

with $Y_N = \sum_{n=1}^{N}\log(T_n/(1+r))$. According to the assumptions, the variables $X_j^N = \log(T_j/(1+r))$ are valued in $\{-\sigma\sqrt{T/N}, \sigma\sqrt{T/N}\}$ and are IID under probability $\mathbb{P}^*$. Moreover,

$$\mathbb{E}^*(X_j^N) = (1 - 2p)\frac{\sigma\sqrt{T}}{\sqrt{N}} = \frac{2 - e^{\sigma\sqrt{T/N}} - e^{-\sigma\sqrt{T/N}}}{e^{\sigma\sqrt{T/N}} - e^{-\sigma\sqrt{T/N}}}\frac{\sigma\sqrt{T}}{\sqrt{N}}.$$

Therefore, the sequence (Y_N) satisfies the conditions of Question 7(a), with $\mu = -\sigma^2 T/2$. If we write $\psi(y) = (Ke^{-RT} - S_0 e^y)_+$, we are able to write

$$|P_0^{(N)} - \mathbb{E}^*(\psi(Y_N))|$$
$$= |\mathbb{E}^*(((1 + RT/N)^{-N}K - S_0\exp(Y_N))_+$$
$$- (Ke^{-RT} - S_0\exp(Y_N))_+)|$$
$$\leq K|(1 + RT/N)^{-N} - e^{-RT}|.$$

Since ψ is a bounded[3], continuous function and because the sequence (Y_N) converges in law, we conclude that

$$\lim_{n\to\infty}P_0^{(N)} = \lim_{N\to\infty}\mathbb{E}^*(\psi(Y_N))$$

$$= \frac{1}{\sqrt{2\pi}}\int_{-\infty}^{+\infty}(Ke^{-RT} - S_0 e^{-\sigma^2 T/2 + \sigma\sqrt{T}y})_+ e^{-y^2/2}dy.$$

---

[3]It is precisely to be able to work with a bounded function that we studied the put first.

The integral can be expressed easily in terms of the cumulative normal distribution function Φ, so that

$$\lim_{n \to \infty} P_0^{(N)} = Ke^{-RT}\Phi(-d_2) - S_0\Phi(-d_1),$$

where $d_1 = (\log(S_0/K) + RT + \sigma^2 T/2)/\sigma\sqrt{T}$, $d_2 = d_1 - \sigma\sqrt{T}$ and

$$\Phi(d) = \frac{1}{\sqrt{2\pi}} \int_{-\infty}^{d} e^{-x^2/2}dx.$$

The price of the call follows easily from put/call parity: $\lim_{N \to \infty} C_0^{(N)} = S_0\Phi(d_1) - Ke^{-RT}\Phi(d_2)$.

Remark 1.4.1. We note that the only non-directly observable parameter is σ. Its interpretation as a variance suggests that it should be estimated by statistical methods. However, we shall resume this question in Chapter 4.

Notes: We have assumed throughout this chapter that the risky assets were not offering any dividend. Actually, Huang and Litzenberger (1988) apply the same ideas to answer the same questions when the stock is carrying dividends. The theorem of characterization of complete markets can also be proved with infinite probability spaces (cf. Dalang et al. (1990) and Morton (1989)). A proof is sketched in Exercise 1 below, based on Rogers (1994). In continuous time, the problem is much more tricky (cf. Harrison and Kreps (1979), Stricker (1990), Delbaen and Schachermayer (1994) and the recent monograph by Delbaen and Schachermayer (2006)). The theory of complete markets in continuous time was developed by Harrison and Pliska (1981), and Harrison and Pliska (1983). An elementary presentation of the Cox-Ross-Rubinstein model is given in the book by Cox and Rubinstein (1985). A generalization of the model (which makes the market incomplete) is sudied in Exercise 2 below. More material on discrete models can be found in Föllmer and Schied (2004) or Pliska (1997).

1.5   Exercises

Exercise 1   Let $X = (X_1, \ldots, X_d)$ be a random vector with values in $\mathbb{R}^d$, defined on some finite probability space $(\Omega, \mathscr{F}, \mathbb{P})$. Denote by $F : \mathbb{R}^d \to \mathbb{R}$ the function defined by $F(\theta) = \mathbb{E}\, e^{\theta.X}$.

1. Show that if $\mathbb{P}(X = 0) < 1$, one can find a subsequence $(X_{i_1}, \ldots, X_{i_k})$ such that each component X_i of X is (almost surely) a linear combination of the random variables $X_{i_1}, \ldots, X_{i_k}$ and, for any $(u_1, \ldots, u_k) \in \mathbb{R}^k$ with at least one nonzero coordinate, $\mathbb{P}(\sum_{j=1}^{k} u_j X_{i_j} = 0) < 1$.

2. Show that if F achieves a minimum at $\theta^* \in \mathbb{R}^d$, we have $\mathbb{E}Xe^{\theta^*.X} = 0$.

3. The random vector X is said to be *arbitrage-free* if for all $\theta \in \mathbb{R}^d$ such that $\mathbb{P}(\theta.X \geq 0) = 1$, we have $\mathbb{P}(\theta.X = 0) = 1$. We want to prove that if X is arbitrage-free, F has a minimum.

   (a) Consider a sequence $(u_n)_{n \in \mathbb{N}}$ of vectors in $\mathbb{R}^d$, with $||u_n|| = 1$, and a sequence $(\lambda_n)_{n \in \mathbb{N}}$ of real numbers such that $\lim_{n \to \infty} F(\lambda_n u_n) = \inf_{\theta \in \mathbb{R}^d} F(\theta)$. Prove that if X is arbitrage free and satisfies $\mathbb{P}(u.X = 0) < 1$ for every nonzero vector $u \in \mathbb{R}^d$, the sequence $(\lambda_n)_{n \in \mathbb{N}}$ is bounded.

   (b) Conclude, using Question 1.

4. Let $\mathscr{B}_1$ and $\mathscr{B}_2$ be two σ-algebras, with $\mathscr{B}_1 \subset \mathscr{B}_2 \subset \mathscr{F}$. Assume that for every $\mathscr{B}_1$-measurable random vector θ with values in $\mathbb{R}^d$, satisfying $\mathbb{P}(\theta.X \geq 0) = 1$, we have $\mathbb{P}(\theta.X = 0) = 1$. Prove that there exists a $\mathscr{B}_2$-measurable random variable U such that $\mathbb{P}(U > 0) = 1$, $\mathbb{E}(U \mid \mathscr{B}_1) = 1$ and $\mathbb{E}(XU) = 0$. Hint: use the previous questions and the atoms of $\mathscr{B}_1$.

5. Use the above to work out another proof of Theorem 1.2.6 (part (b2)).

Exercise 2   Super-replication in an incomplete market. We consider, here, an extended version of the Cox-Ross-Rubinstein model allowing the asset price to take three different values at each time step.

As for the Cox-Ross-Rubinstein model, let S_n be the price at time n of the risky asset, let r be the riskless return over one period of time and $S_n^0 = (1+r)^n$ be the price of the riskless asset. Between two successive periods the relative price change can be a, b or c, with $-1 < a < b < c$:

$$S_{n+1} = \begin{cases} S_n(1+a) \\ S_n(1+b) \\ S_n(1+c). \end{cases}$$

The initial stock price is denoted by S_0. The set of possible states is $\Omega = \{1+a, 1+b, 1+c\}^N$, where each N-tuple represents the values of S_{n+1}/S_n, $n = 0, 1, \ldots, N-1$. We also assume that, for $n = 1, \ldots, N$, $\mathscr{F}_n = \sigma(S_1, \ldots, S_n)$ is the σ-field generated by the random variables $S_1, \ldots, S_n$. We assume that $\mathbb{P}$ gives to each singleton in Ω a strictly positive probability. This assumption defines $\mathbb{P}$ up to an equivalent change of probability.

Part I: Viability and completeness

1. At which condition on a, b, c and r is this model viable? We assume, in the sequel, that this assumption is fulfilled.

2. Assuming that $N = 1$ and $r = 0$, show, by constructing a contingent claim that cannot be replicated, that the model is incomplete

We will now prove that we are able to construct a super-replicating portfolio for every contingent claim with payoff $f(S_N)$, f being convex.

More pecisely, a *self-financing* strategy $\phi = ((H_n^0, H_n), 0 \leq n \leq N)$ is a *super-replicating strategy* for the contingent claim with payoff $f(S_N)$ if and only if, by definition, its value $V_n(\phi) = H_n^0 S_n^0 + H_n S_n$ satisfies $V_N(\phi) \geq f(S_N)$ almost surely.

When such a super-replicating strategy exists, the *super-replication price* of the contingent claim is the smallest initial value of a super-replicating strategy, if such a minimal strategy exists.

Part II: A lower bound for the super-replication price   We assume that there exists a super-replicating strategy $\phi = ((H_n^0, H_n), 0 \leq n \leq N)$ whose value at time n is given by $V_n(\phi)$.

1. Show that if $\tilde{\mathbb{P}}$ is a probability equivalent to $\mathbb{P}$, under which $(\tilde{S}_n = S_n / S_n^0, 0 \leq n \leq N)$ is a martingale, then $V_0(\phi) \geq \tilde{\mathbb{E}} \left(\frac{f(S_N)}{(1+r)^N} \right)$.

2. Let $T_n = S_n / S_{n-1}$. We denote by $\mathbb{P}^{p_1, p_2, p_3}$ the probability on Ω, such that $(T_n, 0 \leq n \leq N)$ is a sequence of independent random variables with

$$\mathbb{P}^{p_1, p_2, p_3}(T_n = 1 + a) = p_1, \qquad (1.5.1)$$

$$\mathbb{P}^{p_1, p_2, p_3}(T_n = 1 + b) = p_2, \qquad (1.5.2)$$

$$\mathbb{P}^{p_1, p_2, p_3}(T_n = 1 + c) = p_3, \qquad (1.5.3)$$

p_1, p_2, p_3 being positive real numbers such that $p_1 + p_2 + p_3 = 1$.

Prove that $(\tilde{S}_n, 0 \leq n \leq N)$ is a martingale under $\mathbb{P}^{p_1, p_2, p_3}$ if and only if $p_1 a + p_2 b + p_3 c = r$.

Under which condition on p_1, p_2, p_3 is this probability equivalent to the initial probability $\mathbb{P}$?

3. Prove that the super-replication price V_0 is greater than

$$\bar{V}_0 = \sup_{\substack{p_1 > 0, p_2 > 0, p_3 > 0 \\ p_1 + p_2 + p_3 = 1 \\ p_1 a + p_2 b + p_3 c = r}} \mathbb{E}^{p_1, p_2, p_3} \left(\frac{f(S_N)}{(1+r)^N} \right).$$

4. Prove that

$$\bar{V}_0 \geq V_{\text{CRR}} = \mathbb{E}^{p^*, 0, 1-p^*} \left(\frac{f(S_N)}{(1+r)^N} \right),$$

where p^* is such that $p^* a + (1 - p^*)c = r$ (note that, using equations $p_1 + p_2 + p_3 = 1$ and $p_1 a + p_2 b + p_3 c = r$, we can express p_1 as $\alpha(p_2)$ and p_3 as $\beta(p_2)$ and that $\mathbb{E}^{\alpha(p_2), p_2, \beta(p_2)}(f(S_N)))$ is a continuous function of p_2).

Give an interpretation for p^* and for V_{CRR} in a Cox-Ross-Rubinstein model with $d = 1 + a$ and $u = 1 + c$.

Part III: Computation of a super-replication strategy We will now show that we can construct a super-replication strategy with initial value V_{CRR} for a contingent claim with a convex payoff function f.

Let $v(n, x)$ be the price in the Cox-Ross-Rubinstein model with parameters $d = 1 + a$ and $u = 1 + c$, at time n and for a current value x of the asset. This price satisfies the recursive equations

$$
\begin{cases}
v(N, x) = f(x), x \in \mathbb{R}^+ \\
v(n, x) = \dfrac{p^* v(n+1, xd) + (1 - p^*) v(n+1, xu)}{1 + r} \\
x \in \mathbb{R}^+, 0 \leq n < N.
\end{cases}
\tag{1.5.4}
$$

Let $\Delta(n+1, x)$ be the corresponding hedge between times n and $n+1$, for a value x, defined by

$$
\Delta(n+1, x) = \frac{v(n+1, xu) - v(n+1, xd)}{x(u - d)}.
$$

Let V_n be the value of the unique self-financing strategy with a quantity $\Delta(n, S_n)$ of risky asset at time n and initial value $V_0 = V_{\text{CRR}} = v(0, S_0)$.

1. Assuming that f is a convex function, prove that if v is solution of (1.5.4), then, for all n, $v(n, .)$ is convex.

2. Let $\tilde{V}_n = V_n/(1 + r)^n$ and $\tilde{S}_n = S_n/(1 + r)^n$. Prove that

$$
\tilde{V}_{n+1} - \tilde{V}_n = \Delta(n+1, S_n) \left(\tilde{S}_{n+1} - \tilde{S}_n \right).
$$

3. Show, using equation (1.5.4), that for $\alpha = a$ or c,

$$
\frac{v(n+1, x(1 + \alpha))}{1 + r} = v(n, x) + \Delta(n+1, x) \left(\frac{x(1 + \alpha)}{1 + r} - x \right),
$$

and deduce, using the convexity of v, that

$$
\frac{v(n+1, x(1 + b))}{1 + r} \leq v(n, x) + \Delta(n+1, x) \left(\frac{x(1 + b)}{1 + r} - x \right).
$$

4. Prove, by induction, that for all $0 \leq n \leq N$, $V_n \geq v(n, S_n)$ and deduce that V_{CRR} is the initial value of a super-replication strategy.

Exercise 3 We consider a discrete-time financial model as in Section 1.1. A *numéraire* is an adapted sequence $W = (W_n)_{n=0,\ldots,N}$ such that $W_0 = 1$, $W_n > 0$ for $n = 1, \ldots, N$, and $W_n = V_n(\theta)$ $(n = 0, \ldots, N)$ for some admissible strategy θ. In this exercise, we consider a given numéraire W, and denote by S^W the W-*discounted* vector price process, defined by $S_n^W = S_n/W_n$, $n = 0, \ldots, N$.

1. Prove that a predictable sequence $\phi = (\phi_n)_{n=0,\dots,N}$, with values in $\mathbb{R}^{d+1}$, is a self-financing strategy if and only if we have

$$V_n^W(\phi) = V_0(\phi) + \sum_{j=1}^{n} \phi_j . \Delta S_j^W, \quad n = 1, \dots, N,$$

with the notation $V_n^W(\phi) = V_n(\phi)/W_n$.

2. Prove that for $n = 1, \dots, N$, $\sum_{j=1}^{n} \theta_j . \Delta S_j^W = 0$.

3. Prove that for any predictable sequence $\phi = (\phi_n)_{n=0,\dots,N}$, with values in $\mathbb{R}^{d+1}$ and any real number V_0, there exists a self-financing strategy $\hat{\phi}$ such that

$$\hat{\phi}_n . S_n^W = V_0 + \sum_{j=1}^{n} \phi_j . \Delta S_j^W, \quad n = 0, \dots, N.$$

4. Prove that the market is viable if and only if there exists a probability $\mathbb{P}^W$, equivalent to $\mathbb{P}$, such that S^W is a $\mathbb{P}^W$-martingale.

5. Prove that, in a viable market, there is at most one deterministic numéraire.

6. Assume that the market is viable and complete, and denote by $\mathbb{P}^*$ the unique equivalent probability measure with respect to which $\tilde{S}$ is a martingale.

   (a) Prove that the probability measure $\mathbb{P}^W$ introduced in Question 4 is unique, and satisfies $d\mathbb{P}^W/d\mathbb{P}^* = W_N/S_N^0$.

   (b) Prove that the fair price at time n of a European contingent claim h is given by
   $$W_n \mathbb{E}^W(h/W_N \mid \mathscr{F}_n).$$

Chapter 2

Optimal stopping problem and American options

The purpose of this chapter is to address the pricing and hedging of American options and to establish the link between these questions and the optimal stopping problem. To do so, we will need to define the notion of stopping time, which will enable us to model exercise strategies for American options. We will also define the Snell envelope, which is the fundamental concept used to solve the optimal stopping problem. The application of these concepts to American options will be described in Section 2.5.

2.1   Stopping time

The buyer of an American option can exercise his or her right at any time until maturity. The decision to exercise or not at time n will be made according to the information available at time n. In a discrete-time model built on a finite filtered probability space $(\Omega, \mathscr{F}, (\mathscr{F}_n)_{0 \le n \le N}, \mathbb{P})$, the exercise date is described by a random variable called a stopping time.

Definition 2.1.1.   A random variable ν taking values in $\{0, 1, 2, \ldots, N\}$ is a stopping time if, for any $n \in \{0, 1, \cdots, N\}$,

$$\{\nu = n\} \in \mathscr{F}_n.$$

Remark 2.1.2.   As in the previous chapter, we assume that $\mathscr{F} = \mathscr{P}(\Omega)$ and $\mathbb{P}(\{\omega\}) > 0, \forall \omega \in \Omega$. This hypothesis is nonetheless not essential: if it does not hold, the results presented in this chapter remain true almost surely. However, we will not assume $\mathscr{F}_0 = \{\emptyset, \Omega\}$ and $\mathscr{F}_N = \mathscr{F}$, except in Section 2.5, dedicated to finance.

Remark 2.1.3.   The reader can verify, as an exercise, that ν is a stopping time if and only if, for any $n \in 0, 1, \ldots, N$,

$$\{\nu \leq n\} \in \mathscr{F}_n.$$

We will use an equivalent definition to generalize the concept of stopping time to the continuous-time setting.

Let us introduce now the concept of a *sequence stopped at a stopping time*. Let $(X_n)_{0 \leq n \leq N}$ be a sequence adapted to the filtration $(\mathscr{F}_n)_{0, \leq n \leq N}$ and let ν be a stopping time. The sequence stopped at time ν is defined as

$$X_n^\nu(\omega) = X_{\nu(\omega) \wedge n(\omega)},$$

i.e. on the set $\nu = j$ we have

$$X_n^\nu = \begin{cases} X_j \text{ if } j \leq n \\ X_n \text{ if } j > n. \end{cases}$$

Note that $X_N^\nu(\omega) = X_{\nu(\omega)}(\omega) (= X_j$ on $\{\nu = j\})$.

Proposition 2.1.4. *Let (X_n) be an adapted sequence and ν be a stopping time. The stopped sequence $(X_n^\nu)_{0 \leq n \leq N}$ is adapted. Moreover, if (X_n) is a martingale (resp. a supermartingale), then (X_n^ν) is a martingale (resp. a supermartingale).*

Proof. We see that, for $n \geq 1$, we have

$$X_{\nu \wedge n} = X_0 + \sum_{j=1}^{n} \phi_j (X_j - X_{j-1}),$$

where $\phi_j = \mathbf{1}_{\{j \leq \nu\}}$. Since $\{j \leq \nu\}$ is the complement of the set $\{\nu < j\} = \{\nu \leq j-1\}$, the process $(\phi_n)_{0 \leq n \leq N}$ is predictable.

It is clear then that $(X_{\nu \wedge n})_{0 \leq n \leq N}$ is adapted to the filtration $(\mathscr{F}_n)_{0 \leq n \leq N}$. Furthermore, if (X_n) is a martingale, $(X_{\nu \wedge n})$ is also a martingale with respect to $(\mathscr{F}_n)$, since it is the martingale transform of (X_n). Similarly, we can show that if the sequence (X_n) is a supermartingale (resp. a submartingale), the stopped sequence is still a supermartingale (resp. a submartingale) using the predictability and the non-negativity of $(\phi_j)_{0 \leq j \leq N}$. $\square$

2.2   The Snell envelope

In this section, we consider an adapted sequence $(Z_n)_{0 \leq n \leq N}$ and define the sequence $(U_n)_{0 \leq n \leq N}$ as follows:

$$\begin{cases} U_N = Z_N \\ U_n = \max(Z_n, \mathbb{E}(U_{n+1} | \mathscr{F}_n)), \quad n = 0, \ldots, N-1. \end{cases}$$

The study of this sequence is motivated by our first approach of American options (see Section 1.3.3). We already know, by Proposition 1.3.6, that $(U_n)_{0\le n\le N}$ is the smallest supermartingale that dominates the sequence $(Z_n)_{0\le n\le N}$. We call it the Snell envelope of the sequence $(Z_n)_{0\le n\le N}$.

By definition, U_n is greater than Z_n (with equality for $n=N$) and in the case of a strict inequality, $U_n = \mathbb{E}(U_{n+1}|\mathscr{F}_n)$. This suggests that, by stopping adequately the sequence (U_n), it is possible to obtain a martingale, as the following proposition shows.

Proposition 2.2.1. *The random variable defined by*

$$\nu_0 = \inf\{n \ge 0 \,|\, U_n = Z_n\} \tag{2.1}$$

is a stopping time and the stopped sequence $(U_{n\wedge\nu_0})_{0\le n\le N}$ is a martingale.

Proof. Since $U_N = Z_N$, ν_0 is a well-defined element of $\{0, 1, \dots, N\}$ and we have

$$\{\nu_0 = 0\} = \{U_0 = Z_0\} \in \mathscr{F}_0,$$

and, for $k \ge 1$,

$$\{\nu_0 = k\} = \{U_0 > Z_0\} \cap \cdots \cap \{U_{k-1} > Z_{k-1}\} \cap \{U_k = Z_k\} \in \mathscr{F}_k.$$

To demonstrate that $(U_n^{\nu_0})$ is a martingale, we write as in the proof of Proposition 2.1.4:

$$U_n^{\nu_0} = U_{n\wedge\nu_0} = U_0 + \sum_{j=1}^n \phi_j \Delta U_j,$$

where $\phi_j = \mathbf{1}_{\{\nu_0 \ge j\}}$. So that, for $n \in \{0, 1, \dots, N-1\}$,

$$U_{n+1}^{\nu_0} - U_n^{\nu_0} = \phi_{n+1}(U_{n+1} - U_n)$$
$$= \mathbf{1}_{\{n+1\le\nu_0\}}(U_{n+1} - U_n).$$

By definition, $U_n = \max(Z_n, \mathbb{E}(U_{n+1}|\mathscr{F}_n))$ and on the set $\{n+1 \le \nu_0\}$, $U_n > Z_n$. Consequently

$$U_n = \mathbb{E}(U_{n+1}|\mathscr{F}_n)$$

and we deduce

$$U_{n+1}^{\nu_0} - U_n^{\nu_0} = \mathbf{1}_{\{n+1\le\nu_0\}}(U_{n+1} - \mathbb{E}(U_{n+1}|F_n))$$

and taking the conditional expectation on both sides of the equality

$$\mathbb{E}((U_{n+1}^{\nu_0} - U_n^{\nu_0})|\mathscr{F}_n) = \mathbf{1}_{\{n+1\le\nu_0\}}\mathbb{E}((U_{n+1} - \mathbb{E}(U_{n+1}|\mathscr{F}_n))|\mathscr{F}_n)$$

because $\{n+1 \le \nu_0\} \in \mathscr{F}_n$ (since the complement of $\{n+1 \le \nu_0\}$ is $\{\nu_0 \le n\}$). Hence

$$\mathbb{E}((U_{n+1}^{\nu_0} - U_n^{\nu_0})|\mathscr{F}_n) = 0$$

which proves that U^{ν_0} is a martingale.   □

In the remainder, we shall denote by $\mathcal{T}_{n,N}$ the set of stopping times taking values in $\{n, n+1, \ldots, N\}$. Notice that $\mathcal{T}_{n,N}$ is a finite set since Ω is assumed to be finite. The martingale property of the sequence U^{ν_0} gives the following result, which relates the concept of Snell envelope to the optimal stopping problem.

Corollary 2.2.2.   *The stopping time ν_0 satisfies*

$$U_0 = \mathbb{E}(Z_{\nu_0}|\mathcal{F}_0) = \sup_{\nu \in \mathcal{T}_{0,N}} \mathbb{E}(Z_\nu|\mathcal{F}_0).$$

If we think of Z_n as the total winnings of a gambler after n games, we see that stopping at time ν_0 maximises the expected gain given $\mathcal{F}_0$.

Proof.   Since U^{ν_0} is a martingale, we have

$$U_0 = U_0^{\nu_0} = \mathbb{E}(U_N^{\nu_0}|\mathcal{F}_0) = \mathbb{E}(U_{\nu_0}|\mathcal{F}_0) = \mathbb{E}(Z_{\nu_0}|\mathcal{F}_0).$$

On the other hand, if $\nu \in \mathcal{T}_{0,N}$, the stopped sequence U^ν is a supermartingale, so that

$$\begin{aligned} U_0 &\geq \mathbb{E}(U_N^\nu|\mathcal{F}_0) = \mathbb{E}(U_\nu|\mathcal{F}_0) \\ &\geq \mathbb{E}(Z_\nu|\mathcal{F}_0), \end{aligned}$$

which yields the result.   □

Remark 2.2.3.   An immediate generalization of Corollary 2.2.2 gives

$$\begin{aligned} U_n &= \sup_{\nu \in \mathcal{T}_{n,N}} \mathbb{E}(Z_\nu|\mathcal{F}_n) \\ &= \mathbb{E}(Z_{\nu_n}|\mathcal{F}_n), \end{aligned}$$

where $\nu_n = \inf\{j \geq n|U_j = Z_j\}$.

Definition 2.2.4.   A stopping time ν^* is called optimal for the sequence $(Z_n)_{0 \leq n \leq N}$ if

$$\mathbb{E}(Z_{\nu^*}|\mathcal{F}_0) = \sup_{\nu \in \mathcal{T}_{0,N}} \mathbb{E}(Z_\nu|\mathcal{F}_0).$$

We can see that ν_0 is optimal. The following result gives a characterization of optimal stopping times that shows that ν_0 is the smallest optimal stopping time.

Theorem 2.2.5.   *A stopping time ν is optimal if and only if*

$$\begin{cases} Z_\nu = U_\nu \\ and \ (U_{\nu \wedge n})_{0 \leq n \leq N} \ is \ a \ martingale. \end{cases} \tag{2.2}$$

Proof. If the stopped sequence U^ν is a martingale, $U_0 = \mathbb{E}(U_\nu|\mathscr{F}_0)$ and consequently, if (2.2) holds, $U_0 = \mathbb{E}(Z_\nu|\mathscr{F}_0)$. Optimality of ν is then ensured by Corollary 2.2.2.

Conversely, if ν is optimal, we have

$$U_0 = \mathbb{E}(Z_\nu|\mathscr{F}_0) \leq \mathbb{E}(U_\nu|\mathscr{F}_0).$$

But, since U^ν is a supermatingale,

$$\mathbb{E}(U_\nu|\mathscr{F}_0) \leq U_0.$$

Therefore

$$\mathbb{E}(U_\nu|\mathscr{F}_0) = \mathbb{E}(Z_\nu|\mathscr{F}_0)$$

and since $U_\nu \geq Z_\nu, U_\nu = Z_\nu$.

Since $\mathbb{E}(U_\nu|\mathscr{F}_0) = U_0$ and from the inequalities

$$U_0 \geq \mathbb{E}(U_{\nu \wedge n}|\mathscr{F}_0) \geq \mathbb{E}(U_\nu|\mathscr{F}_0)$$

(based on the supermartingale property of (U_n^ν)) we get

$$\mathbb{E}(U_{\nu \wedge n}|\mathscr{F}_0) = \mathbb{E}(U_\nu|\mathscr{F}_0) = \mathbb{E}(\mathbb{E}(U_\nu|\mathscr{F}_n)|\mathscr{F}_0).$$

But we have $U_{\nu \wedge n} \geq \mathbb{E}(U_\nu|\mathscr{F}_n)$, therefore $U_{\nu \wedge n} = \mathbb{E}(U_\nu|\mathscr{F}_n)$, which proves that (U_n^ν) is a martingale. $\square$

2.3 Decomposition of supermartingales

The following decomposition (commonly called the 'Doob decomposition') is used in viable complete market models to associate any supermartingale with a trading strategy for which consumption is allowed (see Exercise 8 for that matter).

Proposition 2.3.1. *Every supermartingale* $(U_n)_{0 \leq n \leq N}$ *has the unique following decomposition:*

$$U_n = M_n - A_n,$$

where (M_n) *is a martingale and* (A_n) *is a non-decreasing, predictable process, null at 0.*

Proof. It is clearly seen that the only solution for $n = 0$ is $M_0 = U_0$ and $A_0 = 0$. Then we must have

$$U_{n+1} - U_n = M_{n+1} - M_n - (A_{n+1} - A_n),$$

so that, conditioning both sides with respect to $\mathscr{F}_n$ and using the properties of M and A,

$$-(A_{n+1} - A_n) = \mathbb{E}(U_{n+1}|\mathscr{F}_n) - U_n$$

and
$$M_{n+1} - M_n = U_{n+1} - \mathbb{E}(U_{n+1}|\mathscr{F}_n).$$

(M_n) and (A_n) are entirely determined using the previous equations and we see that (M_n) is a martingale and that (A_n) is predictable and non-decreasing (because (U_n) is a supermartingale).

$\square$

Suppose now that (U_n) is the Snell envelope of an adapted sequence (Z_n). We can then give a characterization of the largest optimal stopping time for (Z_n) using the non-decreasing process (A_n) of the Doob decomposition of (U_n):

Proposition 2.3.2. *The largest optimal stopping time for (Z_n) is given by*

$$\nu_{\max} = \begin{cases} N & if A_N = 0 \\ \inf\{n, A_{n+1} \neq 0\} & if A_N \neq 0. \end{cases}$$

Proof. It is straightforward to see that $\nu_{\max}$ is a stopping time using the fact that $(A_n)_{0 \leq n \leq N}$ is predictable. From $U_n = M_n - A_n$ and because $A_j = 0$, for $j \leq \nu_{\max}$, we deduce that $U^{\nu_{\max}} = M^{\nu_{\max}}$ and conclude that $U^{\nu_{\max}}$ is a martingale. To show the opimality of $\nu_{\max}$, it is sufficient to prove

$$U_{\nu_{\max}} = Z_{\nu_{\max}}.$$

We note that

$$U_{\nu_{\max}} = \sum_{j=0}^{N-1} \mathbf{1}_{\{\nu_{\max}=j\}} U_j + \mathbf{1}_{\{\nu_{\max}=N\}} U_N$$

$$= \sum_{j=0}^{N-1} \mathbf{1}_{\{\nu_{\max}=j\}} \max(Z_j, \mathbb{E}(U_{j+1}|\mathscr{F}_j)) + \mathbf{1}_{\{\nu_{\max}=N\}} Z_N,$$

We have $\mathbb{E}(U_{j+1}|\mathscr{F}_j) = M_j - A_{j+1}$ and, on the set $\{\nu_{\max} = j\}$, $A_j = 0$ and $A_{j+1} > 0$, so $U_j = M_j$ and $\mathbb{E}(U_{j+1}|\mathscr{F}_j) = M_j - A_{j+1} < U_j$. It follows that $U_j = \max(Z_j, \mathbb{E}(U_{j+1}|\mathscr{F}_j)) = Z_j$, so that finally

$$U_{\nu_{\max}} = Z_{\nu_{\max}}.$$

It remains to show that it is the greatest optimal stopping time. If ν is a stopping time such that $\nu \geq \nu_{\max}$ and $\mathbb{P}(\nu > \nu_{\max}) > 0$, then

$$\mathbb{E}(U_\nu) = \mathbb{E}(M_\nu) - \mathbb{E}(A_\nu) = \mathbb{E}(U_0) - \mathbb{E}(A_\nu) < \mathbb{E}(U_0)$$

and U^ν cannot be a martingale, which establishes the claim. $\square$

2.4   Snell envelope and Markov chains

The aim of this section is to compute Snell envelopes in a Markovian setting. A sequence $(X_n)_{n \geq 0}$ of random variables taking their values in a finite set E is called a Markov chain if, for any integer $n \geq 1$ and any elements $x_0, x_1, \ldots, x_{n-1}, x, y$ of E, we have

$$\mathbb{P}(X_{n+1} = y \mid X_0 = x_0, \ldots, X_{n-1} = x_{n-1}, X_n = x) = \mathbb{P}(X_{n+1} = y \mid X_n = x).$$

The chain is said to be homogeneous if the value $P(x, y) = \mathbb{P}(X_{n+1} = y \mid X_n = x)$ does not depend on n. The matrix $P = (P(x, y))_{(x,y) \in E \times E}$, indexed by $E \times E$, is then called the transition matrix of the chain. The matrix P has non-negative entries and satisfies $\sum_{y \in E} P(x, y) = 1$ for all $x \in E$; it is said to be a stochastic matrix. On a filtered probability space $(\Omega, \mathscr{F}, (\mathscr{F}_n)_{0 \leq n \leq N}, \mathbb{P})$, we can define the notion of a Markov chain with respect to the filtration:

Definition 2.4.1. A sequence $(X_n)_{0 \leq n \leq N}$ of random variables taking values in E is a homogeneous Markov chain with respect to the filtration $(\mathscr{F}_n)_{0 \leq n \leq N}$, with transition matrix P, if (X_n) is adapted and if for any real-valued function f on E, we have

$$\mathbb{E}(f(X_{n+1}) \mid \mathscr{F}_n) = Pf(X_n),$$

where Pf represents the function that maps $x \in E$ to

$$Pf(x) = \sum_{y \in E} P(x, y) f(y).$$

Note that, if one interprets real-valued functions on E as matrices with a single column indexed by E, then Pf is indeed the product of the two matrices P and f. It can also be easily seeen that a Markov chain, as defined at the beginning of the section, is a Markov chain with respect to its natural filtration, defined by $\mathscr{F}_n = \sigma(X_0, \ldots, X_n)$.

The following proposition is an immediate consequence of the latter definition and the definition of a Snell envelope. It is the basis for the effective computation of American option prices in discrete models (see Exercise 7).

Proposition 2.4.2. Let (Z_n) be an adapted sequence defined by $Z_n = \psi(n, X_n)$, where (X_n) is a homogeneous Markov chain with transition matrix P, taking values in E, and ψ is a function from $\mathbb{N} \times E$ to $\mathbb{R}$. Then, the Snell envelope (U_n) of the sequence (Z_n) is given by $U_n = u(n, X_n)$, where the function u is defined by

$$u(N, x) = \psi(N, x) \quad \forall x \in E$$

and, for $n \leq N - 1$,

$$u(n, \cdot) = \max(\psi(n, \cdot), Pu(n + 1, \cdot)).$$

2.5 Application to American options

>From now on, we will work in a viable complete market. The modeling will be based on the filtered space $(\Omega, \mathscr{F}, (\mathscr{F}_n)_{0 \leq n \leq N}, \mathbb{P})$ and, as in Sections 1.3.1

and 1.3.3, we will denote by $\mathbb{P}^*$ the unique probability under which the discounted asset prices are martingales.

2.5.1 Hedging American options

In Section 1.3.3, we defined the value process (U_n) of an American option described by the sequence (Z_n) by the system

$$\begin{cases} U_N = Z_N \\ U_n = \max(Z_n, S_n^0 \mathbb{E}^*(U_{n+1}/S_{n+1}^0|\mathscr{F}_n)) & \forall n \leq N-1. \end{cases}$$

Thus, the sequence $(\tilde{U}_n)$ defined by $\tilde{U}_n = U_n/S_n^0$ (discounted price of the option) is the Snell envelope, under $\mathbb{P}^*$, of the sequence $(\tilde{Z}_n)$. We deduce from the above Section 2.2 that

$$\tilde{U}_n = \sup_{\nu \in \mathscr{T}_{n,N}} \mathbb{E}^*(\tilde{Z}_\nu|\mathscr{F}_n)$$

and consequently

$$U_n = S_n^0 \sup_{\nu \in \mathscr{T}_{n,N}} \mathbb{E}^*\left(\frac{Z_\nu}{S_\nu^0}|\mathscr{F}_n\right).$$

>From Section 2.3, we can write

$$\tilde{U}_n = \tilde{M}_n - \tilde{A}_n,$$

where $\tilde{M}_n$ is a $\mathbb{P}^*$-martingale and $(\tilde{A}_n)$ is an increasing predictable process, null at 0. Since the market is complete, there is a self-financing strategy ϕ such that

$$V_N(\phi) = S_N^0 \tilde{M}_N,$$

i.e. $\tilde{V}_N(\phi) = \tilde{M}_N$. Since the sequence $(\tilde{V}_n(\phi))$ is a $\mathbb{P}^*$-martingale, we have

$$\begin{aligned} \tilde{V}_n(\phi) &= \mathbb{E}^*(\tilde{V}_N(\phi)|\mathscr{F}_n) \\ &= \mathbb{E}^*(\tilde{M}_N|\mathscr{F}_n) \\ &= \tilde{M}_n, \end{aligned}$$

and consequently

$$\tilde{U}_n = \tilde{V}_n(\phi) - \tilde{A}_n.$$

Therefore

$$U_n = V_n(\phi) - A_n,$$

where $A_n = S_n^0 \tilde{A}_n$. From the previous equality, it is obvious that the writer of the option can hedge himself perfectly: once he receives the premium $U_0 = V_0(\phi)$, he can generate a wealth equal to $V_n(\phi)$ at time n which is bigger than U_n and *a fortiori* Z_n.

What is the optimal date to exercise the option? The date of exercise is to be chosen among all the stopping times. For the buyer of the option, there is no point in exercising at time n when $U_n > Z_n$, because he would trade an asset worth U_n (the option) for an amount (Z_n) (by exercising the option). Thus an optimal date τ of exercise is such that $U_\tau = Z_\tau$. On the other hand, there is no point in exercising after the time

$$\nu_{\max} = \inf\{j, A_{j+1} \neq 0\}$$

(which is equal to $\inf \{j, \tilde{A}_{j+1} \neq 0\}$) because, at that time, selling the option provides the holder with a wealth $U_{\nu_{\max}} = V_{\nu_{\max}}(\phi)$ and, following the strategy ϕ from that time, he creates a portfolio whose value is strictly bigger than the option's at times $\nu_{\max} + 1, \nu_{\max} + 2, \ldots, N$. Therefore we set, as a second condition, $\tau \leq \nu_{\max}$, which allows us to say that U^τ is a martingale. As a result, optimal dates of exercise are optimal stopping times for the sequence $(\tilde{Z}_n)$, under probability $\mathbb{P}^*$. To make this point clear, let us consider the writer's point of view. If he hedges himself using the strategy ϕ as defined above, and if the buyer exercises at a non-optimal time τ, then $U_\tau > Z_\tau$ or $A_\tau > 0$. In both cases, the writer makes a profit $V_\tau(\phi) - Z_\tau = U_\tau + A_\tau - Z_\tau$, which is positive.

2.5.2   American options and European options

Proposition 2.5.1. *Let C_n be the value at time n of an American option described by an adapted sequence $(Z_n)_{0 \leq n \leq N}$ and let c_n be the value at time n of the European option defined by the $\mathscr{F}_N$-measurable random variable $h = Z_N$. Then, we have $C_n \geq c_n$.*
   Moreover, if $c_n \geq Z_n$ for any n, then

$$c_n = C_n \quad \forall n \in \{0, 1, \ldots, N\}.$$

The inequality $C_n \geq c_n$ makes sense since the American option entitles the holder to more rights than its European counterpart.

Proof.   Since the discounted value $(\tilde{C}_n)$ is a supermartingale under $\mathbb{P}^*$, we have
$$\tilde{C}_n \geq \mathbb{E}^*(\tilde{C}_N | \mathscr{F}_n) = \mathbb{E}^*(\tilde{c}_N | \mathscr{F}_n) = \tilde{c}_n.$$
Hence $C_n \geq c_n$.
   If $c_n \geq Z_n$ for any n, then the sequence $(\tilde{c}_n)$, which is a martingale under $\mathbb{P}^*$, appears to be a supermartingale (under $\mathbb{P}^*$) and an upper bound for the sequence $(\tilde{Z}_n)$ and consequently

$$\tilde{C}_n \leq \tilde{c}_n \quad \forall n \in \{0, 1, \ldots, N\}.$$

$\square$

Remark 2.5.2. One checks readily that if the relationships of Proposition 2.5.1 did not hold, there would be some arbitrage opportunities by trading the options.

To illustrate the last proposition, let us consider the case of a market with a single risky asset, with price S_n at time n and a constant riskless interest rate, equal to $r \geq 0$ on each period, so that $S_n^0 = (1+r)^n$. Then, with the notations of Proposition 2.5.1, if we take $Z_n = (S_n - K)_+$, c_n is the price at time n of a European call with maturity N and strike price K on one unit of the risky asset and C_n is the price of the corresponding American call. We have

$$
\begin{aligned}
\tilde{c}_n &= (1+r)^{-N} \mathbb{E}^*((S_N - K)_+ | \mathscr{F}_n) \\
&\geq \mathbb{E}^*(\tilde{S}_n - K(1+r)^{-N} | \mathscr{F}_n) \\
&= \tilde{S}_n - K(1+r)^{-N},
\end{aligned}
$$

using the martingale property of $(\tilde{S}_n)$. Hence, $c_n \geq S_n - K(1+r)^{-(N-n)} \geq S_n - K$, for $r \geq 0$. As $c_n \geq 0$, we also have $c_n \geq (S_n - K)_+$ and, by Proposition 2.5.1, $C_n = c_n$. There is equality between the price of the European call and the price of the corresponding American call.

This property does not hold for the put, nor in the case of calls on currencies or dividend paying stocks.

Notes: For further discussions on the Snell envelope and optimal stopping, one may consult Neveu (1972), Chapter VI, and Dacunha-Castelle and Duflo (1986a), Chapter 5, Section 1. For the theory of optimal stopping in continuous time, see El Karoui (1981), Shiryayev (1978) and Peskir and Shiryaev (2006).

2.6   Exercises

Exercise 4   Let ν be a stopping time with respect to a filtration $(\mathscr{F}_n)_{0 \leq n \leq N}$. We denote by $\mathscr{F}_\nu$ the set of events A such that $A \cap \{\nu = n\} \in \mathscr{F}_n$ for any $n \in \{0, \dots, N\}$.

1. Show that $\mathscr{F}_\nu$ is a sub-σ-algebra of $\mathscr{F}_N$. $\mathscr{F}_\nu$ is often called 'σ-algebra of events determined prior to the stopping time ν'.

2. Show that the random variable ν is $\mathscr{F}_\nu$-measurable.

3. Let X be a real-valued random variable. Prove the equality

$$
\mathbb{E}(X | \mathscr{F}_\nu) = \sum_{j=0}^{N} \mathbf{1}_{\{\nu = j\}} \mathbb{E}(X | \mathscr{F}_j).
$$

4. Let τ be a stopping time such that $\tau \geq \nu$. Show that $\mathscr{F}_\nu \subset \mathscr{F}_\tau$.

5. Under the same hypothesis, show that if (M_n) is a martingale, we have

$$
M_\nu = \mathbb{E}(M_\tau | \mathscr{F}_\nu).
$$

(Hint: first consider the case $\tau = N$.)

Exercise 5 Let (U_n) be the Snell envelope of an adapted sequence (Z_n). Without assuming that $\mathcal{F}_0$ is trivial, show that

$$\mathbb{E}(U_0) = \sup_{\nu \in \mathcal{T}_{0,N}} \mathbb{E}(Z_\nu),$$

and more generally

$$\mathbb{E}(U_n) = \sup_{\nu \in \mathcal{T}_{n,N}} \mathbb{E}(Z_\nu).$$

Exercise 6 Show that ν is optimal according to Definition 2.2.4 if and only if

$$\mathbb{E}(Z_\nu) = \sup_{\tau \in \mathcal{F}_{0,N}} \mathbb{E}(Z_\tau).$$

Exercise 7 The purpose of this exercise is to study the American put in the model of Cox-Ross-Rubinstein. Notations are those of Chapter 1.

1. Show that the price $\mathcal{P}_n$, at time n, of an American put on a share with maturity N and strike price K can be writen as

$$\mathcal{P}_n = P_{am}(n, S_n),$$

where $P_{am}(n,x)$ is defined by $P_{am}(N,x) = (K-x)_+$ and, for $n \leq N-1$,

$$P_{am}(n,x) = \max\left((K-x)_+, \frac{f(n+1,x)}{1+r}\right),$$

with

$$f(n+1,x) = pP_{am}(n+1,xd) + (1-p)P_{am}(n+1,xu)$$

and $p = (u-1-r)/(u-d)$.

2. Show that the function $P_{am}(0,.)$ can be expressed as

$$P_{am}(0,x) = \sup_{\nu, \in \mathcal{T}_{0,N}} \mathbb{E}^*((1+r)^\tau (K - xV_\nu)_+),$$

where the sequence of random variables $(V_n)_{0 \leq n \leq N}$ is defined by $V_0 = 1$ and for $n \geq 1, V_n = \prod_{i=1}^n U_i$, where the U_i's are some random variables. Give their joint distribution under $\mathbb{P}^*$.

3. From the last formula, show that the function $x \mapsto P_{am}(0,x)$ is convex and non-increasing.

4. We assume $d < 1$. Show that there is a real number $x^* \in [0, K]$ such that, for $x \leq x^*$, $P_{am}(0,x) = (K-x)_+$ and, for $x^* < x < K/d^N$,

$$P_{am}(0,x) > (K-x)_+.$$

5. An agent holds the American put at time 0. For which values of the spot S_0 would he rather exercise his option immediately?

6. Show that the hedging strategy of the American put is determined by a quantity $H_n = \Delta(n, S_{n-1})$ of the risky asset to be held at time n, where Δ can be written as a function of P_{am}.

Exercise 8 Consumption strategies. The self-financing strategies defined in Chapter 1 ruled out any consumption. Consumption strategies can be introduced in the following way: at time n, once the new prices $S_n^0, \ldots, S_n^d$ are quoted, the investor readjusts his positions from ϕ_n to ϕ_{n+1} and selects the wealth γ_{n+1} to be consumed at time $n+1$. With any endowment being excluded and the new positions being decided given prices at time n, we deduce

$$\phi_{n+1}.S_n = \phi_n.S_n - \gamma_{n+1}. \tag{2.3}$$

So, a trading strategy with consumption will be defined as a pair (ϕ, γ), where ϕ is a predictable process taking values in $\mathbb{R}^{d+1}$, representing the numbers of assets held in the portfolio, and $\gamma = (\gamma_n)_{1 \le n \le N}$ is a predictable process taking values in $\mathbb{R}^+$, representing the wealth consumed at any time. Equation (2.3) gives the relationship between the processes ϕ and γ and replaces the self-financing condition of Chapter 1.

1. Let ϕ be a predictable process taking values in $\mathbb{R}^{d+1}$ and let γ be a predictable process taking values in $\mathbb{R}^+$. We set $V_n(\phi) = \phi_n.S_n$ and $\tilde{V}_n(\phi) = \phi_n.\tilde{S}_n$. Show the equivalence between the following conditions:

   (a) The pair (ϕ, γ) defines a trading strategy with consumption.

   (b) For any $n \in \{1, \ldots, N\}$,

   $$V_n(\phi) = V_0(\phi) + \sum_{j=1}^{n} \phi_j.\Delta S_j - \sum_{j=1}^{n} \gamma_j.$$

   (c) For any $n \in \{1, \ldots, N\}$,

   $$\tilde{V}_n(\phi) = V_0(\phi) + \sum_{j=1}^{n} \phi_j.\Delta \tilde{S}_j - \sum_{j=1}^{n} \gamma_j/S_{j-1}^0.$$

2. In the remainder, we assume that the market is viable and complete and we denote by $\mathbb{P}^*$ the unique probability under which the discounted asset prices are martingales. Show that if the pair (ϕ, γ) defines a trading strategy with consumption, then $(\tilde{V}_n(\phi))$ is a supermartingale under $\mathbb{P}^*$.

3. Let (U_n) be an adapted sequence such that $(\tilde{U}_n)$ is a supermartingale under $\mathbb{P}^*$. Using the Doob decomposition, show that there is a trading strategy with consumption (ϕ, γ) such that $V_n(\phi) = U_n$ for any $n \in \{0, \ldots, N\}$.

4. Let (Z_n) be an adapted sequence. We say that a trading strategy with consumption (ϕ, γ) hedges the American option defined by (Z_n) if $V_n(\phi) \geq Z_n$ for any $n \in \{0, 1, \ldots, N\}$. Show that there is at least one trading strategy with consumption that hedges (Z_n), whose value is precisely the value (U_n) of the American option. Also, prove that any trading strategy with consumption (ϕ, γ) hedging (Z_n) satisfies $V_n(\phi) \geq U_n$, for any $n \in \{0, 1, \ldots, N\}$.

5. Let x be a non-negative number representing the investor's endowment and let $\gamma = (\gamma_n)_{1 \leq n \leq N}$ be a predictable strategy taking values in $\mathbb{R}^+$. The consumption process (γ_n) is said to be budget-feasible from endowment x if there is a predictable process ϕ taking values in $\mathbb{R}^{d+1}$, such that the pair (ϕ, γ) defines a trading strategy with consumption satisfying $V_0(\phi) = x$ and $V_n(\phi) \geq 0$, for any $n \in \{0, \ldots, N\}$. Show that (γ_n) is budget-feasible from endowment x if and only if $\mathbb{E}^* \left(\sum_{j=1}^{N} \gamma_j / S_{j-1}^0 \right) \leq x$.

Chapter 3

Brownian motion and stochastic differential equations

The first two chapters of this book dealt with discrete-time models. We had the opportunity to see the importance of the concepts of martingales, self-financing strategy and the Snell envelope. In this chapter, we are going to elaborate on these ideas in a continuous-time framework. In particular, we shall introduce the mathematical tools needed to model financial assets and to price options. In continuous-time, the technical aspects are more advanced and more difficult to handle than in discrete-time, but the main ideas are fundamentally the same.

Why do we consider continuous-time models? The primary motivation comes from the nature of the processes that we want to model. In practice, the price changes in the market are actually so frequent that a discrete-time model can barely follow the moves. On the other hand, continuous-time models lead to more explicit computations, even if numerical methods are sometimes required. Indeed, the most widely used model is the continuous-time Black-Scholes model, which leads to an extremely simple formula. As we mentioned in the Introduction, the connections between stochastic processes and finance are not recent. Bachelier (1900), in his dissertation called *Théorie de la spéculation*, is not only among the first to look at the properties of Brownian motion, but he also derived option pricing formulae.

We will be giving a few mathematical definitions in order to understand continuous-time models. In particular, we shall define the Brownian motion since it is the core concept of the Black-Scholes model and appears in most financial asset models. Then we shall state the concept of martingales in a continuous-time set up, and, finally, we shall construct the stochastic integral and introduce the differential calculus associated with it, namely the Itô calculus.

It is advisable that, upon first reading, the reader passes over the proofs in small print, as they are rather technical.

3.1   General comments on continuous-time processes

What do we exactly mean by *continuous-time processes?*

Definition 3.1.1. Let $(E, \mathscr{E})$ be a measurable space. A continuous-time stochastic process with state space $(E, \mathscr{E})$ is a family $(X_t)_{t \in \mathbb{R}^+}$ of random variables defined on a probability space $(\Omega, \mathscr{A}, \mathbb{P})$ with values in $(E, \mathscr{E})$.

Remark 3.1.2.

- In practice, the index t stands for the time.

- A process can also be considered as a random map: for each ω in Ω we associate the map from $\mathbb{R}^+$ to E: $t \to X_t(\omega)$, called a *path* of the process.

- A process can be considered as a map from $\mathbb{R}^+ \times \Omega$ into E. We will always assume that this map is measurable when we endow the product set $\mathbb{R}^+ \times \Omega$ with the product σ-algebra $\mathscr{B}(\mathbb{R}^+) \times \mathscr{A}$ and when the set E is endowed with $\mathscr{E}$. In other words, we will deal only with *measurable* processes (see Karatzas and Shreve (1988) for technical details).

- We will only work with processes that are indexed on a finite time interval $[0, T]$.

As in discrete-time, we introduce the concept of *filtration.*

Definition 3.1.3. A *filtration* on a probability space $(\Omega, \mathscr{A}, \mathbb{P})$ is an increasing family $(\mathscr{F}_t)_{t \geq 0}$ of σ-algebras included in $\mathscr{A}$.

The σ-algebra $\mathscr{F}_t$ represents the information available at time t. We say that a process $(X_t)_{t \geq 0}$ is *adapted* to $(\mathscr{F}_t)_{t \geq 0}$, if, for any t, X_t is $\mathscr{F}_t$-measurable.

Remark 3.1.4. >From now on, we will be working with filtrations that have the following property:

$$\text{If } A' \subset A \in \mathscr{A} \text{ and if } \mathbb{P}(A) = 0, \text{ then for any } t, A' \in \mathscr{F}_t.$$

In other words $\mathscr{F}_t$ contains all the $\mathbb{P}$-null sets of $\mathscr{A}$. A useful consequence of this technical assumption is that if $X = Y$ $\mathbb{P}$ a.s. and Y is $\mathscr{F}_t$-measurable, then X is also $\mathscr{F}_t$-measurable.

We can build a filtration generated by a process $(X_t)_{t \geq 0}$ and we write $\mathscr{F}_t = \sigma(X_s, s \leq t)$. In general, this filtration does not satisfy the previous condition. However, if we replace $\mathscr{F}_t$ by $\tilde{\mathscr{F}}_t$, which is the σ-algebra generated by both $\mathscr{F}_t$ and $\mathscr{N}$ (the σ-algebra generated by all the $\mathbb{P}$-null sets of $\mathscr{A}$), we obtain a proper filtration satisfying the desired condition. We call it the *natural filtration* of the process $(X_t)_{t\geq 0}$. When we talk about a filtration without mentioning anything, it is assumed that we are dealing with the natural filtration of the process that we are considering. Obviously, a process is adapted to its natural filtration.

As in discrete-time, the concept of *stopping time* will be useful. A stopping time is a random time that depends on the underlying process in a non-anticipative way. In other words, at a given time t, we *know* if the stopping time is less than or equal to t. Formally, the definition is the following:

Definition 3.1.5. A stopping time with respect to the filtration $(\mathscr{F}_t)_{t\geq 0}$ is a random variable τ, with values in $\mathbb{R}^+ \cup \{+\infty\}$, such that for any $t \geq 0$,

$$\{\tau \leq t\} \in \mathscr{F}_t.$$

The σ-algebra associated with τ is defined as

$$\mathscr{F}_\tau = \{A \in \mathscr{A}, \text{ for any } t \geq 0, A \cap \{\tau \leq t\} \in \mathscr{F}_t\}.$$

This σ-algebra represents the information available before the random time τ. One can prove that (refer to Exercises 11, 12, 13, 14 and 17):

Proposition 3.1.6.

- If S is a stopping time, S is $\mathscr{F}_S$-measurable.

- If S is a stopping time, finite almost surely, and $(X_t)_{t\geq 0}$ is a continuous, adapted process, then X_S is $\mathscr{F}_S$-measurable.

- If S and T are two stopping times such that $S \leq T$ $\mathbb{P}$ a.s., then $\mathscr{F}_S \subset \mathscr{F}_T$.

- If S and T are two stopping times, then $S \wedge T = \inf(S,T)$ is a stopping time. In particular, if S is a stopping time and t is a deterministic time, $S \wedge t$ is a stopping time.

3.2 Brownian motion

A particularly important example of a stochastic process is the *Brownian motion*. It will be the core of most financial models, whether we consider stocks, currencies or interest rates.

Definition 3.2.1. A Brownian motion is a real-valued, continuous stochastic process $(X_t)_{t\geq 0}$, with independent and stationary increments. In other words:

- continuity: $\mathbb{P}$ a.s. the map $s \mapsto X_s(\omega)$ is continuous.

- independent increments: if $s \leq t, X_t - X_s$ is independent of

$$\mathscr{F}_s = \sigma(X_u, u \leq s).$$

- stationary increments: if $s \leq t, X_t - X_s$ and $X_{t-s} - X_0$ have the same probability law.

This definition induces the distribution of the process X_t, but the result is difficult to prove and the reader ought to consult the book by Gikhman and Skorokhod (1969) for a proof of the following theorem.

Theorem 3.2.2. *If $(X_t)_{t \geq 0}$ is a Brownian motion, then $X_t - X_0$ is a normal random variable with mean rt and variance $\sigma^2 t$, where r and σ are constant real numbers.*

Remark 3.2.3. A Brownian motion is *standard* if

$$X_0 = 0 \; \mathbb{P} \text{ a.s.} \qquad \mathbb{E}(X_t) = 0, \qquad \mathbb{E}(X_t^2) = t.$$

>From now on, a Brownian motion is assumed to be standard if nothing else is mentioned. In that case, the distribution of X_t is the following:

$$\frac{1}{\sqrt{2\pi t}} \exp\left(-\frac{x^2}{2t}\right) dx,$$

where dx is the Lebesgue measure on $\mathbb{R}$.

The following theorem emphasises the Gaussian property of the Brownian motion. We have just seen that for any t, X_t is a normal random variable. A stronger result is the following:

Theorem 3.2.4. *If $(X_t)_{t \geq 0}$ is a Brownian motion and if $0 \leq t_1 < \cdots < t_n$, then $(X_{t_1}, \ldots, X_{t_n})$ is a Gaussian vector.*

The reader ought to consult the Appendix, page 237, to recall some properties of Gaussian vectors.

Proof. Consider $0 \leq t_1 < \cdots < t_n$. Then the random vector

$$(X_{t_1}, X_{t_2} - X_{t_1}, \ldots, X_{t_n} - X_{t_{n-1}})$$

is composed of normal, independent random variables (by Theorem 3.2.2 and by definition of the Brownian motion). Therefore, this vector is Gaussian and so is $(X_{t_1}, \ldots, X_{t_n})$. □

We shall also need a definition of a Brownian motion with respect to a filtration $(\mathscr{F}_t)$.

Definition 3.2.5. A real-valued, continuous stochastic process is an $(\mathscr{F}_t)$-Brownian motion if it satisfies:

- For any $t \geq 0, X_t$ is $\mathscr{F}_t$-measurable.

- If $s \leq t, X_t - X_s$ is independent of the σ-algebra $\mathscr{F}_s$.

- If $s \leq t, X_t - X_s$ and $X_{t-s} - X_0$ have the same law.

Remark 3.2.6. The first point of this definition shows that $\sigma(X_u, u \leq t) \subset \mathscr{F}_t$. Moreover, it is easy to check that an $\mathscr{F}_t$-Brownian motion is also a Brownian motion with respect to its natural filtration.

3.3 Continuous-time martingales

As in discrete-time models, the concept of martingales is a crucial tool to explain the notion of arbitrage. The following definition is an extension of the one in discrete-time.

Definition 3.3.1. Let us consider a probability space $(\Omega, \mathscr{A}, \mathbb{P})$ and a filtration $(\mathscr{F}_t)_{t \geq 0}$ on this space. An adapted family $(M_t)_{t \geq 0}$ of integrable random variables, i.e. $\mathbb{E}(|M_t|) < +\infty$, for any t is

- a martingale if, for any $s \leq t, \mathbb{E}(M_t|\mathscr{F}_s) = M_s$;

- a supermartingale if, for any $s \leq t, \mathbb{E}(M_t|\mathscr{F}_s) \leq M_s$;

- a submartingale if, for any $s \leq t, \mathbb{E}(M_t|\mathscr{F}_s) \geq M_s$.

Remark 3.3.2. It follows from this definition that, if $(M_t)_{t \geq 0}$ is a martingale, then $\mathbb{E}(M_t) = \mathbb{E}(M_0)$ for any t.

Here are some examples of martingales.

Proposition 3.3.3. *If $(X_t)_{t \geq 0}$ is a standard $\mathscr{F}_t$-Brownian motion, then:*

1. *X_t is an $\mathscr{F}_t$-martingale.*

2. *$X_t^2 - t$ is an $\mathscr{F}_t$-martingale.*

3. *$\exp(\sigma X_t - (\sigma^2/2)t)$ is an $\mathscr{F}_t$-martingale, for every $\sigma \in \mathbb{R}$.*

Proof. If $s \leq t$, then $X_t - X_s$ is independent of the σ-algebra $\mathscr{F}_s$. Thus $\mathbb{E}(X_t - X_s|\mathscr{F}_s) = \mathbb{E}(X_t - X_s)$. Since a standard Brownian motion has an expectation equal to zero, we have $\mathbb{E}(X_t - X_s) = 0$. Hence the first assertion is proved. To show the second one, we remark that

$$\mathbb{E}(X_t^2 - X_s^2|\mathscr{F}_s) = \mathbb{E}((X_t - X_s)^2 + 2X_s(X_t - X_s)|\mathscr{F}_s)$$
$$= \mathbb{E}((X_t - X_s)^2|\mathscr{F}_s) + 2X_s\mathbb{E}(X_t - X_s|\mathscr{F}_s),$$

and since $(X_t)_{t \geq 0}$ is a martingale, $\mathbb{E}(X_t - X_s|\mathscr{F}_s) = 0$, whence

$$\mathbb{E}(X_t^2 - X_s^2|\mathscr{F}_s) = \mathbb{E}((X_t - X_s)^2|\mathscr{F}_s).$$

Because the Brownian motion has independent and stationary increments, it follows that

$$\mathbb{E}((X_t - X_s)^2|\mathscr{F}_s) = \mathbb{E}(X_{t-s}^2)$$
$$= t - s.$$

The last equality is due to the fact that X_t has a normal distribution with mean zero and variance t. That yields $\mathbb{E}(X_t^2 - t|\mathscr{F}_s) = X_s^2 - s$, if $s < t$.

Finally, let us recall that if g is a standard normal random variable, we have

$$\mathbb{E}(e^{\lambda g}) = \int_{-\infty}^{+\infty} e^{\lambda x} e^{-x^2/2} \frac{dx}{\sqrt{2\pi}} = e^{\lambda^2/2}.$$

On the other hand, if $s < t$,

$$\mathbb{E}(e^{\sigma X_t - \sigma^2 t/2}|\mathscr{F}_s) = e^{\sigma X_s - \sigma^2 t/2}\mathbb{E}(e^{\sigma(X_t - X_s)}|\mathscr{F}_s)$$

because X_s is $\mathscr{F}_s$-measurable. Since $X_t - X_s$ is independent of $\mathscr{F}_s$, it turns out that

$$\mathbb{E}(e^{\sigma(X_t - X_s)}|\mathscr{F}_s) = \mathbb{E}(e^{\sigma(X_t - X_s)})$$
$$= \mathbb{E}(e^{\sigma X_{t-s}})$$
$$= \mathbb{E}(e^{\sigma g\sqrt{t-s}})$$
$$= \exp\left(\frac{1}{2}\sigma^2(t-s)\right).$$

This completes the proof.   □

If $(M_t)_{t\geq 0}$ is a martingale, the property $\mathbb{E}(M_t|\mathscr{F}_s) = M_s$ is also true if t and s are *bounded stopping times*. This result is actually an adaptation of Exercise 4 in Chapter 2 to the continuous case and is called the *optional sampling theorem*. We will not prove this theorem, but the reader ought to refer to Karatzas and Shreve (1988), page 19.

Theorem 3.3.4 (optional sampling theorem) *If $(M_t)_{t\geq 0}$ is a continuous martingale with respect to the filtration $(\mathscr{F}_t)_{t\geq 0}$, and if τ_1 and τ_2 are two stopping times such that $\tau_1 \leq \tau_2 \leq K$, where K is a finite real number, then M_{τ_2} is integrable and*

$$\mathbb{E}(M_{\tau_2}|\mathscr{F}_{\tau_1}) = M_{\tau_1} \quad \mathbb{P} \ a.s.$$

Remark 3.3.5.

- This result implies that if τ is a bounded stopping time, then $\mathbb{E}(M_\tau) = \mathbb{E}(M_0)$ (apply the theorem with $\tau_1 = 0, \tau_2 = \tau$ and take the expectation on both sides).

- If M_t is a submartingale, the same theorem is true if we replace the previous equality by

$$\mathbb{E}(M_{\tau_2}|\mathscr{F}_{\tau_1}) \geq M_{\tau_1} \quad \mathbb{P} \ a.s.$$

We shall now apply that result to study the properties of the *hitting time* of a point by a Brownian motion.

Proposition 3.3.6. *Let $(X_t)_{t \geq 0}$ be an $\mathscr{F}_t$-Brownian motion. For any real number a, let $T_a = \inf\{s \geq 0, X_s = a\}$ or $+\infty$ if that set is empty. Then, T_a is a stopping time, finite almost surely, and its distribution is characterized by its Laplace transform,*

$$\mathbb{E}(e^{-\lambda T_a}) = e^{-\sqrt{2\lambda}|a|}, \quad \lambda \geq 0.$$

Proof. We will assume that $a \geq 0$. First, we show that T_a is a stopping time. Indeed, since X_s is continuous,

$$\{T_a \leq t\} = \bigcap_{\varepsilon \in \mathbb{Q}^{+*}} \left\{ \sup_{s \leq t} X_s > a - \varepsilon \right\} = \bigcap_{\varepsilon \in \mathbb{Q}^{+*}} \bigcap_{s \in \mathbb{Q}^+, s \leq t} \{X_s > a - \varepsilon\}.$$

The last set belongs to $\mathscr{F}_t$, and therefore the result is proved. In the following, we write $x \wedge y = \inf(x, y)$.

Let us apply the sampling theorem to the martingale $M_t = \exp(\sigma X_t - (\sigma^2/2)t)$, with $\sigma > 0$. We cannot apply the theorem to T_a, which is not necessarily bounded; however, if n is a positive integer, $T_a \wedge n$ is a bounded stopping time (see Proposition 3.1.6), and from the optional sampling theorem,

$$\mathbb{E}(M_{T_a \wedge n}) = 1.$$

Note that $M_{T_a \wedge n} = \exp(\sigma X_{T_a \wedge n} - \sigma^2(T_a \wedge n)/2) \leq \exp(\sigma a)$. On the other hand, if $T_a < +\infty$,

$$\lim_{n \to +\infty} M_{T_a \wedge n} = M_{T_a},$$

and if $T_a = +\infty$, $X_t \leq a$ at any t, therefore $\lim_{n \to +\infty} M_{T_a \wedge n} = 0$. The Lebesgue theorem implies that

$$\mathbb{E}\left(1_{\{T_a < +\infty\}} M_{T_a}\right) = 1.$$

It follows, since $X_{T_a} = a$ when $T_a < +\infty$, that

$$\mathbb{E}\left(1_{\{T_a < +\infty\}} \exp\left(-\frac{\sigma^2}{2} T_a\right)\right) = e^{-\sigma a}.$$

By letting σ converge to 0, we get $\mathbb{P}(T_a < +\infty) = 1$ (which means that the Brownian motion reaches the level a almost surely). Also

$$\mathbb{E}\left(\exp\left(-\frac{\sigma^2}{2} T_a\right)\right) = e^{-\sigma a}.$$

The case $a < 0$ is easily solved if we notice that

$$T_a = \inf\{s \geq 0, -X_s = -a\},$$

where $(-X_t)_{t\geq 0}$ is an $\mathscr{F}_t$-Brownian motion because it is a continuous stochastic process with zero mean and variance t and with stationary, independent increments.   □

The optional sampling theorem is also very useful to compute expectations involving the running maximum of a martingale. If M_t is a square integrable martingale, we can show that the second-order moment of $\sup_{0\leq t\leq T}|M_t|$ can be bounded. This is known as the *Doob inequality*.

Theorem 3.3.7 (Doob's inequality)   *If $(M_t)_{0\leq t\leq T}$ is a continuous martingale, we have*

$$\mathbb{E}\left(\sup_{0\leq t\leq T}|M_t|^2\right) \leq 4\mathbb{E}(|M_T|^2).$$

The proof of this theorem is the purpose of Exercise 16.

3.4   Stochastic integral and Itô calculus

In a discrete-time model, if we follow a self-financing strategy, the discounted value at time n of the portfolio with initial wealth V_0 is

$$V_0 + \sum_{j=1}^{n} H_j(\tilde{S}_j - \tilde{S}_{j-1}),$$

where H_j is the number of units of the risky asset held at time j, assuming for simplicity that there is only one risky asset. That wealth appears to be a *martingale transform* under a certain probability measure such that the discounted price of the stock is a martingale. As far as continuous-time models are concerned, integrals of the form $\int_0^t H_s d\tilde{S}_s$ will help us to describe the same idea.

However, the processes modelling stock prices are normally functions of one or several Brownian motions. But one of the most important properties of a Brownian motion is that, almost surely, its paths are not differentiable at any point. In other words, if (X_t) is a Brownian motion, it can be proved that for almost every $\omega \in \Omega$, there is no time t in $\mathbb{R}^+$ such that dX_t/dt exists at t. As a result, we are not able to define the integral above as

$$\int_0^t f(s)dX_s = \int_0^t f(s)\frac{dX_s}{ds}ds.$$

Nevertheless, we are able to define this type of integral with respect to a Brownian motion, and we shall call them *stochastic integrals*. That is the whole purpose of this section.

3.4.1 Construction of the stochastic integral

Suppose that $(W_t)_{t\geq 0}$ is a standard $\mathscr{F}_t$-Brownian motion defined on a filtered probability space

$$(\Omega, \mathscr{A}, (\mathscr{F}_t)_{t\geq 0}, \mathbb{P}).$$

We are about to give a meaning to the expression $\int_0^t f(s,\omega)dW_s$ for a certain class of processes $f(s,\omega)$ adapted to the filtration $(\mathscr{F}_t)_{t\geq 0}$. To start with, we shall construct this stochastic integral for a set of processes called *simple processes*. Throughout the text, T will be a fixed strictly positive, finite real number.

Definition 3.4.1. $(H_t)_{0\leq t\leq T}$ is called a simple process if it can be written as

$$H_t(\omega) = \sum_{i=1}^p \phi_i(\omega)\mathbf{1}_{(t_{i-1},t_i]}(t),$$

where $0 = t_0 < t_1 < \cdots < t_p = T$ and ϕ_i is $\mathscr{F}_{t_{i-1}}$-measurable and bounded.

Then, by definition, the stochastic integral of a simple process H is the continuous process $(I(H)_t)_{0\leq t\leq T}$ defined for any $t \in (t_k, t_{k+1}]$ as

$$I(H)_t = \sum_{1\leq i\leq k} \phi_i(W_{t_i} - W_{t_{i-1}}) + \phi_{k+1}(W_t - W_{t_k}).$$

Note that $I(H)_t$ can be written as

$$I(H)_t = \sum_{1\leq i\leq p} \phi_i(W_{t_i\wedge t} - W_{t_{i-1}\wedge t}),$$

which proves the continuity of $t \mapsto I(H)_t$. We shall write $\int_0^t H_s dW_s$ for $I(H)_t$. The following proposition is fundamental.

Proposition 3.4.2. *If $(H_t)_{0\leq t\leq T}$ is a simple process, then:*

- $(\int_0^t H_s dW_s)_{0\leq t\leq T}$ *is a continuous $\mathscr{F}_t$-martingale.*

- $\mathbb{E}\left(\left(\int_0^t H_s dW_s\right)^2\right) = \mathbb{E}\left(\int_0^t H_s^2 ds\right).$

- $\mathbb{E}\left(\sup_{t\leq T}\left|\int_0^t H_s dW_s\right|^2\right) \leq 4\mathbb{E}\left(\int_0^T H_s^2 ds\right).$

Proof. In order to prove this proposition, we are going to use discrete-time processes. Indeed, to show that $(\int_0^t H_s dW_s)$ is a martingale, we just need to check that, for any $t > s$,

$$\mathbb{E}\left(\int_0^t H_u dW_u \middle| \mathscr{F}_s\right) = \int_0^s H_u dW_u.$$

If we include s and t in the subdivision $t_0 = 0 < t_1 < \cdots < t_p = T$, and if we call $M_n = \int_0^{t_n} H_s dW_s$ and $\mathscr{G}_n = \mathscr{F}_{t_n}$ for $0 \le n \le p$, we want to show that M_n is a $\mathscr{G}_n$-martingale. To prove it, we notice that

$$M_n = \int_0^{t_n} H_s dW_s = \sum_{i=1}^{n} \phi_i (W_{t_i} - W_{t_{i-i}})$$

with ϕ_i $\mathscr{G}_{i-1}$-measurable. Moreover, $X_n = W_{t_n}$ is a $\mathscr{G}_n$-martingale since $(W_t)_{t \ge 0}$ is a Brownian motion. $(M_n)_{n=0,\dots,p}$ turns out to be a martingale transform of $(X_n)_{n=0,\dots,p}$. Proposition 1.2.3 allows us to conclude that $(M_n)_{n=0,\dots,p}$ is a martingale. The second assertion comes from the fact that

$$\mathbb{E}(M_n^2) = \mathbb{E}\left(\left(\sum_{i=1}^{n} \phi_i (X_i - X_{i-1}) \right)^2 \right)$$

$$= \sum_{i=1}^{n} \sum_{j=1}^{n} \mathbb{E}(\phi_i \phi_j (X_i - X_{i-1})(X_j - X_{j-1})). \tag{3.1}$$

Also, if $i < j$, we have

$$\mathbb{E}(\phi_i \phi_j (X_i - X_{i-1})(X_j - X_{j-1}))$$
$$= \mathbb{E}(\mathbb{E}(\phi_i \phi_j (X_i - X_{i-1})(X_j - X_{j-1})|\mathscr{G}_{j-1}))$$
$$= \mathbb{E}(\phi_i \phi_j (X_i - X_{i-1}) \mathbb{E}(X_j - X_{j-1}|\mathscr{G}_{j-1})).$$

Since X_j is a martingale, $\mathbb{E}(X_j - X_{j-1}|\mathscr{G}_{j-1}) = 0$. Therefore, if $i < j$, $\mathbb{E}(\phi_i \phi_j (X_i - X_{i-1})(X_j - X_{j-1})) = 0$. If $j > i$, we get the same thing. Finally, if $i = j$,

$$\mathbb{E}(\phi_i^2 (X_i - X_{i-1})^2) = \mathbb{E}(\mathbb{E}(\phi_i^2 (X_i - X_{i-1})^2|\mathscr{G}_{i-1}))$$
$$= \mathbb{E}(\phi_i^2 \mathbb{E}((X_i - X_{i-1})^2|\mathscr{G}_{i-1})),$$

as a result

$$\mathbb{E}((X_i - X_{i-1})^2|\mathscr{G}_{i-1}) = E((W_{t_i} - W_{t_{i-1}})^2) = t_i - t_{i-1}. \tag{3.2}$$

From (3.1) and (3.2) we conclude that

$$E\left(\left(\sum_{i=1}^{n} \phi_i (X_i - X_{i-1}) \right)^2 \right) = \mathbb{E}\left(\sum_{i=1}^{n} \phi_i^2 (t_i - t_{i-1}) \right) = \mathbb{E}\left(\int_0^t H_s^2 ds \right).$$

The continuity of $t \to \int_0^t H_s dW_s$ is clear from the definition. The third assertion is just a consequence of Doob's inequality (cf. Theorem 3.3.7) applied to the continuous martingale $\left(\int_0^t H_s dW_s \right)_{t \ge 0}$. $\qquad \square$

Remark 3.4.3. We write by definition

$$\int_t^T H_s dW_s = \int_0^T H_s dW_s - \int_0^t H_s dW_s.$$

If $t \le T$, and if $A \in \mathscr{F}_t$, then $s \to \mathbf{1}_A \mathbf{1}_{\{t < s\}} H_s$ is still a simple process and it is easy to check from the definition of the integral that

$$\int_0^T 1_A H_s 1_{\{t<s\}} dW_s = 1_A \int_t^T H_s dW_s. \tag{3.3}$$

Now that we have defined the stochastic integral for simple processes and stated some of its properties, we are going to extend the concept to the following larger class of adapted processes $\mathcal{H}$:

$$\mathcal{H} = \left\{ (H_t)_{0 \le t \le T}, (\mathcal{F}_t)_{t \ge 0} - \text{adapted process}, \mathbb{E}\left(\int_0^T H_s^2 ds \right) < +\infty \right\}.$$

Proposition 3.4.4. *Consider* $(W_t)_{t \ge 0}$ *an* $\mathcal{F}_t$-*Brownian motion. There exists a unique linear mapping* J *from* $\mathcal{H}$ *to the space of continuous* $\mathcal{F}_t$-*martingales defined on* $[0, T]$, *such that:*

1. *If* $(H_t)_{t \le T}$ *is a simple process,* $\mathbb{P}$ *a.s. for any* $0 \le t \le T, J(H)_t = I(H)_t$.

2. *If* $t \le T, \mathbb{E}(J(H)_t^2) = \mathbb{E}\left(\int_0^t H_s^2 ds \right)$.

This linear mapping is unique in the following sense: if both J *and* J' *satisfy the previous properties, then*

$$\mathbb{P} \ a.s. \quad \forall 0 \le t \le T, J(H)_t = J'(H)_t.$$

We denote, for $H \in \mathcal{H}$, $\int_0^t H_s dW_s \doteq J(H)_t$.

On top of that, the stochastic integral satisfies the following properties:

Proposition 3.4.5. *If* $(H_t)_{0 \le t \le T}$ *belongs to* $\mathcal{H}$, *then:*

1. *We have*

$$\mathbb{E}\left(\sup_{t \le T} \left| \int_0^t H_s dW_s \right|^2 \right) \le 4\mathbb{E}\left(\int_0^T H_s^2 ds \right). \tag{3.4}$$

2. *If* τ *is an* $\mathcal{F}_t$-*stopping time,*

$$\mathbb{P} \ a.s. \int_0^\tau H_s dW_s = \int_0^T 1_{\{s \le \tau\}} H_s dW_s. \tag{3.5}$$

Proof. We shall use the fact that if $(H_s)_{s \le T}$ is in $\mathcal{H}$, there exists a sequence $(H_s^n)_{s \le T}$ of simple processes such that

$$\lim_{n \to +\infty} \mathbb{E}\left(\int_0^T |H_s - H_s^n|^2 ds \right) = 0.$$

A proof of this result can be found in Karatzas and Shreve (1988) (page 134, problem 2.5).

If $H \in \mathscr{H}$ and $(H^n)_{n \geq 0}$ is a sequence of simple processes converging to H in the previous sense, we have

$$\mathbb{E}\left(\sup_{t \leq T} |I(H^{n+p})_t - I(H^n)_t|^2\right) \leq 4\mathbb{E}\left(\int_0^T |H_s^{n+p} - H_s^n|^2 ds\right). \qquad (3.6)$$

Therefore, there exists a subsequence $H^{\phi(n)}$ such that

$$\mathbb{E}\left(\sup_{t \leq T} |I(H^{\phi(n+1)})_t - I(H^{\phi(n)})_t|^2\right) \leq \frac{1}{2^n}.$$

Thus, the series whose general term is $I(H^{\phi(n+1)}) - I(H^{\phi(n)})$ is uniformly convergent, almost surely. Consequently, $I(H^{\phi(n)})_t$ converges towards a continuous function that will be, by definition, the map $t \mapsto J(H)_t$. Taking the limit in (3.6), we obtain

$$\mathbb{E}\left(\sup_{t \leq T} |J(H)_t - I(H^n)_t|^2\right) \leq 4\mathbb{E}\left(\int_0^T |H_s - H_s^n|^2 ds\right). \qquad (3.7)$$

This implies that $(J(H)_t)_{0 \leq t \leq T}$ does not depend on the approximating sequence. On the other hand, $(J(H)_t)_{0 \leq t \leq T}$ is a martingale. Indeed,

$$\mathbb{E}(I(H^n)_t | \mathscr{F}_s) = I(H^n)_s,$$

and, for any t, $\lim_{n \to +\infty} I(H^n)_t = J(H)_t$ in $L^2(\Omega, \mathbb{P})$-norm and, because the conditional expectation is continuous in $L^2(\Omega, \mathbb{P})$, we can conclude.

From (3.7) and from the fact that

$$\mathbb{E}(I(H^n)_t^2) = \mathbb{E}\left(\int_0^T |H_s^n|^2 ds\right),$$

it follows that $\mathbb{E}\left(J(H)_t^2\right) = \mathbb{E}\left(\int_0^T |H_s|^2 ds\right)$. In the same way, from (3.7) and since $\mathbb{E}(\sup_{t \leq T} I(H^n)_t^2) \leq 4\mathbb{E}\left(\int_0^T |H_s^n|^2 ds\right)$, we prove (3.4).

The uniqueness of the extension results from the fact that the set of simple processes is dense in $\mathscr{H}$.

We now prove (3.5). First of all, we notice that (3.3) is still true if $H \in \mathscr{H}$. This is justified by the fact that the simple processes are dense in $\mathscr{H}$ and by (3.7). We first consider stopping times of the form $\tau = \sum_{1 \leq i \leq n} t_i \mathbf{1}_{A_i}$, where $0 < t_1 < \cdots < t_n = T$ and the A_i's are disjoint and $\mathscr{F}_{t_i}$-measurable, and we prove (3.5) in that case. First

$$\int_0^T \mathbf{1}_{\{s > \tau\}} H_s dW_s = \int_0^T \left(\sum_{1 \leq i \leq n} \mathbf{1}_{A_i} \mathbf{1}_{\{s > t_i\}}\right) H_s dW_s.$$

Also, each $\mathbf{1}_{\{s > t_i\}} \mathbf{1}_{A_i} H_s$ is adapted because this process is zero if $s \leq t_i$ and is equal to $\mathbf{1}_{A_i} H_s$ otherwise, therefore it belongs to $\mathscr{H}$. It follows that

$$\int_0^T \mathbf{1}_{\{s > \tau\}} H_s dW_s = \sum_{1 \leq i \leq n} \int_0^T \mathbf{1}_{A_i} \mathbf{1}_{\{s > t_i\}} H_s dW_s$$

$$= \sum_{1 \leq i \leq n} \mathbf{1}_{A_i} \int_{t_i}^T H_s dW_s = \int_\tau^T H_s dW_s,$$

and then $\int_0^T \mathbf{1}_{\{s \le \tau\}} H_s dW_s = \int_0^\tau H_s dW_s$.

In order to prove this result for an arbitrary stopping time τ, we observe that τ can be approximated by a decreasing sequence of stopping times of the previous form. Indeed, let

$$\tau_n = \sum_{0 \le i \le 2^n} \frac{(k+1)T}{2^n} \mathbf{1}_{\left\{\frac{kT}{2^n} \le \tau < \frac{(k+1)T}{2^n}\right\}}.$$

Then, τ_n converges almost surely to τ. By continuity of the map $t \mapsto \int_0^t H_s dW_s$ we can assert that, almost surely, $\int_0^{\tau_n} H_s dW_s$ converges to $\int_0^\tau H_s dW_s$. On the other hand,

$$\mathbb{E}\left(\left|\int_0^T \mathbf{1}_{\{s \le \tau\}} H_s dW_s - \int_0^T \mathbf{1}_{\{s \le \tau_n\}} H_s dW_s\right|^2\right) = \mathbb{E}\left(\int_0^T \mathbf{1}_{\{\tau < s \le \tau_n\}} H_s^2 ds\right).$$

This last term converges to 0 by dominated convergence, therefore

$$\int_0^T \mathbf{1}_{\{s \le \tau_n\}} H_s dW_s$$

converges to

$$\int_0^T \mathbf{1}_{\{s \le \tau\}} H_s dW_s$$

in $L^2(\Omega, \mathbb{P})$ (and, so, a subsequence converges almost surely). That completes the proof of (3.5) for an arbitrary stopping time. □

In the modelling, we shall need processes that only satisfy a weaker integrability condition than the processes in $\mathscr{H}$, which is why we define

$$\tilde{\mathscr{H}} = \left\{(H_s)_{0 \le s \le T} \quad (\mathscr{F}_t)_{t \ge 0} - \text{adapted process}, \quad \int_0^T H_s^2 ds < +\infty \ \mathbb{P} \text{ a.s.}\right\}.$$

The following proposition defines an extension of the stochastic integral from $\mathscr{H}$ to $\tilde{\mathscr{H}}$.

Proposition 3.4.6. *There exists a unique linear mapping $\tilde{J}$ from $\tilde{\mathscr{H}}$ into the vector space of continuous processes defined on $[0, T]$, such that:*

1. *Extension property: If $(H_t)_{0 \le t \le T}$ is a simple process, then*

$$\mathbb{P} \text{ a.s. } \forall 0 \le t \le T, \tilde{J}(H)_t = I(H)_t.$$

2. *Continuity property: If $(H^n)_{n \ge 0}$ is a sequence of proceses in $\tilde{H}$ such that $\int_0^T (H_s^n)^2 \ ds$ converges to 0 in probability, then $\sup_{t \le T} |\tilde{J}(H^n)_t|$ converges to 0 in probability.*

Consistently, we write $\int_0^t H_s dW_s$ for $\tilde{J}(H)_t$.

Remark 3.4.7. It is crucial to notice that in this case $(\int_0^t H_s dW_s)_{0 \le t \le T}$ is *not necessarily a martingale.*

Proof. It is easy to deduce from the extension property and from the continuity property that if $H \in \mathscr{H}$, then $\mathbb{P}$ a.s., $\forall t \leq T$, $\tilde{J}(H)_t = J(H)_t$. We now construct $\tilde{J}$.

Let $H \in \mathscr{\tilde{H}}$, and define $T_n = \inf\{0 \leq s \leq T, \int_0^s H_u^2 du \geq n\}$ ($+\infty$ if that set is empty), and $H_s^{(n)} = H_s \mathbf{1}_{\{s \leq T_n\}}$.

Firstly, we show that T_n is a stopping time. Since $\{T_n \leq t\} = \{\int_0^t H_u^2 du \geq n\}$, we just need to prove that $\int_0^t H_u^2 du$ is $\mathscr{F}_t$-measurable. This result is true if H is a simple process and, by density, it is true if $H \in \mathscr{H}$. Finally, if $H \in \mathscr{\tilde{H}}$, $\int_0^t H_u^2 du$ is also $\mathscr{F}_t$-measurable because it is the limit of $\int_0^t (H_u \wedge K)^2 du$ almost surely as K tends to infinity. Then, we see immediately that the processes $H_s^{(n)}$ are adapted and bounded, hence they belong to $\mathscr{H}$. Moreover,

$$\int_0^t H_s^{(n)} dW_s = \int_0^t \mathbf{1}_{\{s \leq T_n\}} H_s^{(n+1)} dW_s,$$

and relation (3.5) implies that

$$\int_0^t H_s^{(n)} dW_s = \int_0^{t \wedge T_n} H_s^{n+1} dW_s.$$

Thus, on the set $\{\int_0^T H_u^2 du < n\}$, for any $t \leq T$, $J(H^{(n)})_t = J(H^{(n+1)})_t$. Since

$$\bigcup_{n \geq 0} \left\{ \int_0^T H_u^2 du < n \right\} = \left\{ \int_0^T H_u^2 du < +\infty \right\},$$

we can define almost surely a process $\tilde{J}(H)_t$ by: on the set $\{\int_0^t H_u^2 du < n\}$,

$$\forall t \leq T \quad \tilde{J}(H)_t = J(H^{(n)})_t.$$

The process $t \mapsto \tilde{J}(H)_t$ is almost surely continuous, by definition. The extension property is satisfied by construction. We just need to prove the continuity property of $\tilde{J}$. To do so, we first notice that

$$\mathbb{P}\left(\sup_{t \leq T} |\tilde{J}(H)_t| \geq \varepsilon \right) \leq \mathbb{P}\left(\int_0^T H_s^2 ds \geq 1/N \right)$$

$$+ \mathbb{P}\left(\mathbf{1}_{\{\int_0^T H_u^2 du < 1/N\}} \sup_{t \leq T} |\tilde{J}(H)_t| \geq \varepsilon \right).$$

If we call $\tau_N = \inf\{s \leq T, \int_0^s H_u^2 du \geq 1/N\}$ ($+\infty$ if this set is empty), then on the set $\left\{ \int_0^T H_u^2 du < 1/N \right\}$, it follows from (3.5) that, for any $t \leq T$,

$$\int_0^t H_s dW_s = \tilde{J}(H)_t = J(H^1)_t = \int_0^t H_s^1 \mathbf{1}_{\{s \leq \tau_N\}} dW_s = \int_0^t H_s \mathbf{1}_{\{s \leq \tau_N\}} dW_s,$$

whence, by applying (3.4) to the process $s \mapsto H_s \mathbf{1}_{\{s \leq \tau_N\}}$ we get

$$\mathbb{P}\left(\sup_{t \leq T} \left| \tilde{J}(H)_t \right| \geq \varepsilon \right) \leq \mathbb{P}\left(\int_0^T H_s^2 ds \geq \frac{1}{N} \right)$$

$$+ \frac{4}{\varepsilon^2} \mathbb{E}\left(\int_0^T H_s^2 \mathbf{1}_{\{s \leq \tau_N\}} ds \right)$$

$$\leq \mathbb{P}\left(\int_0^T H_s^2 ds \geq \frac{1}{N} \right) + \frac{4}{N\varepsilon^2}.$$

As a result, if $\int_0^T (H_s^n)^2 ds$ converges to 0 in probability, then $\sup_{t \leq T} |\tilde{J}(H^n)_t|$ converges to 0 in probability.

In order to prove the linearity of $\tilde{J}$, let us consider two processes belonging to $\mathscr{H}$, called H and K, and the two sequences H_t^n and K_t^n defined at the beginning of the proof such that $\int_0^T (H_s^n - H_s)^2 ds$ and $\int_0^T (K_s^n - K_s)^2 ds$ converge to 0 in probability. By continuity of $\tilde{J}$ we can take the limit in the equality $J(\lambda H^n + \mu K^n)_t = \lambda J(H^n)_t + \mu J(K^n)_t$ to prove the linearity of $\tilde{J}$.

Finally, the fact that if $H \in \mathscr{H}$ then $\int_0^T (H_t - H_t^n)^2 dt$ converges to 0 in probability and the continuity property yield the uniqueness of the extension. $\square$

We are about to summarise the conditions needed to define the stochastic integral with respect to a Brownian motion and we want to specify the assumptions that make it a martingale.

Summary:

Let us consider an $\mathscr{F}_t$-Brownian motion $(W_t)_{t \geq 0}$ and an $\mathscr{F}_t$-adapted process $(H_t)_{0 \leq t \leq T}$. We are able to define the stochastic integral $(\int_0^t H_s dW_s)_{0 \leq t \leq T}$ as soon as $\int_0^T H_s^2 ds < +\infty$ $\mathbb{P}$ a.s. By construction, the process $(\int_0^t H_s dW_s)_{0 \leq t \leq T}$ is a martingale *if* $\mathbb{E}\left(\int_0^T H_s^2 ds\right) < +\infty$. This condition is not necessary. Indeed, the inequality $\mathbb{E}\left(\int_0^T H_s^2 ds\right) < +\infty$ is satisfied if and only if

$$\mathbb{E}\left(\sup_{0 \leq t \leq T} \left(\int_0^t H_s dW_s\right)^2\right) < +\infty,$$

in which case we have

$$\mathbb{E}\left[\left(\int_0^T H_t dW_t\right)^2\right] = \mathbb{E}\left(\int_0^T H_t^2\right).$$

This is proved in Exercise 18.

3.4.2 Itô calculus

It is now time to introduce a differential calculus based on this stochastic integral. It will be called the *Itô calculus* and the main ingredient is the famous *Itô formula*.

In particular, the Itô formula allows us to differentiate such a function as $t \mapsto f(W_t)$ if f is twice continuously differentiable. The following example will simply show that a naive extension of the classical differential calculus is bound to fail. Let us try to *differentiate* the function $t \to W_t^2$ in terms of 'dW_t'. Typically, for a differentiable function $f(t)$ null at the origin, we have $f(t)^2 = 2 \int_0^t f(s) \dot{f}(s) ds = 2 \int_0^t f(s) df(s)$. In the Brownian case, it is impossible to have a similar formula $W_t^2 = 2 \int_0^t W_s dW_s$. Indeed, from the previous section we know that $\int_0^t W_s dW_s$ is a martingale (because $\mathbb{E}\left(\int_0^t W_s^2 ds\right) < +\infty$), null at zero. If it were equal to $W_t^2/2$, it would be non-negative, and a non-negative martingale vanishing at zero can only be identically zero.

We shall define precisely the class of processes for which the Itô formula is applicable.

Definition 3.4.8. Let $(\Omega, \mathscr{F}, (\mathscr{F}_t)_{t\geq 0}, \mathbb{P})$ be a filtered probability space and $(W_t)_{t\geq 0}$ an $\mathscr{F}_t$-Brownian motion. $(X_t)_{0\leq t\leq T}$ is an $\mathbb{R}$-valued Itô process if it can be written as

$$\mathbb{P} \text{ a.s. } \forall t \leq T \quad X_t = X_0 + \int_0^t K_s ds + \int_0^t H_s dW_s,$$

where

- X_0 is $\mathscr{F}_0$-measurable.

- $(K_t)_{0\leq t\leq T}$ and $(H_t)_{0\leq t\leq T}$ are $\mathscr{F}_t$-adapted processes.

- $\int_0^T |K_s| ds < +\infty$ $\mathbb{P}$ a.s.

- $\int_0^T |H_s|^2 ds < +\infty$ $\mathbb{P}$ a.s.

We can prove the following proposition (see Exercise 19), which clarifies the uniqueness of the previous decomposition.

Proposition 3.4.9. *If $(M_t)_{0\leq t\leq T}$ is a continuous martingale such that*

$$M_t = \int_0^t K_s ds, \quad with \quad \mathbb{P} \text{ a.s. } \int_0^T |K_s| ds < +\infty,$$

then

$$\mathbb{P} \text{ a.s. } \forall t \leq T, \ M_t = 0.$$

This implies that:

- *An Itô process decomposition is unique. This means that if*

$$X_t = X_0 + \int_0^t K_s ds + \int_0^t H_s dW_s = X_0' + \int_0^t K_s' ds + \int_0^t H_s' dW_s,$$

   then

$$X_0 = X_0' \ d\mathbb{P} \text{ a.s.} \quad H_s = H_s' \ ds \times d\mathbb{P} \text{ a.e.} \quad K_s = K_s' \ ds \times d\mathbb{P} \text{ a.e.}$$

- *If $(X_t)_{0\leq t\leq T}$ is a martingale of the form $X_0 + \int_0^t K_s ds + \int_0^t H_s dW_s$, then $K_t = 0 \ dt \times d\mathbb{P} \text{ a.e.}$*

We shall state the Itô formula for continuous martingales. The interested reader should refer to Bouleau (1988) for an elementary proof in the Brownian case, i.e. when (W_t) is a standard Brownian motion, or to Karatzas and Shreve (1988) for a complete proof.

Theorem 3.4.10. Let $(X_t)_{0 \leq t \leq T}$ be an Itô process,

$$X_t = X_0 + \int_0^t K_s ds + \int_0^t H_s dW_s,$$

and f be a twice continuously differentiable function. Then

$$f(X_t) = f(X_0) + \int_0^t f'(X_s)dX_s + \frac{1}{2}\int_0^t f''(X_s)d\langle X, X\rangle_s$$

where, by definition,

$$\langle X, X\rangle_t = \int_0^t H_s^2 ds$$

and

$$\int_0^t f'(X_s)dX_s = \int_0^t f'(X_s)K_s ds + \int_0^t f'(X_s)H_s dW_s.$$

Likewise, if $(t, x) \rightarrow f(t, x)$ is a function that is twice differentiable with respect to x and once with respect to t, and if these partial derivatives are continuous with respect to (t, x) (i.e. f is a function of class $C^{1,2}$), the Itô formula becomes

$$f(t, X_t) = f(0, X_0) + \int_0^t f'_s(s, X_s)ds$$

$$+ \int_0^t f'_x(s, X_s)dX_s + \frac{1}{2}\int_0^t f''_{xx}(s, X_s)d\langle X, X\rangle_s.$$

3.4.3 Examples: Itô formula in practice

Let us start by giving an elementary example. If $f(x) = x^2$ and $X_t = W_t$, we identify $K_s = 0$ and $H_s = 1$. Thus

$$W_t^2 = 2\int_0^t W_s dW_s + \frac{1}{2}\int_0^t 2ds.$$

It turns out that

$$W_t^2 - t = 2\int_0^t W_s dW_s.$$

Since $\mathbb{E}\left(\int_0^t W_s^2 ds\right) < +\infty$, it confirms the fact that $W_t^2 - t$ is a martingale.

We now want to tackle the problem of finding the solutions $(S_t)_{t \geq 0}$ of

$$S_t = x_0 + \int_0^t S_s(\mu ds + \sigma dW_s). \tag{3.8}$$

This is often written in the symbolic form

$$dS_t = S_t(\mu dt + \sigma dW_t), \qquad S_0 = x_0. \tag{3.9}$$

We are actually looking for an adapted process $(S_t)_{t \geq 0}$ such that the integrals $\int_0^t S_s ds$ and $\int_0^t S_s dW_s$ exist and at any time t

$$\mathbb{P} \text{ a.s. } S_t = x_0 + \int_0^t \mu S_s ds + \int_0^t \mu S_s ds + \int_0^t \sigma S_s dW_s.$$

To put it in a simple way, let us do a formal calculation. We write $Y_t = \log(S_t)$, where S_t is a solution of (3.8). S_t is an Itô process with $K_s = \mu S_s$ and $H_s = \sigma S_s$. Assuming that S_t is non-negative, we apply the Itô formula to $f(x) = \log(x)$ (at least formally, because $f(x)$ is not a C^2 function on the real line!), and we obtain

$$\log(S_t) = \log(S_0) + \int_0^t \frac{dS_s}{S_s} + \frac{1}{2} \int_0^t \left(\frac{-1}{S_s^2} \right) \sigma^2 S_s^2 ds.$$

Using (3.9), we get

$$Y_t = Y_0 + \int_0^t (\mu - \sigma^2/2) dt + \int_0^t \sigma dW_t,$$

and finally

$$Y_t = \log(S_t) = \log(S_0) + (\mu - \sigma^2/2)t + \sigma W_t.$$

Taking this into account, it seems that

$$S_t = x_0 \exp((\mu - \sigma^2/2)t + \sigma W_t)$$

is a solution of equation (3.8). We must check that conjecture rigorously. We have $S_t = f(t, W_t)$ with

$$f(t, x) = x_0 \exp((\mu - \sigma^2/2)t + \sigma x).$$

The Itô formula is now applicable and yields

$$\begin{aligned} S_t &= f(t, W_t) \\ &= f(0, W_0) + \int_0^t f'_s(s, W_s) ds \\ &\quad + \int_0^t f'_x(s, W_s) dW_s + \frac{1}{2} \int_0^t f''_{xx}(s, W_s) d\langle W, W \rangle_s. \end{aligned}$$

Furthermore, since $\langle W, W \rangle_t = t$, we can write

$$S_t = X_0 + \int_0^t S_s(\mu - \sigma^2/2) ds + \int_0^t S_s \sigma dW_s + \frac{1}{2} \int_0^t S_s \sigma^2 ds.$$

In conclusion,

$$S_t = X_0 + \int_0^t S_s \mu ds + \int_0^t S_s \sigma dW_s.$$

Remark 3.4.11. We could have obtained the same result (exercise) by applying the Itô formula to $S_t = \phi(Z_t)$, with $Z_t = (\mu - \sigma^2/2)t + \sigma W_t$ (which is an Itô process) and $\phi(x) = x_0 \exp(x)$.

We have just proved the existence of a solution to equation (3.8). we are about to prove its uniqueness. To do that, we shall use the *integration by parts* formula.

Proposition 3.4.12 (integration by parts formula) Let X_t and Y_t be two Itô processes,

$$X_t = X_0 + \int_0^t K_s ds + \int_0^t H_s dW_s$$

and

$$Y_t = Y_0 + \int_0^t K'_s ds + \int_0^t H'_s dW_s.$$

Then

$$X_t Y_t = X_0 Y_0 + \int_0^t X_s dY_s + \int_0^t Y_s dX_s + \langle X, Y \rangle_t$$

with the following convention:

$$\langle X, Y \rangle_t = \int_0^t H_s H'_s ds.$$

Proof. By the Itô formula,

$$(X_t + Y_t)^2 = (X_0 + Y_0)^2$$
$$+ 2\int_0^t (X_s + Y_s) d(X_s + Y_s)$$
$$+ \int_0^t (H_s + H'_s)^2 ds$$
$$X_t^2 = X_0^2 + 2\int_0^t X_s dX_s + \int_0^t H_s^2 ds$$
$$Y_t^2 = Y_0^2 + 2\int_0^t Y_s dY_s + \int_0^t H'^2_s ds.$$

By subtracting the last two equalities from the first one, we get

$$X_t Y_t = X_0 Y_0 + \int_0^t X_s dY_s + \int_0^t Y_s dX_s + \int_0^t H_s H'_s ds.$$

$\square$

We now have the tools to show that equation (3.8) has a unique solution. Recall that
$$S_t = x_0 \exp((\mu - \sigma^2/2)t + \sigma W_t)$$
is a solution of (3.8) and assume that $(X_t)_{t \geq 0}$ is another one. We attempt to compute the *stochastic differential* of the quantity $X_t S_t^{-1}$. Define

$$Z_t = \frac{S_0}{S_t} = \exp((-\mu + \sigma^2/2)t - \sigma W_t),$$

$\mu' = -\mu + \sigma^2$ and $\sigma' = -\sigma$. Then $Z_t = \exp((\mu' - \sigma'^2/2)t + \sigma'W_t)$ and the verification that we have just done shows that

$$Z_t = 1 + \int_0^t Z_s(\mu'ds + \sigma'dW_s) = 1 + \int_0^t Z_s((-\mu + \sigma^2)ds - \sigma dW_s).$$

>From the integration by parts formula, we can compute the *differential* of $X_t Z_t$,

$$d(X_t Z_t) = X_t dZ_t + Z_t dX_t + d\langle X, Z \rangle_t.$$

In this case, we have

$$\langle X, Z \rangle_t - \langle \int_0^{\cdot} X_s \sigma dW_s, - \int_0^{\cdot} Z_s \sigma dW_s \rangle_t = - \int_0^t \sigma^2 X_s Z_s ds.$$

Therefore

$$\begin{aligned} d(X_t Z_t) &= X_t Z_t((-\mu + \sigma^2)dt - \sigma dW_t) \\ &\quad + X_t Z_t(\mu dt + \sigma dW_t) - X_t Z_t \sigma^2 dt \\ &= 0. \end{aligned}$$

Hence, $X_t Z_t$ is equal to $X_0 Z_0$, which implies that

$$\forall t \geq 0, \qquad \mathbb{P} \text{ a.s. } X_t = x_0 Z_t^{-1} = S_t.$$

The processes X_t and Z_t being continuous, this proves that

$$\mathbb{P} \text{ a.s. } \forall t \geq 0, \quad X_t = x_0 Z_t^{-1} = S_t.$$

We have just proved the following theorem:

Theorem 3.4.13. *If we consider two real numbers σ, μ and a Brownian motion $(W_t)_{t \geq 0}$ and a strictly positive constant T, there exists a unique Itô process $(S_t)_{0 \leq t \leq T}$ that satisfies, for any $t \leq T$,*

$$S_t = x_0 + \int_0^t S_s(\mu ds + \sigma dW_s).$$

This process is given by

$$S_t = X_0 \exp((\mu - \sigma^2/2)t + \sigma W_t).$$

Remark 3.4.14.

- The process (S_t) that we just studied will model the evolution of a stock price in the Black-Scholes model.

- When $\mu = 0$, (S_t) is actually a martingale (see Proposition 3.3.3) called the *exponential martingale of Brownian motion*.

Remark 3.4.15. Let Θ be an open set in $\mathbb{R}$ and $(X_t)_{0 \leq t \leq T}$ an Itô process that stays in Θ at all times. If we consider a function f from Θ to $\mathbb{R}$ that is twice continuously differentiable, we can derive an extension of the Itô formula in that case:

$$f(X_t) = f(X_0) + \int_0^t f'(X_s)dX_s + \frac{1}{2}\int_0^t f''(X_s)d\langle X, X \rangle_s.$$

This result allows us to apply the Itô formula to $\log(X_t)$, for instance, if X_t is a strictly positive process.

3.4.4 Multidimensional Itô formula

We apply a multidimensional version of the Itô formula when f is a function of several Itô processes that are themselves defined with several Brownian motions. This version will prove to be very useful when we model complex interest rate structures, for instance.

Definition 3.4.16. We call a standard p-dimensional $\mathscr{F}_t$-Brownian motion an $\mathbb{R}^p$-valued process

$$(W_t = (W_t^1, \ldots, W_t^p))_{t \geq 0}$$

adapted to $\mathscr{F}_t$, where all the $(W_t^i)_{t \geq 0}$ are independent standard $\mathscr{F}_t$-Brownian motions.

It can be proved that if $(W_t)_{t \geq 0}$ is a standard p-dimensional $\mathscr{F}_t$-Brownian motion, the vector $W_t - W_s$ is independent of $\mathscr{F}_s$, for $t \geq s \geq 0$.

Definition 3.4.17. An Itô process with respect to $(W_t, \mathscr{F}_t)$, where $(W_t)_{t \geq 0}$ is a standard p-dimensional $\mathscr{F}_t$-Brownian motion, is a stochastic process $(X_t)_{0 \leq t \leq T}$ of the form

$$X_t = X_0 + \int_0^t K_s ds + \sum_{i=1}^p \int_0^t H_s^i dW_s^i,$$

where:

- K_t and all the processes (H_t^i) are adapted to $(\mathscr{F}_t)$.

- $\int_0^T |K_s| ds < +\infty$ $\mathbb{P}$ *a.s.*

- $\int_0^T (H_s^i)^2 ds < +\infty$ $\mathbb{P}$ *a.s.*

The Itô formula becomes:

Proposition 3.4.18. Let $(X_t^1), \ldots, (X_t^n)$ be n Itô processes:

$$X_t^i = X_0^i + \int_0^t K_s^i ds + \sum_{j=1}^p \int_0^t H_s^{i,j} dW_s^j.$$

Then, if f is twice differentiable with respect to x and once differentiable with respect to t, with continuous partial derivatives in (t, x),

$$f(t, X_t^1, \ldots, X_t^n) = f(0, X_0^1, \ldots, X_0^n) + \int_0^t \frac{\partial f}{\partial s}(s, X_s^1, \ldots, X_s^n)ds$$

$$+ \sum_{i=1}^n \int_0^t \frac{\partial f}{\partial x_i}(s, X_s^1, \ldots, X_s^n)dX_s^i$$

$$+ \frac{1}{2} \sum_{i,j=1}^n \int_0^t \frac{\partial^2 f}{\partial x_i x_j}(s, X_s^1, \ldots, X_s^n)d\langle X^i, X^j \rangle_s,$$

with:

- $dX_s^i = K_s^i ds + \sum_{j=1}^p H_s^{i,j} dW_s^j,$

- $d\langle X^i, X^j \rangle_s = \sum_{m=1}^p H_s^{i,m} H_s^{j,m} ds.$

Remark 3.4.19. If $(X_s)_{0 \le t \le T}$ and $(Y_s)_{0 \le t \le T}$ are two Itô processes, we can define formally the *cross-variation* of X and Y (denoted by $\langle X, Y \rangle_s$) through the following properties:

1. $\langle X, Y \rangle_t$ is bilinear and symmetric.

2. $\langle \int_0^{\cdot} K_s ds, X. \rangle_t = 0$ if $(X_t)_{0 \le t \le T}$ is an Itô process.

3. $\langle \int_0^{\cdot} H_s dW_t^i, \int_0^{\cdot} H_s' dW_t^j \rangle_t = 0$ if $i \ne j$.

4. $\langle \int_0^{\cdot} H_s dW_t^i, \int_0^{\cdot} H_s' dW_t^i \rangle_t = \int_0^t H_s H_s' ds.$

This definition leads to the cross-variation stated in the previous proposition.

3.5   Stochastic differential equations

In Section 3.4.2, we studied in detail the solutions to the equation

$$X_t = x + \int_0^t X_s(\mu ds + \sigma dW_s).$$

We can now consider some more general equations of the type

$$X_t = Z + \int_0^t b(s, X_s)ds + \int_0^t \sigma(s, X_s)dW_s. \tag{3.10}$$

These equations are called *stochastic differential equations* and a solution of equation (3.10) is called a *diffusion*. These equations are useful to model most financial assets, whether we are speaking about stocks or interest rate processes. Let us first study some properties of the solutions to these equations.

3.5.1 Itô theorem

What do we mean by *a solution of (3.10)*?

Definition 3.5.1. We consider a probability space $(\Omega, \mathscr{A}, \mathbb{P})$ equipped with a filtration $(\mathscr{F}_t)_{t \geq 0}$. We also have functions $b : \mathbb{R}^+ \times \mathbb{R} \to \mathbb{R}$, $\sigma : \mathbb{R}^+ \times \mathbb{R} \to \mathbb{R}$; an $\mathscr{F}_0$-measurable random variable Z and finally a standard $\mathscr{F}_t$-Brownian motion $(W_t)_{t \geq 0}$. A solution to equation (3.10) is an $\mathscr{F}_t$-adapted continuous stochastic process $(X_t)_{t \geq 0}$ that satisfies:

1. For any $t \geq 0$, the integrals $\int_0^t b(s, X_s)ds$ and $\int_0^t \sigma(s, X_s)dW_s$ exist:

$$\int_0^t |b(s, X_s)|ds < +\infty \text{ and } \int_0^t |\sigma(s, X_s)|^2 ds < +\infty \ \mathbb{P} \text{ a.s.}$$

2. $(X_t)_{t \geq 0}$ satisfies (3.10), i.e.

$$\forall t \geq 0 \ \mathbb{P} \text{ a.s. } X_t = Z + \int_0^t b(s, X_s)ds + \int_0^t \sigma(s, X_s)dW_s.$$

Remark 3.5.2. Formally, we often write equation (3.10) as

$$\begin{cases} dX_t &= b(t, X_t)dt + \sigma(t, X_t)dW_t \\ X_0 &= Z. \end{cases}$$

The following theorem gives sufficient conditions on b and σ to guarantee the existence and uniqueness of a solution of equation (3.10).

Theorem 3.5.3. *If b and σ are continuous functions, and if there exists a constant $K < +\infty$ such that*

1. $|b(t, x) - b(t, y)| + |\sigma(t, x) - \sigma(t, y)| \leq K|x - y|$

2. $|b(t, x)| + |\sigma(t, x)| \leq K(1 + |x|)$

3. $\mathbb{E}(Z^2) < +\infty$

then, for any $T \geq 0$, (3.10) admits a unique solution in the interval $[0, T]$. Moreover, this solution $(X_s)_{0 \leq s \leq T}$ satisfies

$$\mathbb{E}\left(\sup_{0 \leq s \leq T} |X_s|^2 \right) < +\infty.$$

The uniqueness of the solution means that if $(X_t)_{0 \leq t \leq T}$ and $(Y_t)_{0 \leq t \leq T}$ are two solutions of (3.10), then $\mathbb{P}$ a.s. $\forall 0 \leq t \leq T$, $X_t = Y_t$.

Proof. We define the set

$$\mathscr{E} = \left\{ (X_s)_{0 \leq s \leq T}, \ \mathscr{F}_t\text{-adapted continuous processes,} \right.$$

$$\left. \text{such that } \mathbb{E}\left(\sup_{s \leq T} |X_s|^2 \right) < +\infty \right\}.$$

Together with the norm $\|X\| = (\mathbb{E}(\sup_{0 \leq s \leq T} |X_s|^2))^{1/2}$, $\mathscr{E}$ is a complete normed vector space. In order to show the existence of a solution, we are going to use the theorem of existence of a fixed point for a contracting mapping. Let Φ be the function that maps a process $(X_s)_{0 \leq s \leq T}$ into a process $(\Phi(X)_s)_{0 \leq s \leq T}$ defined by

$$\Phi(X)_t = Z + \int_0^t b(s, X_s)ds + \int_0^t \sigma(s, X_s)dW_s.$$

If X belongs to $\mathscr{E}$, $\Phi(X)$ is well defined and, furthermore, if X and Y are both in $\mathscr{E}$, we can use the fact that $(a+b)^2 \leq 2(a^2 + b^2)$ and so write that

$$|\Phi(X)_t - \Phi(Y)_t|^2 \leq 2 \left(\sup_{0 \leq t \leq T} \left| \int_0^t (b(s, X_s) - b(s, Y_s))ds \right|^2 \right.$$
$$\left. + \sup_{0 \leq t \leq T} \left| \int_0^t (\sigma(s, X_s) - \sigma(s, Y_s))dW_s \right|^2 \right),$$

and therefore by (3.4)

$$\mathbb{E} \left(\sup_{s \leq T} |\Phi(X)_t - \Phi(Y)_t|^2 \right)$$
$$\leq 2\mathbb{E} \left(\sup_{0 \leq t \leq T} \left(\int_0^t |b(s, X_s) - b(s, Y_s)|ds \right)^2 \right)$$
$$+ 8\mathbb{E} \left(\int_0^T (\sigma(s, X_s - \sigma(s, Y_s))^2 ds \right)$$
$$\leq 2(K^2 T^2 + 4K^2 T)\mathbb{E} \left(\sup_{0 \leq t \leq T} |X_t - Y_t|^2 \right),$$

whence $\|\Phi(X) - \Phi(Y)\| \leq (2(K^2 T^2 + 4K^2 T))^{1/2} \|X - Y\|$. On the other hand, if we denote by 0 the process that is identically equal to zero, and if we notice that $(a+b+c)^2 \leq 3(a^2 + b^2 + c^2)$,

$$|\Phi(0)_t|^2 \leq 3 \left(Z^2 + \sup_{0 \leq t \leq T} \left| \int_0^t b(s, 0)ds \right|^2 + \sup_{0 \leq t \leq T} \left| \int_0^t \sigma(s, 0)dW_s \right|^2 \right).$$

Therefore

$$\mathbb{E} \left(\sup_{0 \leq t \leq T} |\Phi(0)_t|^2 \right) \leq 3(\mathbb{E}(Z^2) + K^2 T^2 + 4K^2 T) < +\infty.$$

We deduce that Φ is a mapping from $\mathscr{E}$ to $\mathscr{E}$ with a Lipschitz norm bounded from above by $k(T) = (2(K^2 T^2 + 4K^2 T))^{1/2}$. If we assume that T is small enough so that $k(T) < 1$, it turns out that Φ is a contraction from $\mathscr{E}$ into $\mathscr{E}$. Thus it has a fixed point in $\mathscr{E}$. Moreover, if X is a fixed point of Φ, it is a solution of (3.10). This completes the proof of the existence for T small enough. On the other hand, a solution of (3.10) that belongs to $\mathscr{E}$ is a fixed point of Φ. That proves the uniqueness of a solution of equation (3.10) in the class $\mathscr{E}$. In order to prove the uniqueness in the whole class of Itô processes, we just need to show that a solution of (3.10) always belongs to $\mathscr{E}$. Let X be a solution of (3.10), and define $T_n = \inf\{s \geq 0, \quad |X_s| > n\}$ and $f^n(t) =$

$\mathbb{E}(\sup_{0 \leq s \leq t \wedge T_n} |X_s|^2)$. It is easy to check that $f^n(t)$ is finite and continuous. Using the same comparison arguments as before, we can say that

$$\mathbb{E}(\sup_{0 \leq u \leq t \wedge T_n} |X_u|^2) \leq 3 \left(\mathbb{E}(Z^2) + \mathbb{E}\left(\int_0^{t \wedge T_n} K(1 + |X_s|)ds \right)^2 \right.$$

$$+ 4\mathbb{E}\left(\int_0^{t \wedge T_n} K^2(1 + |X_s|)^2 ds \right) \Big)$$

$$\leq 3(\mathbb{E}(Z^2) + 2(K^2 T + 4K^2))$$

$$\times \int_0^t (1 + \mathbb{E}(\sup_{0 \leq u \leq s \wedge T_n} |X_u|^2))ds).$$

This yields the following inequality:

$$f^n(t) \leq a + b \int_0^t f^n(s)ds.$$

In order to complete the proof, let us recall the Gronwall lemma.

Lemma 3.5.4 (Gronwall lemma) *If f is a continuous function such that for any $0 \leq t \leq T$, $f(t) \leq a + b \int_0^t f(s)ds$, then $f(T) \leq ae^{bT}$.*

Proof. Let us write $u(t) = e^{-bt} \int_0^t f(s)ds$. Then,

$$u'(t) = e^{-bt}(f(s) - b \int_0^t f(s)ds) \leq ae^{-bt}.$$

By first-order integration we obtain $u(T) \leq a(1 - e^{-bT})/b$ and

$$f(T) \leq a + b \int_0^T f(s)ds = a + be^{bT} u(T) \leq ae^{bT}.$$

In our case, we have $f^n(T) < K < +\infty$, where K is a function of T independent of n. It follows from the Fatou lemma that, for any T,

$$\mathbb{E}\left(\sup_{0 \leq s \leq T} |X_s|^2 \right) < K < +\infty.$$

Therefore X belongs to $\mathscr{E}$ and that completes the proof for small T. For an arbitrary T, we consider a large enough integer n and think successively on the intervals $[0, T/n], [T/n, 2T/n], \ldots, [(n-1)T/n, T]$. $\qquad \square$

3.5.2 The Ornstein-Ulhenbeck process

The Ornstein-Ulhenbeck process is the unique solution of the following equation:

$$\begin{cases} dX_t = -cX_t dt + \sigma dW_t \\ X_0 = x. \end{cases}$$

It can be written explicitly. Indeed, if we consider $Y_t = X_t e^{ct}$ and integrate by parts, it yields

$$dY_t = dX_t e^{ct} + X_t d(e^{ct}) + d\langle X, e^{c\cdot} \rangle_t.$$

Furthermore, because $d(e^{ct}) = ce^{ct}dt$,

$$\langle X, e^{c\cdot} \rangle_t = 0.$$

It follows that $dY_t = \sigma e^{ct} dW_t$ and thus

$$X_t = xe^{-ct} + \sigma e^{-ct} \int_0^t e^{cs} dW_s.$$

This enables us to compute the mean and variance of X_t:

$$\mathbb{E}(X_t) = xe^{-ct} + \sigma e^{-ct} \mathbb{E}\left(\int_0^t e^{cs} dW_s\right) = xe^{-ct}$$

(since $\mathbb{E}(\int_0^t (e^{cs})^2 ds) < +\infty$, $\int_0^t e^{cs} dW_s$ is a martingale null at time 0 and therefore its expectation is zero). Similarly

$$\mathrm{Var}(X_t) = \mathbb{E}((X_t - \mathbb{E}(X_t))^2)$$

$$= \sigma^2 \mathbb{E}\left(e^{-2ct}\left(\int_0^t e^{cs} dW_s\right)^2\right)$$

$$= \sigma^2 e^{-2ct} \mathbb{E}\left(\int_0^t e^{2cs} ds\right)$$

$$= \sigma^2 \frac{1 - e^{-2ct}}{2c}.$$

We can also prove that X_t is a normal random variable, since X_t can be written as $\int_0^t f(s) dW_s$, where $f(.)$ is a deterministic function of time and $\int_0^t f^2(s) ds < +\infty$ (see Exercise 12). More precisely, the *process* $(X_t)_{t \geq 0}$ is Gaussian. This means that if $\lambda_1, \ldots, \lambda_n$ are real numbers and if $0 \leq t_1 < \cdots < t_n$, the random variable $\lambda_1 X_{t_1} + \cdots + \lambda_n X_{t_n}$ is normal. To convince ourselves, we just notice that

$$X_{t_i} = xe^{-ct_i} + \sigma e^{-ct_i} \int_0^{+\infty} \mathbf{1}_{\{s \leq t_i\}} e^{cs} dW_s = m_i + \int_0^t f_i(s) dW_s.$$

Then $\lambda_1 X_{t_1} + \cdots + \lambda_n X_{t_n} = \sum_{i=1}^n \lambda_i m_i + \int_0^t (\sum_{i=1}^n \lambda_i f_i(s)) dW_s$, which is indeed a normal random variable (since it is a stochastic integral of a deterministic function of time).

3.5.3 Multidimensional stochastic differential equations

The analysis of stochastic differential equations can be extended to the case when processes evolve in $\mathbb{R}^n$. This generalization proves to be useful in finance when we want to model baskets of stocks or currencies. We consider

- $W = (W^1, \ldots, W^p)$ an $\mathbb{R}^p$-valued $\mathscr{F}_t$-Brownian motion.

- $b : \mathbb{R}^+ \times \mathbb{R}^n \to \mathbb{R}^n, b(s, x) = (b^1(s, x), \ldots, b^n(s, x)).$

- $\sigma : \mathbb{R}^+ \times \mathbb{R}^n \to \mathbb{R}^{n \times p}$, $\sigma(s, x) = (\sigma_{i,j}(s, x))_{1 \le i \le n, 1 \le j \le p}$.

- $Z = (Z^1, \ldots, Z^n)$ an $\mathscr{F}_0$-measurable random variable with values in $\mathbb{R}^n$.

We are also interested in the following stochastic differential equation:

$$X_t = Z + \int_0^t b(s, X_s)ds + \int_0^t \sigma(s, X_s)dW_s. \tag{3.11}$$

In other words, we are looking for a process $(X_t)_{0 \le t \le T}$ with values in $\mathbb{R}^n$, adapted to the filtration $(\mathscr{F}_t)_{t \ge 0}$ and such that $\mathbb{P}$ a.s., for any t and for any $i \le n$:

$$X_t^i = Z^i + \int_0^t b^i(s, X_s)ds + \sum_{j=1}^p \int_0^t \sigma_{i,j}(s, X_s)dW_s^j.$$

The theorem of existence and uniqueness of a solution of (3.11) can be stated as:

Theorem 3.5.5. *If $x \in \mathbb{R}^n$, we denote by $|x|$ the Euclidean norm of x and if $\sigma \in \mathbb{R}^{n \times p}$*

$$|\sigma|^2 = \sum_{1 \le i \le n, \ 1 \le j \le p} \sigma_{i,j}^2.$$

We assume that

1. $|b(t, x) - b(t, y)| + |\sigma(t, x) - \sigma(t, y)| \le K|x - y|$

2. $|b(t, x)| + |\sigma(t, x)| \le K(1 + |x|)$

3. $E(|Z|^2) < +\infty$

then there exists a unique solution to (3.11). Moreover, this solution satisfies for any T

$$\mathbb{E}\left(\sup_{0 \le s \le T} |X_s|^2 \right) < +\infty.$$

The proof is very similar to the one in the scalar case.

3.5.4 The Markov property of the solution of a stochastic differential equation

The intuitive meaning of the Markov property is that the future behavior of the process $(X_t)_{t \ge 0}$ after t depends only on the value X_t and is not influenced by the history of the process before t. This is a crucial property of the Markovian model and it will have great consequences in the pricing of options. For instance, it will allow us to show that the price of an option on an underlying asset whose price is Markovian depends only on the price of this underlying asset at time t.

Mathematically speaking, an $\mathscr{F}_t$-adapted process $(X_t)_{t\geq 0}$ satisfies the Markov property if, for any bounded Borel function f and for any s and t such that $s \leq t$, we have

$$\mathbb{E}(f(X_t)|\mathscr{F}_s) = \mathbb{E}(f(X_t)|X_s).$$

We are going to state this property for a solution of equation (3.10). We shall denote by $(X_s^{t,x}, s \geq t)$ the solution of equation (3.10) starting from x at time t and by $X^x = X^{0,x}$ the solution starting from x at time 0. For $s \geq t, X_s^{t,x}$ satisfies

$$X_s^{t,x} = x + \int_t^s b(u, X_u^{t,x})du + \int_t^s \sigma(u, X_u^{t,x})dW_u.$$

A priori, $X^{t,x}$ is defined for any (t,x) almost surely. However, under the assumptions of Theorem 3.5.3, we can build a process depending on (t, x, s) that is almost surely continuous with respect to these variables and is a solution of the previous equation. This result is difficult to prove and the interested reader should refer to Rogers and Williams (1987) for the proof.

The Markov property is a consequence of the *flow* property of a solution of a stochastic differential equation that is itself an extension of the flow property of solutions of ordinary differential equations.

Lemma 3.5.6. *Under the assumptions of Theorem 3.5.3, if $s \geq t$,*

$$X_s^{0,x} = X_s^{t,X_t^x} \quad \mathbb{P} \ a.s.$$

Proof. We are only going to sketch the proof of this lemma. For any x, we have

$$\mathbb{P} \ a.s. \ X_s^{t,x} = x + \int_t^s b(u, X_u^{t,x})du + \int_t^s \sigma(u, X_u^{t,x})dW_u.$$

It follows that, $\mathbb{P}$ a.s. for any $y \in \mathbb{R}$,

$$X_s^{t,y} = y + \int_t^s b(u, X_u^{t,y})du + \int_t^s \sigma(u, X_u^{t,y})dW_u,$$

and also

$$X_s^{t,X_t^x} = X_t^x + \int_t^s b\left(u, X_u^{t,X_t^x}\right)du + \int_t^s \sigma\left(u, X_u^{t,X_t^x}\right)dW_u.$$

These results are intuitive, but they can be proved rigorously by using the continuity of $y \mapsto X^{t,y}$ We can also notice that X_s^x is also a solution of the previous equation.

Indeed, if $t \leq s$,

$$X_s^{0,x} = x + \int_0^s b(u, X_u^x)du + \int_0^s \sigma(u, X_u^x)dW_u$$

$$= X_t^x + \int_t^s b(u, X_u^x)du + \int_t^s \sigma(u, X_u^x)dW_u.$$

The uniqueness of the solution to this equation implies that $X_s^{0,x} = X_s^{t,X_t}$ for $t \leq s$.

In this case, the Markov property can be stated as follows: □

Theorem 3.5.7. *Let $(X_t)_{t\geq 0}$ be a solution of (3.10). It is a Markov process with respect to the filtration $(\mathscr{F}_t)_{t\geq 0}$. Furthermore, for any bounded Borel function f we have*

$$\mathbb{P} \text{ a.s. } \mathbb{E}(f(X_t)|\mathscr{F}_s) = \phi(X_s),$$

with $\phi(x) = \mathbb{E}(f(X_t^{s,x}))$.

Remark 3.5.8. The previous equality is often written as

$$\mathbb{E}(f(X_t)|\mathscr{F}_s) = \mathbb{E}(f(X_t^{s,x}))|_{x=X_s}.$$

Proof. Yet again, we shall only sketch the proof of this theorem. For a full proof, the reader ought to refer to Friedman (1975).

The flow property shows that if $s \leq t$, $X_t^x = X_t^{s,X_s^x}$. On the other hand, we can prove that $X_t^{s,x}$ is a measurable function of x and the Brownian increments $(W_{s+u} - W_s, u \geq 0)$ (this result is natural but it is quite tricky to justify (see Friedman (1975)). If we use this result for fixed s and t, we obtain $X_t^{s,x} = \Phi(x, W_{s+u} - W_s; u \geq 0)$ and thus

$$X_t^x = \phi(X_s^x, W_{s+u} - W_s; \ u \geq 0),$$

where X_s^x is $\mathscr{F}_s$-measurable and $(W_{s+u} - W_s)_{u\geq 0}$ is independent of $\mathscr{F}_s$.

If we apply the result of Proposition A.2.5 in the Appendix to X_s, $(W_{s+u} - W_s)_{u\geq 0}$, Φ and $\mathscr{F}_s$, it turns out that

$$\mathbb{E}(f(\Phi(X_s^x, W_{s+u} - W_s; u \geq 0))|\mathscr{F}_s)$$
$$= \mathbb{E}(f(\Phi(x, W_{s+u} - W_s; u \geq 0)))|_{x=X_s^x}$$
$$= \mathbb{E}(f(X_t^{s,x}))|_{x=X_s^x}.$$

The previous result can be extended to the case when we consider a function of the whole path of a diffusion after time s. In particular, the following theorem is useful when we do computations involving interest rate models. □

Theorem 3.5.9. *Let $(X_t)_{t\geq 0}$ be a solution of (3.10) and $r(s,x)$ be a nonnegative measurable function. For $t > s$,*

$$\mathbb{P} \text{ a.s. } \mathbb{E}\left(e^{-\int_s^t r(u,X_u)du} f(X_t)|\mathscr{F}_s\right) = \phi(X_s)$$

with

$$\phi(x) = \mathbb{E}\left(e^{-\int_s^t r(u,X_u^{s,x})du} f(X_t^{s,x})\right).$$

It is also written as

$$\mathbb{E}\left(e^{-\int_s^t r(u,X_u)du} f(X_t)|\mathscr{F}_s\right) = \mathbb{E}\left(e^{-\int_s^t r(u,X_u^{s,x})du} f(X_t^{s,x})\right)\Bigg|_{x=X_s}.$$

Remark 3.5.10. Actually, one can prove a more general result. Without getting into the technicalities, let us just mention that if ϕ is a *function of the whole path* of X_t after time s, the following stronger result is still true:

$$\mathbb{P} \text{ a.s. } \mathbb{E}(\phi(X_t^x, \ t \geq s)|\mathscr{F}_s) = \mathbb{E}(\phi(X_t^{s,x}, t \geq s))|_{x=X_s}.$$

Remark 3.5.11. When b and σ are independent of x (the diffusion is said to be homogeneous), we can show that the law of $X_{s+t}^{s,x}$ is the same as the one of $X_t^{0,x}$, which implies that if f is a bounded measurable function, then

$$\mathbb{E}(f(X_{s+t}^{s,x})) = \mathbb{E}(f(X_t^{0,x})).$$

We can extend this result and show that if r is a function of x only, then

$$\mathbb{E}\left(e^{-\int_s^{s+t} r(X_u^{s,x})du} f(X_{s+t}^{s,x})\right) = \mathbb{E}\left(e^{-\int_0^t r(X_u^{0,x})du} f(X_t^{0,x})\right).$$

In that case, Theorem 3.5.9 yields

$$\mathbb{E}\left(e^{-\int_s^t r(X_u)du} f(X_t)|\mathscr{F}_s\right) = \mathbb{E}\left(e^{-\int_0^{t-s} r(X_u^{0,x})du} f(X_{t-s}^{0,x})\right)\Bigg|_{x=X_s}.$$

3.6   Exercises

Exercise 9   Let $(M_t)_{t\geq 0}$ be a martingale such that for any t, $\mathbb{E}(M_t^2) < +\infty$. Prove that if $s \leq t$,

$$\mathbb{E}((M_t - M_s)^2|\mathscr{F}_s) = \mathbb{E}(M_t^2 - M_s^2|\mathscr{F}_s).$$

Exercise 10   Let X_t be a process with independent stationary increments and zero initial value such that for any t, $\mathbb{E}(X_t^2) < +\infty$. We shall also assume that the map $t \mapsto \mathbb{E}(X_t^2)$ is continuous. Prove that $\mathbb{E}(X_t) = ct$ and that $\mathrm{Var}(X_t) = c't$, where c and c' are two constants.

Exercise 11   Prove that, if τ is a stopping time,

$$\mathscr{F}_\tau = \{A \in \mathscr{A}, \text{ for all } t \geq 0, \ A \cap \{\tau \leq t\} \in \mathscr{F}_t\}$$

is a σ-algebra.

Exercise 12   Let S be a stopping time. Prove that S is $\mathscr{F}_S$-measurable.

Exercise 13   Let S and T be two stopping times such that $S \leq T$ $\mathbb{P}$ a.s. Prove that $\mathscr{F}_S \subset \mathscr{F}_T$.

Exercise 14   Let S be a stopping time almost surely finite, and $(X_t)_{t\geq 0}$ be an adapted process almost surely continuous.

1. Prove that, $\mathbb{P}$ a.s., for any s

$$X_s = \lim_{n \to +\infty} \sum_{k \geq 0} 1_{[k/n,(k+1)/n)}(s) X_{k/n}(\omega).$$

2. Prove that the mapping

$$([0,t] \times \Omega, \mathscr{B}([0,t]) \times \mathscr{F}_t) \mapsto (\mathbb{R}, \mathscr{B}(\mathbb{R}))$$
$$(s,\omega) \qquad\qquad \mapsto \quad X_s(\omega)$$

is measurable.

3. Conclude that if $S \leq t$, X_S is $\mathscr{F}_t$-measurable, and thus that X_S is $\mathscr{F}_S$-measurable.

Exercise 15   This exercise is an introduction to the concept of stochastic integration. We want to build an integral of the form $\int_0^{+\infty} f(s)dX_s$, where $(X_t)_{t \geq 0}$ is an $\mathscr{F}_t$-Brownian motion and $f(s)$ is a measurable function from $(\mathbb{R}^+, \mathscr{B}(\mathbb{R}^+))$ into $(\mathbb{R}, \mathscr{B}(\mathbb{R}))$ such that $\int_0^{+\infty} f^2(s)ds < +\infty$. This type of integral is called a *Wiener integral* and it is a particular case of Itô integral that is studied in Section 3.4.

We recall that the set $\mathscr{H}$ of functions of the form $\sum_{0 \leq i \leq N-1} a_i 1_{(t_i, t_{i+1}]}$, with $a_i \in \mathbb{R}$, and $t_0 = 0 \leq t_1 \leq \cdots \leq t_N$ is dense in the space $L^2(\mathbb{R}^+, dx)$ endowed with the norm $\|f\|_{L^2} = \left(\int_0^{+\infty} f^2(s)ds\right)^{1/2}$.

1. Consider $a_i \in \mathbb{R}$ and $0 = t_0 \leq t_1 \leq \cdots \leq t_N$, and call

$$f = \sum_{0 \leq i \leq N-1} a_i 1_{(t_i, t_{i+1}]}.$$

We define

$$I_e(f) = \sum_{0 \leq i \leq N-1} a_i (X_{t_{i+1}} - X_{t_i}).$$

Prove that $I_e(f)$ is a normal random variable and compute its mean and variance. In particular, show that

$$\mathbb{E}(I_e(f)^2) = \|f\|_{L^2}^2.$$

2. >From this, show that there exists a unique linear mapping I from $L^2(\mathbb{R}^+, dx)$ into $L^2(\Omega, \mathscr{F}, \mathbb{P})$, such that $I(f) = I_e(f)$, when f belongs to $\mathscr{H}$ and $\mathbb{E}(I(f)^2) = \|f\|_{L^2}$, for any f in $L^2(\mathbb{R}^+)$.

3. Prove that if $(X_n)_{n \geq 0}$ is a sequence of normal random variables with zero mean that converges to X in $L^2(\Omega, \mathscr{F}, \mathbb{P})$, then X is also a normal random variable with zero mean. Deduce that if $f \in L^2(\mathbb{R}^+, dx)$, then $I(f)$ is a normal random variable with zero mean and a variance equal to $\int_0^{+\infty} f^2(s)ds$.

4. We consider $f \in L^2(\mathbb{R}^+, dx)$, and we define

$$Z_t = \int_0^t f(s)dX_s = \int \mathbf{1}_{[0,t]}(s)f(s)dX_s.$$

Prove that Z_t is adapted to $\mathscr{F}_t$, and that $Z_t - Z_s$ is independent of $\mathscr{F}_s$. (Hint: begin with the case $f \in H$.)

5. Prove that the processes $Z_t, Z_t^2 - \int_0^t f^2(s)ds, \exp(Z_t - \frac{1}{2}\int_0^t f^2(s)ds)$ are $\mathscr{F}_t$-martingales.

Exercise 16 Let T be a positive real number and $(M_t)_{0 \leq t \leq T}$ be a continuous $\mathscr{F}_t$-martingale. We assume that $\mathbb{E}(M_T^2)$ is finite.

1. Prove that $(|M_t|)_{0 \leq t \leq T}$ is a submartingale.

2. Show that if $M^* = \sup_{0 \leq t \leq T} |M_t|$,

$$\lambda \mathbb{P}(M^* \geq \lambda) \leq \mathbb{E}(|M_T|\mathbf{1}_{\{M^* \geq \lambda\}}).$$

(Hint: apply the optional sampling theorem to the submartingale $(|M_t|)$ between $\tau \wedge T$ and T, where $\tau = \inf\{t \leq T, |M_t| \geq \lambda\}$ (if this set is empty, τ is equal to $+\infty$).)

3. >From the previous result, deduce that for positive A

$$\mathbb{E}((M^* \wedge A)^2) \leq 2\mathbb{E}((M^* \wedge A)|M_T|).$$

(Use the fact that $(M^* \wedge A)^p = \int_0^{M^* \wedge A} px^{p-1}dx$ for $p = 1, 2$.)

4. Prove that $\mathbb{E}(M^*)$ is finite and

$$\mathbb{E}\left(\sup_{0 \leq t \leq T} |M_t|^2\right) \leq 4\mathbb{E}(|M_T|^2).$$

Exercise 17

1. Prove that if S and S' are two $\mathscr{F}_t$-stopping times, then $S \wedge S' = \inf(S, S')$ and $S \vee S' = \sup(S, S')$ are also two $\mathscr{F}_t$-stopping times.

2. By applying the sampling theorem to the stopping time $S \vee s$, prove that

$$\mathbb{E}(M_S \mathbf{1}_{\{s>s\}}|\mathscr{F}_s) = M_s \mathbf{1}_{\{s>s\}}.$$

3. Deduce that for $s \leq t$,

$$\mathbb{E}(M_{S \wedge t} \mathbf{1}_{\{S>s\}}|\mathscr{F}_s) = M_s \mathbf{1}_{\{S>s\}}.$$

4. Remembering that $M_{S \wedge s}$ is $\mathscr{F}_s$-measurable, show that $t \to M_{S \wedge t}$ is an $\mathscr{F}_t$-martingale.

Exercise 18

1. Let $(H_t)_{0 \le t \le T}$ be an adapted measurable process such that $\int_0^T H_t^2 dt < \infty$, a.s. Let $M_t = \int_0^t H_s dW_s$. Show that if $\mathbb{E}(\sup_{0 \le t \le T} M_t^2) < \infty$, then $\mathbb{E}\left(\int_0^T H_t^2 dt\right) < \infty$. (Hint: introduce the sequence of stopping times $\tau_n = \inf\{t \ge 0 | \int_0^t H_s^2 ds = n\}$ and show that $E(M_{t \wedge \tau_n}^2) = \mathbb{E}(\int_0^{T \wedge \tau_n} H_s^2 ds).$)

2. Let $p(t, x) = 1/\sqrt{1-t} \exp(-x^2/2(1-t))$, for $0 \le t < 1$ and $x \in \mathbb{R}$, and $p(1, x) = 0$. Define $M_t = p(t, W_t)$, where $(W_t)_{0 \le t \le 1}$ is standard Brownian motion.

   (a) Prove that
   $$M_t = M_0 + \int_0^t \frac{\partial p}{\partial x}(s, W_s) dW_s.$$

   (b) Let
   $$H_t = \frac{\partial p}{\partial x}(t, W_t).$$
   Prove that $\int_0^1 H_t^2 dt < \infty$, a.s., and $\mathbb{E}\left(\int_0^1 H_t^2 dt\right) = +\infty$.

Exercise 19 Let $(M_t)_{0 \le t \le T}$ be a continuous $\mathscr{F}_t$-martingale equal to $\int_0^t K_s ds$, where $(K_t)_{0 \le t \le T}$ is an $\mathscr{F}_t$-adapted process such that $\int_0^T |K_s| ds < +\infty$ $\mathbb{P}$ a.s.

1. Moreover, we assume that $\mathbb{P}$ a.s. $\int_0^T |K_s| ds \le C < +\infty$. Prove that if we write $t_i^n = Ti/n$ for $0 \le i \le n$, then
   $$\lim_{n \to +\infty} \mathbb{E}\left(\sum_{i=1}^n \left(M_{t_i^n} - M_{t_{i-1}^n}\right)^2\right) = 0.$$

2. Under the same assumptions, prove that
   $$\mathbb{E}\left(\sum_{i=1}^n \left(M_{t_i^n} - M_{t_{i-1}^n}\right)^2\right) = \mathbb{E}(M_T^2 - M_0^2).$$
   Conclude that $M_T = 0$ $\mathbb{P}$ a.s., and thus $\mathbb{P}$ a.s. $\forall t \le T$, $M_t = 0$.

3. $\int_0^T |K_s| ds$ is now assumed to be finite almost surely as opposed to bounded. We shall accept the fact that the random variable $\int_0^t |K_s| ds$ is $\mathscr{F}_t$-measurable. Show that T_n defined by
   $$T_n = \inf\{0 \le s \le T\}, \int_0^t |K_s| ds \ge n\}$$
   (or T if this set is empty) is a stopping time. Prove that $\mathbb{P}$ a.s. $\lim_{n \to +\infty} T_n = T$. Considering the sequence of martingales $(M_{t \wedge T_n})_{t \ge 0}$, prove that
   $$\mathbb{P} \text{ a.s. } \forall t \le T, \ M_t = 0.$$

4. Let M_t be a martingale of the form $\int_0^t H_s dW_s + \int_0^t K_s ds$ with $\int_0^t H_s^2 ds < +\infty$ $\mathbb{P}$ a.s. and $\int_0^t |K_s| ds < +\infty$ $\mathbb{P}$ a.s. Define the sequence of stopping times $T_n = \inf\{t \le T, \int_0^t H_s^2 ds \ge n\}$, in order to prove that $K_t = 0$ $dt \times \mathbb{P}$ a.s.

Exercise 20 Let us call X_t the solution of the following stochastic differential equation:

$$\begin{cases} dX_t = (\mu X_t + \mu')dt + (\sigma X_t + \sigma')dW_t \\ X_0 = 0. \end{cases}$$

We write $S_t = \exp((\mu - \sigma^2/2)t + \sigma W_t)$.

1. Derive the stochastic differential equation satisfied by S_t^{-1}.

2. Prove that
$$d(X_t S_t^{-1}) = S_t^{-1}((\mu' - \sigma\sigma')dt + \sigma' dW_t).$$

3. Obtain the explicit representation of X_t.

Exercise 21 Let $(W_t)_{t \ge 0}$ be an $\mathscr{F}_t$-Brownian motion. The purpose of this exercise is to compute the joint distribution of $(W_t, \sup_{s \le t} W_s)$.

1. Consider S a bounded stopping time. Apply the optional sampling theorem to the martingale $M_t = \exp(izW_t + z^2 t/2)$, where z is a real number to prove that if $0 \le u \le v$, then

$$\mathbb{E}(\exp(iz(W_{v+s} - W_{u+S}))|\mathscr{F}_{u+S}) = \exp(-z^2(v-u)/2).$$

2. Deduce that $W_u^S = W_{u+S} - W_S$ is an $\mathscr{F}_{S+u}$-Brownian motion independent of the σ-algebra $\mathscr{F}_S$.

3. Let $(Y_t)_{t \ge 0}$ be a continuous stochastic process independent of the σ-algebra $\mathscr{B}$ such that
$$\mathbb{E}(\sup_{0 \le s \le K} |Y_s|) < +\infty.$$

Let T be a non-negative $\mathscr{B}$-measurable random variable bounded from above by K. Show that

$$\mathbb{E}(Y_T|\mathscr{B}) = \mathbb{E}(Y_t)|_{t=T}.$$

(Hint: assume first that T can be written as $\sum_{1 \le i \le n} t_i \mathbf{1}_{A_i}$, where $0 < t_1 < \cdots < t_n = K$, and the A_i's are disjoint $\mathscr{B}$-measurable sets.)

4. Let $\tau^\lambda = \inf\{s \ge 0, W_s > \lambda\}$. Prove that if f is a bounded Borel function, we have
$$\mathbb{E}\left(f(W_t)\mathbf{1}_{\{\tau^\lambda \le t\}}\right) = \mathbb{E}\left(\mathbf{1}_{\{\tau^\lambda \le t\}}\phi(t - \tau^\lambda)\right),$$
where $\phi(u) = \mathbb{E}(f(W_u + \lambda))$. Using $\mathbb{E}(f(W_u + \lambda)) = \mathbb{E}(f(-W_u + \lambda))$, deduce that
$$\mathbb{E}(f(W_t)\mathbf{1}_{\{\tau^\lambda \le t\}}) = \mathbb{E}(f(2\lambda - W_t)\mathbf{1}_{\{\tau^\lambda \le t\}}).$$

5. Show that if we write $W_t^* = \sup_{s \le t} W_s$ and if $\lambda \ge 0$,

$$\mathbb{P}(W_t \le \lambda, \ W_t^* \ge \lambda) = \mathbb{P}(W_t \ge \lambda, \ W_t^* \ge \lambda) = \mathbb{P}(W_t \ge \lambda).$$

Conclude that the random variables W_t^* and $|W_t|$ have the same distribution.

6. If $\lambda \ge \mu$ and $\lambda \ge 0$, prove that

$$\mathbb{P}(W_t \le \mu, \ W_t^* \ge \lambda) = \mathbb{P}(W_t \ge 2\lambda - \mu, W_t^* \ge \lambda) = \mathbb{P}(W_t \ge 2\lambda - \mu),$$

and if $\lambda \le \mu$ and $\lambda \ge 0$,

$$\mathbb{P}(W_t \le \mu, W_t^* \ge \lambda) = 2\mathbb{P}(W_t \ge \lambda) - \mathbb{P}(W_t \ge \mu).$$

7. Finally, check that the law of (W_t, W_t^*) is given by

$$\mathbf{1}_{\{0 \le y\}} \mathbf{1}_{\{x \le y\}} \frac{2(2y - x)}{\sqrt{2\pi t^3}} \exp\left(-\frac{(2y - x)^2}{2t}\right) \, dx dy.$$

Chapter 4

The Black-Scholes model

Black and Scholes (1973) tackled the problem of pricing and hedging a European option (call or put) on a non-dividend paying stock. Their method, which is based on similar ideas to those developed in discrete-time in Chapter 1 of this book, leads to some formulae frequently used by practitioners, despite the simplifying character of the model. In this chapter, we give an up-to-date presentation of their work. The case of the American option is investigated and some extensions of the model are presented.

4.1 Description of the model

4.1.1 The behavior of prices

The model suggested by Black and Scholes to describe the behavior of prices is a continuous-time model with one risky asset (a stock with price S_t at time t) and a riskless asset (with price S_t^0 at time t). We suppose the behavior of S_t^0 to be encapsulated by the following (ordinary) differential equation:

$$dS_t^0 = rS_t^0 dt, \tag{4.1}$$

where r is a non-negative constant. Note that r is an instantaneous interest rate and should not be confused with the one-period rate in discrete-time models. We set $S_0^0 = 1$, so that $S_t^0 = e^{rt}$ for $t \geq 0$.

We assume that the behavior of the stock price is determined by the following stochastic differential equation:

$$dS_t = S_t(\mu dt + \sigma dB_t), \tag{4.2}$$

where μ and σ are two constants and (B_t) is a standard Brownian motion. The constant σ is called the *volatility* of the asset.

The model is valid on the interval $[0, T]$, where T is the maturity of the option. As we saw previously (Chapter 3, Section 3.4.3), equation (4.2) has a closed-form solution,

$$S_t = S_0 \exp\left(\mu t - \frac{\sigma^2}{2}t + \sigma B_t\right),$$

where S_0 is the spot price observed at time 0. One particular result from this model is that the law of S_t is lognormal (i.e. its logarithm follows a normal law).

More precisely, we see that the process (S_t) is a solution of an equation of the type (4.2) if and only if the process $(\log(S_t))$ is a Brownian motion (not necessarily standard). According to Definition 3.2.1 of Chapter 3, the process (S_t) has the following properties:

- continuity of the sample paths;

- independence of the relative increments: if $u \leq t$, S_t/S_u or (equivalently) the relative increment $(S_t - S_u)/S_u$ is independent of the σ-algebra $\sigma(S_v, v \leq u)$;

- stationarity of the relative increments: if $u \leq t$, the law of $(S_t - S_u)/S_u$ is identical to the law of $(S_{t-u} - S_0)/S_0$.

These three properties express in concrete terms the hypotheses of Black and Scholes on the behavior of the stock price.

4.1.2   Self-financing strategies

A strategy will be defined as a process $\phi = (\phi_t)_{0 \leq t \leq T} = (H_t^0, H_t)_{0 \leq t \leq T}$ with values in $\mathbb{R}^2$, adapted to the natural filtration $(\mathcal{F}_t)$ of the Brownian motion; the components H_t^0 and H_t are the quantities of riskless asset and risky asset, respectively, held in the portfolio at time t. The value of the portfolio at time t is then given by

$$V_t(\phi) = H_t^0 S_t^0 + H_t S_t.$$

In the discrete-time models, we have characterized self-financing strategies by the equality $V_{n+1}(\phi) - V_n(\phi) = \phi_{n+1}.(S_{n+1} - S_n)$ (see Chapter 1, Remark 1.1.1). This equality is extended to give the self-financing condition in the continuous-time case:

$$dV_t(\phi) = H_t^0 dS_t^0 + H_t dS_t.$$

To give a meaning to this equality, we set the condition

$$\int_0^T |H_t^0| dt < +\infty \text{ a.s.} \quad \text{and} \quad \int_0^T H_t^2 < +\infty \text{ a.s.}$$

Then the integral

$$\int_0^T H_t^0 dS_t^0 = \int_0^T H_t^0 r e^{rt} dt$$

is well defined, as is the stochastic integral

$$\int_0^T H_t^0 dS_t^0 = \int_0^T (H_t S_t \mu) dt + \int_0^T \sigma H_t S_t dB_t,$$

since the map $t \mapsto S_t$ is continuous, and thus bounded on $[0, T]$ almost surely.

Definition 4.1.1. A self-financing strategy is defined by a pair ϕ of adapted processes $(H_t^0)_{0 \leq t \leq T}$ and $(H_t)_{0 \leq t \leq T}$ satisfying:

1. $\displaystyle\int_0^T |H_t^0| dt + \int_0^T H_t^2 dt < +\infty$ a.s..

2. $H_t^0 S_t^0 + H_t S_t = H_0^0 S_0^0 + H_0 S_0 + \displaystyle\int_0^t H_u^0 dS_u^0 + \int_0^t H_u dS_u$ a.s., for all $t \in [0, T]$.

We denote by $\tilde{S}_t = e^{-rt} S_t$ the discounted price of the risky asset. The following proposition is the counterpart of Proposition 1.1.2 of Chapter 1.

Proposition 4.1.2. Let $\phi = (H_t^0, H_t)_{0 \leq t \leq T}$ be an adapted process with values in $\mathbb{R}^2$, satisfying $\int_0^T |H_t^0| dt + \int_0^T H_t^2 dt < +\infty$ a.s.. Let $V_t(\phi) = H_t^0 S_t^0 + H_t S_t$ and $\tilde{V}_t(\phi) = e^{-rt} V_t(\phi)$. Then, ϕ defines a self-financing strategy if and only if

$$\tilde{V}_t(\phi) = V_0(\phi) + \int_0^t H_u d\tilde{S}_u \text{ a.s.,} \tag{4.3}$$

for all $t \in [0, T]$.

Proof. Suppose ϕ is a self-financing strategy. From the equality

$$d\tilde{V}_t(\phi) = -r\tilde{V}_t(\phi) dt + e^{-rt} dV_t(\phi),$$

which results from the differentiation of the product of the processes (e^{-rt}) and $(V_t(\phi))$ (the cross-variation term $d\langle e^{-r\cdot}, V.(\phi) \rangle_t$ is null), we deduce

$$\begin{aligned} d\tilde{V}_t(\phi) &= -re^{-rt}(H_t^0 e^{rt} + H_t S_t) dt + e^{-rt} H_t^0 d(e^{rt}) + e^{-rt} H_t dS_t \\ &= H_t(-re^{-rt} S_t dt + e^{-rt} dS_t) \\ &= H_t d\tilde{S}_t, \end{aligned}$$

which yields (4.3). The converse is justified similarly. □

Remark 4.1.3. We have not imposed any condition of *predictability* on strategies, unlike in Chapter 1. Actually, it is still possible to define a predictable process in continuous-time, but, in the case of the filtration of a Brownian motion, this does not restrict the class of adapted processes significantly (because of the continuity of sample paths of Brownian motion). In our study of complete discrete models, we had to consider at some stage a probability measure equivalent to the initial probability, under which discounted prices of assets are martingales. We were then able to design self-financing strategies replicating the option. The following section provides the tools that allow us to apply these methods in continuous-time.

4.2 Change of probability. Representation of martingales

4.2.1 Equivalent probabilities

Let $(\Omega, \mathscr{A}, \mathbb{P})$ be a probability space. A probability measure $\mathbb{Q}$ on $(\Omega, \mathscr{A})$ is absolutely continuous with respect to $\mathbb{P}$ if

$$\forall A \in \mathscr{A}, \quad \mathbb{P}(A) = 0 \Rightarrow \mathbb{Q}(A) = 0.$$

Theorem 4.2.1. *A probability measure $\mathbb{Q}$ is absolutely continuous with respect to $\mathbb{P}$ if an only if there exists a non-negative random variable Z on $(\Omega, \mathscr{A})$ such that*

$$\forall A \in \mathscr{A}, \quad \mathbb{Q}(A) = \int_A Z(\omega) d\mathbb{P}(\omega).$$

Z is called the density of $\mathbb{Q}$ with respect to $\mathbb{P}$ and denoted by $d\mathbb{Q}/d\mathbb{P}$.

The sufficiency is obvious, and the converse is a version of the Radon-Nikodym theorem (see for example Dacunha-Castelle and Duflo (1986b), or Williams (1991), Section 5.14).

The probabilities $\mathbb{P}$ and $\mathbb{Q}$ are *equivalent* if each of them is absolutely continuous with respect to the other. Note that if $\mathbb{Q}$ is absolutely continuous with respect to $\mathbb{P}$, with density Z, then $\mathbb{P}$ and $\mathbb{Q}$ are equivalent if and only if $\mathbb{P}(Z > 0) = 1$.

4.2.2 The Girsanov theorem

Let $(\Omega, \mathscr{F}, (\mathscr{F}_t)_{0 \le t \le T}, \mathbb{P})$ be a filtered probability space and $(B_t)_{0 \le t \le T}$ an $(\mathscr{F}_t)$-standard Brownian motion. The following theorem is known as the Girsanov theorem (for a proof, see Karatzas and Shreve (1988), or Dacunha-Castelle and Duflo (1986a), Chapter 8).

Theorem 4.2.2. *Let $(\theta_t)_{0 \le t \le T}$ be an adapted process satisfying $\int_0^T \theta_s^2 ds < \infty$ a.s. and such that the process $(L_t)_{0 \le t \le T}$ defined by*

$$L_t = \exp\left(-\int_0^t \theta_s dB_s - \frac{1}{2} \int_0^t \theta_s^2 ds \right)$$

is a martingale. Then, under the probability $\mathbb{P}^{(L)}$ with density L_T with respect to $\mathbb{P}$, the process $(W_t)_{0 \le t \le T}$ defined by $W_t = B_t + \int_0^t \theta_s ds$ is an $(\mathscr{F}_t)$-standard Brownian motion.

Remark 4.2.3. A sufficient condition for $(L_t)_{0 \le t \le T}$ to be a martingale is

$$\mathbb{E}\left(\exp\left(\frac{1}{2}\int_0^T \theta_t^2 \, dt\right)\right) < \infty,$$

This is known as Novikov's criterion (see Karatzas and Shreve (1988), Dacunha-Castelle and Duflo (1986a)). The proof of Girsanov's theorem when (θ_t) is constant is the purpose of Exercise 22.

4.2.3 Representation of Brownian martingales

Let $(B_t)_{0 \le t \le T}$ be a standard Brownian motion built on a probability space $(\Omega, \mathscr{F}, \mathbb{P})$ and let $(\mathscr{F}_t)_{0 \le t \le T}$ be its *natural filtration*. Recall (see Chapter 3, Proposition 3.4.4) that if $(H_t)_{0 \le t \le T}$ is an adapted process such that $\mathbb{E}(\int_0^T H_t^2 \, dt) < \infty$, the process $(\int_0^t H_s \, dB_s)$ is a square-integrable martingale, null at 0. The following theorem shows that any Brownian martingale can be represented in terms of a stochastic integral.

Theorem 4.2.4. Let $(M_t)_{0 \le t \le T}$ be a square-integrable martingale, with respect to the filtration $(\mathscr{F}_t)_{0 \le t \le T}$. There exists an adapted process $(H_t)_{0 \le t \le T}$ such that $\mathbb{E}(\int_0^T H_s^2 \, ds) < +\infty$ and

$$\forall t \in [0, T], \quad M_t = M_0 + \int_0^t H_s \, dB_s \quad a.s. \tag{4.4}$$

Note that this representation only applies to martingales relative to the *natural* filtration of the Brownian motion (cf. Exercise 29).

>From this theorem, it follows that if U is an $\mathscr{F}_T$-measurable, square-integrable random variable, it can be written as

$$U = \mathbb{E}(U) + \int_0^T H_s \, dB_s \quad a.s.,$$

where (H_t) is an adapted process such that $\mathbb{E}(\int_0^T H_t^2 \, ds) < +\infty$. To prove this, consider the martingale $M_t = \mathbb{E}(U | \mathscr{F}_t)$. It can also be shown (see, for example, Karatzas and Shreve (1988)) that if $(M_t)_{0 \le t \le T}$ is a martingale (not necessarily square-integrable), there is a representation similar to (4.4) with a process satisfying only $\int_0^T H_t^2 \, ds < \infty$, a.s. This result will be used in Chapter 6.

4.3 Pricing and hedging options in the Black-Scholes model

4.3.1 A probability under which $(\tilde{S}_t)$ is a martingale

We now consider the model introduced in Section 4.1. We will prove that there exists a probability equivalent to $\mathbb{P}$, under which the discounted stock price

$\tilde{S}_t = e^{-rt} S_t$ is a martingale. From the stochastic differential equation satisfied by (S_t), we have

$$d\tilde{S}_t = -re^{-rt} S_t dt + e^{-rt} dS_t$$
$$= \tilde{S}_t((\mu - r)dt + \sigma dB_t).$$

Consequently, if we set $W_t = B_t + (\mu - r)t/\sigma$,

$$d\tilde{S}_t = \tilde{S}_t \sigma dW_t. \tag{4.5}$$

>From Theorem 4.2.2, applied with $\theta_t = (\mu - r)/\sigma$, there exists a probability $\mathbb{P}^*$ equivalent to $\mathbb{P}$, under which $(W_t)_{0 \le t \le T}$ is a standard Brownian motion. It can be proved (see Exercise 28) that the definition of the stochastic integral is invariant by a change of equivalent probability. Then, under the probability $\mathbb{P}^*$, we deduce from (4.5) that $(\tilde{S}_t)$ is a martingale and that

$$\tilde{S}_t = \tilde{S}_0 \exp(\sigma W_t - \sigma^2 t/2).$$

4.3.2 Pricing

In this section, we will focus on European options. A European option will be defined by a non-negative, $\mathscr{F}_T$-measurable, random variable h. Quite often, h can be written as $f(S_T)$ ($f(x) = (x - K)_+$ in the case of a call, $f(x) = (K - x)_+$ in the case of a put). As in the discrete-time setting, we will define the option value by a replication argument. For technical reasons, we will limit our study to the following admissible strategies:

Definition 4.3.1. A strategy $\phi = (H_t^0, H_t)_{0 \le t \le T}$ is admissible if it is self-financing and if the discounted value $\tilde{V}_t(\phi) = H_t^0 + H_t \tilde{S}_t$ of the corresponding portfolio is, for all t, non-negative, and such that $\sup_{t \in [0,T]} \tilde{V}_t$ is square-integrable under $\mathbb{P}^*$.

An option is said to be replicable if its payoff at maturity is equal to the final value of an admissible strategy. It is clear that, for the option defined by h to be replicable, it is necessary that h should be square-integrable under $\mathbb{P}^*$. In the case of a call ($h = (S_T - K)_+$), this property indeed holds since $\mathbb{E}^*(S_T^2) < \infty$; note that in the case of a put, h is even bounded.

Theorem 4.3.2. In the Black-Scholes model, any option defined by a non-negative, $\mathscr{F}_T$-measurable random variable h, which is square-integrable under the probability $\mathbb{P}^*$, is replicable and the value at time t of any replicating portfolio is given by

$$V_t = \mathbb{E}^* \left(e^{-r(T-t)} h | \mathscr{F}_t \right).$$

Thus, the option value at time t can be naturally defined by the expression $\mathbb{E}^*(e^{-r(T-t)} h | \mathscr{F}_t)$.

Proof. First, assume that there is an admissible strategy (H^0, H), replicating the option. The value at time t of the portfolio (H_t^0, H_t) is given by

$$V_t = H_t^0 S_t^0 + H_t S_t,$$

and, by assumption, we have $V_T = h$. Let $\tilde{V}_t = V_t e^{-rt}$ be the discounted value

$$\tilde{V}_t = H_t^0 + H_t \tilde{S}_t.$$

Since the strategy is self-financing, we get from Proposition 4.1.2 and equality (4.5)

$$\tilde{V}_t = V_0 + \int_0^t H_u d\tilde{S}_u$$

$$= V_0 + \int_0^t H_u \sigma \tilde{S}_u dW_u.$$

Under the probability $\mathbb{P}^*$, $\sup_{t \in [0,T]} \tilde{V}_t$ is square-integrable, by definition of admissible strategies. Furthermore, the preceding equality shows that the process $(\tilde{V}_t)$ is a stochastic integral relative to (W_t). It follows (cf. Chapter 3, Proposition 3.4.4 and Exercise 18) that $(\tilde{V}_t)$ is a square-integrable martingale under $\mathbb{P}^*$. Hence

$$\tilde{V}_t = \mathbb{E}^* \left(\tilde{V}_T \mid \mathscr{F}_t \right),$$

and consequently

$$V_t = \mathbb{E}^* \left(e^{-r(T-t)} h \mid \mathscr{F}_t \right). \tag{4.6}$$

So we have proved that if a portfolio (H^0, H) replicates the option defined by h, its value is given by (4.6). To complete the proof of the theorem, it remains to show that the option is indeed replicable, i.e. to find some processes (H_t^0) and (H_t) defining an admissible strategy, such that

$$H_t^0 S_t^0 + H_t S_t = \mathbb{E}^* \left(e^{-r(T-t)} h \mid \mathscr{F}_t \right).$$

Under the probability $\mathbb{P}^*$, the process defined by $M_t = \mathbb{E}^*(e^{-rT} h \mid \mathscr{F}_t)$ is a square-integrable martingale. The filtration $(\mathscr{F}_t)$, which is the natural filtration of (B_t), is also the natural filtration of (W_t), and, from the Martingale Representation Theorem (cf. Theorem 4.2.4), there exists an adapted process $(K_t)_{0 \leq t \leq T}$ such that $\mathbb{E}^*(\int_0^T K_s^2 ds) < +\infty$ and

$$\forall t \in [0, T], \quad M_t = M_0 + \int_0^t K_s dW_s \text{ a.s.}$$

The strategy $\phi = (H^0, H)$, with $H_t = K_t/(\sigma \tilde{S}_t)$ and $H_t^0 = M_t - H_t \tilde{S}_t$, is then, from Proposition 4.1.2 and equality (4.5), a self-financing strategy; its value at time t is given by

$$V_t(\phi) = e^{rt} M_t = \mathbb{E}^* \left(e^{-r(T-t)} h \mid \mathscr{F}_t \right).$$

This expression clearly shows that $V_t(\phi)$ is a non-negative random variable, with $\sup_{0 \le t \le} V_t(\phi)$ square-integrable under $\mathbb{P}^*$ and that $V_T(\phi) = h$. We have found an admissible strategy replicating h. □

Remark 4.3.3. When the random variable h can be written as $h = f(S_T)$, we can express the option value V_t at time t as a function of t and S_t. We have indeed

$$V_t = \mathbb{E}^* \left(e^{-r(T-t)} f(S_T) \mid \mathscr{F}_t \right)$$
$$= \mathbb{E}^* \left(e^{-r(T-t)} f \left(S_t e^{r(T-t)} e^{\sigma(W_T - W_t) - (\sigma^2/2)(T-t)} \right) \mid \mathscr{F}_t \right).$$

The random variable S_t is $\mathscr{F}_t$-measurable and, under $\mathbb{P}^*$, $W_T - W_t$ is independent of $\mathscr{F}_t$. Therefore, from Proposition A.2.5 of the Appendix, we deduce

$$V_t = F(t, S_t),$$

where

$$F(t, x) = \mathbb{E}^* \left(e^{-r(T-t)} f \left(x e^{r(T-t)} e^{\sigma(W_T - W_t) - (\sigma^2/2)(T-t)} \right) \right). \qquad (4.7)$$

Since, under $\mathbb{P}^*$, $W_T - W_t$ is a zero-mean normal variable with variance $T - t$,

$$F(t, x) = e^{-r(T-t)} \int_{-\infty}^{+\infty} f \left(x e^{(r-\sigma^2/2)(T-t) + \sigma y \sqrt{T-t}} \right) \frac{e^{-y^2/2} dy}{\sqrt{2\pi}}.$$

The function F can be computed explicitly for calls and puts. If we take the case of the call, where $f(x) = (x - K)_+$, we have, from equality (4.7),

$$F(t, x) = \mathbb{E}^* \left(e^{-r(T-t)} \left(x e^{(r-\sigma^2/2)(T-t) + \sigma(W_T - W_t)} - K \right)_+ \right)$$
$$= \mathbb{E} \left(x e^{\sigma \sqrt{\theta} g - \sigma^2 \theta/2} - K e^{-r\theta} \right)_+,$$

where g is a standard Gaussian variable and $\theta = T - t$.
　　Let us set

$$d_1 = \frac{\log(x/K) + (r + \sigma^2/2)\theta}{\sigma \sqrt{\theta}} \quad \text{and} \quad d_2 = d_1 - \sigma \sqrt{\theta}.$$

Using these notations, we have

$$F(t, x) = \mathbb{E} \left[\left(x e^{\sigma \sqrt{\theta} g - \sigma^2 \theta/2} - K e^{-r\theta} \right) 1_{\{g + d_2 \ge 0\}} \right]$$
$$= \int_{-d_2}^{+\infty} \left(x e^{\sigma \sqrt{\theta} y - \sigma^2 \theta/2} - K e^{-r\theta} \right) \frac{e^{-y^2/2}}{\sqrt{2\pi}} dy$$
$$= \int_{-\infty}^{d_2} \left(x e^{\sigma \sqrt{\theta} y - \sigma^2 \theta/2} - K e^{-r\theta} \right) \frac{e^{-y^2/2}}{\sqrt{2\pi}} dy.$$

Writing this expression as the difference of two integrals and in the first one using the change of variable $z = y + \sigma\sqrt{\theta}$, we obtain

$$F(t, x) = xN(d_1) - Ke^{-r\theta}N(d_2), \tag{4.8}$$

where

$$N(d) = \frac{1}{\sqrt{2\pi}} \int_{-\infty}^{d} e^{-x^2/2} dx.$$

Using identical notations and through similar calculations, the price of the put is equal to

$$F(t, x) = Ke^{-r\theta}N(-d_2) - xN(-d_1). \tag{4.9}$$

The reader will find efficient methods to compute $N(d)$ in the Appendix.

4.3.3 Hedging calls and puts

In the proof of Theorem 4.3.2, we referred to the Martingale Representation Theorem to show the existence of a replicating portfolio. In practice, an existence theorem is not satisfactory and it is essential to be able to build a real replicating portfolio to hedge an option.

When the option is defined by a random variable $h = f(S_T)$, we show that it is possible to find an explicit hedging portfolio. A replicating portfolio must have, at any time t, a discounted value equal to

$$\tilde{V}_t = e^{-rt}F(t, S_t),$$

where F is the function defined by equality (4.7). Under large hypothesis on f (and, in particular, in the case of calls and puts where we have the closed-form solutions of Remark 4.3.3), we see that the function F is of class C^∞ on $[0, T) \times \mathbb{R}$. If we set

$$\tilde{F}(t, x) = e^{-rt}F(t, xe^{rt}),$$

we have $\tilde{V}_t = \tilde{F}(t, \tilde{S}_t)$ and, for $t < T$, from Itô's formula,

$$\tilde{F}(t, \tilde{S}_t) = \tilde{F}(0, \tilde{S}_0) + \int_0^t \frac{\partial \tilde{F}}{\partial x}(u, \tilde{S}_u) d\tilde{S}_u$$

$$+ \int_0^t \frac{\partial \tilde{F}}{\partial t}(u, \tilde{S}_u) du + \int_0^t \frac{1}{2} \frac{\partial^2 \tilde{F}}{\partial x^2}(u, \tilde{S}_u) d\langle \tilde{S}, \tilde{S}\rangle_u.$$

>From equality $d\tilde{S}_t = \sigma\tilde{S}_t dW_t$, we deduce

$$d\langle \tilde{S}, \tilde{S}\rangle_u = \sigma^2 \tilde{S}_u^2 du,$$

so that $\tilde{F}(t, \tilde{S}_t)$ can be written as

$$\tilde{F}(t, \tilde{S}_t) = \tilde{F}(0, \tilde{S}_0) + \int_0^t \sigma \frac{\partial \tilde{F}}{\partial x}(u, \tilde{S}_u)\tilde{S}_u dW_u + \int_0^t K_u du.$$

Since $\tilde{F}(t, \tilde{S}_t)$ is a martingale under $\mathbb{P}^*$, the process K_u is necessarily null (cf. Chapter 3, Exercise 19). Hence

$$\tilde{F}(t, \tilde{S}_t) = \tilde{F}(0, \tilde{S}_0) + \int_0^t \sigma \frac{\partial \tilde{F}}{\partial x}(u, \tilde{S}_u)\tilde{S}_u dW_u.$$
$$= \tilde{F}(0, \tilde{S}_0) + \int_0^t \frac{\partial \tilde{F}}{\partial x}(u, \tilde{S}_u)d\tilde{S}_u.$$

The natural candidate for the hedging process H_t is then

$$H_t = \frac{\partial \tilde{F}}{\partial x}(t, \tilde{S}_t) = \frac{\partial F}{\partial x}(t, S_t).$$

If we set $H_t^0 = \tilde{F}(t, \tilde{S}_t) - H_t\tilde{S}_t$, the portfolio (H_t^0, H_t) is self-financing and its discounted value is indeed $\tilde{V}_t = \tilde{F}(t, \tilde{S}_t)$.

Remark 4.3.4. The preceding argument shows that it is not absolutely necessary to use the Martingale Representation Theorem to deal with options of the form $f(S_T)$.

Remark 4.3.5. In the case of the call, we have, using the same notations as in Remark 4.3.3,

$$\frac{\partial F}{\partial x}(t, x) = N(d_1),$$

and, in the case of a put,

$$\frac{\partial F}{\partial x}(t, x) = -N(-d_1).$$

This is left as an exercise (the easiest way is to differentiate under the expectation sign). This quantity is often called the 'delta' of the option by practitioners. More generally, when the value at time t of a portfolio can be expressed as $\Psi(t, S_t)$, the quantity $(\partial \Psi/\partial x)(t, S_t)$, which measures the sensitivity of the portfolio with respect to the variations of the asset price at time t, is called the 'delta' of the portfolio. Similarly, 'gamma' refers to the second-order derivative $(\partial^2 \Psi/\partial x^2)(t, S_t)$, 'theta' to the derivative with respect to time and 'vega' to the derivative of Ψ with respect to the volatility σ.

4.4   American options

4.4.1   Pricing American options

We have seen in Chapter 2 how the pricing of American options and the optimal stopping problem are related in a discrete-time setting. The theory of optimal stopping in continuous-time is based on the same ideas as in discrete-time but is far more complex technically speaking. The approach we proposed in Section 1.3.3 of Chapter 1, based on an induction argument, cannot be used directly to price American options. Exercise 8 in Chapter 2 shows that, in a discrete model, it is possible to associate with any American option a hedging scheme with consumption.

Definition 4.4.1. A trading strategy with consumption is defined as an adapted process $\phi = (H_t^0, H_t)_{0 \leq t \leq T}$, with values in $\mathbb{R}^2$, satisfying the following properties:

1. $\int_0^T |H_t^0| dt + \int_0^T H_t^2 dt < +\infty \quad a.s.$

2. $H_t^0 S_t^0 + H_t S_t = H_0^0 S_0^0 + H_0 S_0 + \int_0^t H_u^0 dS_u^0 + \int_0^t H_u dS_u - C_t$ for all $t \in [0, T]$, where $(C_t)_{0 \leq t \leq T}$ is an adapted, continuous, non-decreasing process null at $t = 0$; C_t corresponds to the cumulative consumption up to time t.

An American option is naturally defined by an adapted non-negative process $(h_t)_{0 \leq t \leq T}$. For the sake of simplicity, we will only consider payoff processes of the form $h_t = \psi(S_t)$, where ψ is a continuous function from $\mathbb{R}^+$ to $\mathbb{R}^+$, satisfying $\psi(x) \leq A + Bx, \forall x \in \mathbb{R}^+$, for some non-negative constants A and B. For a call, we have $\psi(x) = (x - K)_+$ and for a put $\psi(x) = (K - x)_+$.

The trading strategy with consumption $\phi = (H_t^0, H_t)_{0 \leq t \leq T}$ is said to hedge the American option defined by $h_t = \psi(S_t)$ if, setting $V_t(\phi) = H_t^0 S_t^0 + H_t S_t$, we have

$$\forall t \in [0, T], \quad V_t(\phi) \geq \psi(S_t) \quad a.s.$$

Denote by Φ^ψ the set of all trading strategies with consumption hedging the American option defined by $h_t = \psi(S_t)$. If the writer of the option follows a strategy $\phi \in \Phi^\psi$, he or she possesses at any time t a wealth at least equal to $\psi(S_t)$, which is precisely the payoff if the option is exercised at time t. The following theorem introduces the minimal value of a hedging strategy for an American option.

Theorem 4.4.2. Let u be the map from $[0, T] \times \mathbb{R}^+$ to $\mathbb{R}$ defined by

$$u(t, x) = \sup_{\tau \in \mathscr{T}_{t,T}} \mathbb{E}^*[e^{-r(\tau - t)} \psi(x \exp((r - (\sigma^2/2))(\tau - t) + \sigma(W_\tau - W_t)))],$$

where $\mathscr{T}_{t,T}$ is the set of all stopping times with values in $[t, T]$. There exists a strategy $\tilde{\phi} \in \Phi^\psi$, such that $V_t(\tilde{\phi}) = u(t, S_t)$, for all $t \in [0, T]$. Moreover, for any strategy $\phi \in \Phi^\psi$, we have $V_t(\phi) \geq u(t, S_t)$, for all $t \in [0, T]$.

In order to avoid technical difficulties, we only sketch the proof (see Karatzas and Shreve (1988) for details). First, one shows that the process $(e^{-rt} u(t, S_t))$ is the Snell envelope of the process $(e^{-rt} \psi(S_t))$, i.e. the smallest right-continuous $\mathbb{P}^*$-supermartingale that dominates it. As it can be proved that the discounted value of a trading strategy with consumption is a supermartingale under $\mathbb{P}^*$, we deduce the inequality $V_t(\phi) \geq u(t, S_t)$, for any strategy $\phi \in \Phi^\psi$.

To show the existence of a strategy $\bar{\phi}$ such that $V_t(\bar{\phi}) = u(t, S_t)$, we have to use a decomposition theorem for supermartingales similar to Proposition 2.3.1 of Chapter 2 as well as the Representation Theorem for Brownian martingales.

It is natural to consider $u(t, S_t)$ as a price for the American option at time t, since it is the minimal value of a strategy hedging the option.

Remark 4.4.3. Let τ be a stopping time with values in $[0, T]$. The value at time 0 of an admissible strategy in the sense of Definition 4.3.1 with value $\psi(S_\tau)$ at time τ is given by $\mathbb{E}^*(e^{-r\tau}\psi(S_\tau))$, since the discounted value of any admissible strategy is a martingale under $\mathbb{P}^*$. Thus the quantity $u(0, S_0) = \sup_{\tau \in \mathscr{T}_{0,T}} \mathbb{E}^*(e^{-r\tau}\psi(S_\tau))$ is the minimal initial wealth that hedges the whole range of possible exercises.

As in discrete models, we notice that the American call price (on a non-dividend paying stock) is equal to the European call price:

Proposition 4.4.4. *If in Theorem 4.4.2, ψ is given by $\psi(x) = (x - K)_+$, for all real x, then we have*
$$u(t, x) = F(t, x),$$
where F is the function defined by equation (4.8) corresponding to the European call price.

Proof. We assume here that $t = 0$ (the proof is the same for $t > 0$). Then it is sufficient to show that, for any stopping time τ,
$$\mathbb{E}^*(e^{r\tau}(S_\tau - K)_+) \leq \mathbb{E}^*(e^{-rT}(S_T - K)_+) = \mathbb{E}^*(\tilde{S}_T - e^{-rT}K)_+.$$

On the other hand, we have
$$\mathbb{E}^*\left((\tilde{S}_T - e^{-rT}K)_+ \mid \mathscr{F}_\tau\right) \geq \mathbb{E}^*\left((\tilde{S}_T - e^{-rT}K) \mid \mathscr{F}_\tau\right) = \tilde{S}_\tau - e^{-rT}K$$

since $(\tilde{S}_t)$ is a martigale under $\mathbb{P}^*$. Hence
$$\mathbb{E}^*\left((\tilde{S}_T - e^{-rT}K)_+ \mid \mathscr{F}_\tau\right) \geq \tilde{S}_\tau - e^{-r\tau}K$$

since $r \geq 0$ and by non-negativity of the left-hand term,
$$\mathbb{E}^*\left((\tilde{S}_T - e^{-rT}K)_+ \mid \mathscr{F}_\tau\right) \geq (\tilde{S}_\tau - e^{-r\tau}K)_+.$$

We obtain the desired inequality by taking expectations.   □

4.4.2 Perpetual puts, critical price

In the case of the put, the American option price is not equal to the European one and there is no closed-form solution for the function u. One has to use numerical methods; we present some of them in Chapter 5. In this section we will only use the formula

$$u(t,x) = \sup_{\tau \in \mathscr{T}_{t,T}} \mathbb{E}^*(Ke^{-r(\tau-t)} - x \exp(-\sigma^2(\tau - t)/2 + \sigma(W_\tau - W_t)))_+$$

$$(4.10)$$

to derive some qualitative properties of the function u. To make our point clearer, we assume $t = 0$. In fact, it is always possible to come down to this case by replacing T with $T - t$. Equation (4.10) becomes

$$u(0,x) = \sup_{\tau \in \mathscr{T}_{0,T}} \mathbb{E}^*(Ke^{-r\tau} - x \exp(\sigma W_\tau - \sigma^2 \tau/2))_+. \qquad (4.11)$$

Let us consider a probability space $(\Omega, \mathscr{F}, \mathbb{P})$, and let $(B_t)_{0 \le t < \infty}$ be a standard Brownian motion with infinite time horizon defined on this space. Then, we get

$$u(0,x) = \sup_{\tau \in \mathscr{T}_{0,T}} \mathbb{E}(Ke^{-r\tau} - x \exp(\sigma B_\tau - (\sigma^2 \tau/2)))_+$$

$$\le \sup_{\tau \in \mathscr{T}_{0,\infty}} \mathbb{E}\left[(Ke^{-r\tau} - x \exp(\sigma B_\tau - (\sigma^2 \tau/2)))_+ \mathbf{1}_{\{\tau < \infty\}}\right], \quad (4.12)$$

where $\mathscr{T}_{0,\infty}$ denotes the set of all stopping times of the natural filtration of $(B_t)_{t \ge 0}$ and $\mathscr{T}_{0,T}$ the set of all elements of $\mathscr{T}_{0,\infty}$ with values in $[0,T]$. The right-hand side in inequality (4.12) can be interpreted naturally as the value of a 'perpetual' put (i.e. it can be exercised at any time). The following proposition gives an explicit expression for the upper bound in (4.12).

Proposition 4.4.5. *The function*

$$u^\infty(x) = \sup_{\tau \in \mathscr{T}_{0,\infty}} \mathbb{E}\left[(Ke^{-r\tau} - x \exp(\sigma B_\tau - (\sigma^2 \tau/2)))_+ \mathbf{1}_{\{\tau < \infty\}}\right] \qquad (4.13)$$

is given by the formulae

$$u^\infty(x) = \begin{cases} K - x, & \text{for } x \le x^* \\ (K - x^*)\left(\frac{x}{x^*}\right)^{-\gamma}, & \text{for } x > x^* \end{cases}$$

with $x^* = K\gamma/(1+\gamma)$ *and* $\gamma = 2r/\sigma^2$.

Proof. From formula (4.13) we deduce that the function u^∞ is convex, decreasing on $[0, \infty)$ and satisfies $u^\infty(x) \ge (K - x)_+$, and, for any $T > 0$, $u^\infty(x) \ge \mathbb{E}(Ke^{-rT} - x \exp(\sigma B_T - (\sigma^2 T/2)))_+$, which implies $u^\infty(x) > 0$, for all $x \ge 0$. Now let $x^* = \sup\{x \ge 0 \mid u^\infty(x) = K - x\}$. From the properties of u^∞ we have just stated, it follows that

$$\forall x \le x^*, \quad u^\infty(x) = K - x \quad \text{and} \quad \forall x > x^* \quad u^\infty(x) > (K - x)_+. \qquad (4.14)$$

Now, given $x \in [0, +\infty)$, denote by $(X_t^x)_{t \geq 0}$ the process defined by $X_t^x = x \exp((r - \sigma^2/2)t + \sigma B_t)$. The theory of the Snell envelope in continuous-time (cf. El Karoui (1981), Kushner (1977), as well as Chapter 5) enables us to show

$$u^\infty(x) = \mathbb{E}\left[(Ke^{-r\tau_x} - x\exp(\sigma B_{\tau_x} - (\sigma^2 \tau_x/2)))_+ 1_{\{\tau_x < \infty\}}\right],$$

where τ_x is the stopping time defined by

$$\tau_x = \inf\{t \geq 0 \,|\, e^{-rt} u^\infty(X_t^x) = e^{-rt}(K - X_t^x)_+\} \quad \text{(with } \inf \emptyset = +\infty\text{)}.$$

The stopping time τ_x is therefore an optimal stopping time (note the analogy with the results in Chapter 2).

It follows from (4.14) that

$$\tau_x = \inf\{t \geq 0 \,|\, X_t^x \leq x^*\} = \inf\{t \geq 0 \,|\, (r - \sigma^2/2)t + \sigma B_t \leq \log(x^*/x)\}.$$

Introduce, for any $z \in \mathbb{R}_+$, the stopping time $\tau_{x,z}$ defined by

$$\tau_{x,z} = \inf\{t \geq 0 \,|\, X_t^x \leq z\}.$$

With these notations, the optimal stopping time is given by $\tau_x = \tau_{x,x^*}$. We fix x and denote by ϕ the function of z defined by

$$\phi(z) = \mathbb{E}\left(e^{-r\tau_{x,z}} 1_{\{\tau_{x,z} < \infty\}}\left(K - X_{\tau_{x,z}}^x\right)_+\right).$$

Since τ_{x,x^*} is optimal, the function ϕ attains its maximum at point $z = x^*$. We are going to calculate ϕ explicitly, then we will maximize it to determine x^* and $u^\infty(x) = \phi(x^*)$.

If $z > x$, it is obvious that $\tau_{x,z} = 0$ and $\phi(z) = (K - x)_+$. If $z \leq x$, we have, by the continuity of the paths of $(X_t^x)_{t \geq 0}$,

$$\tau_{x,z} = \inf\{t \geq 0 \,|\, X_t^x = x\}$$

and consequently

$$\phi(z) = (K - z)_+ \mathbb{E}\left(e^{-r\tau_{x,z}} 1_{\{\tau_{x,z} < \infty\}}\right)$$
$$= (K - z)_+ \mathbb{E}(e^{-r\tau_{x,z}})$$

with, by convention, $e^{-r\infty} = 0$. Using the expression of X_t^x in terms of B_t, we see that, for $z \leq x$,

$$\tau_{x,z} = \inf\left\{t \geq 0 \,\Big|\, \left(r - \frac{\sigma^2}{2}\right)t + \sigma B_t = \log(z/x)\right\}$$
$$= \inf\left\{t \geq 0 \,\Big|\, \mu t + B_t = \frac{1}{\sigma}\log(z/x)\right\},$$

with $\mu = r/\sigma - \sigma/2$. Thus, if we introduce, for any real number b,

$$T_b = \inf\{t \geq 0 \,|\, \mu t + B_t = b\},$$

we get

$$\phi(z) = \begin{cases} (K-x)_+ & \text{if } z > x \\ (K-z)\mathbb{E}(\exp(-rT_{\log(z/x)/\sigma})) & \text{if } z \in [0,x] \cap [0,K] \\ 0 & \text{if } z \in [0,x] \cap [K,+\infty). \end{cases}$$

The maximum of ϕ is attained on the interval $[0,x] \cap [0,K]$. Using the following formula (proved in Exercise 27):

$$\mathbb{E}(e^{-\alpha T_b}) = \exp\left(\mu b - |b|\sqrt{\mu^2 + 2\alpha}\right),$$

it can be seen that

$$\forall z \in [0,x] \cap [0,K] \quad \phi(z) = (K-z)\left(\frac{z}{x}\right)^\gamma,$$

where $\gamma = 2r/\sigma^2$. The derivative of this function is given by

$$\phi'(z) = \frac{z^{\gamma-1}}{x^\gamma}(K\gamma - (\gamma+1)z).$$

It results that if $x \leq K\gamma/(\gamma+1)$,

$$\max_z \phi(z) = \phi(x) = K - x$$

and if $x > K\gamma/(\gamma+1)$,

$$\max_z \phi(z) = \phi(K\gamma/(\gamma+1)),$$

and we recognize the required expressions. □

Remark 4.4.6. Let us go back to the American put with finite maturity T. Following the same arguments as in the beginning of the proof of Proposition 4.4.5, we see that, for any $t \in [0,T)$, there exists a real number $s(t) \in [0,K]$ satisfying

$$\forall x \leq s(t), \quad u(t,x) = K - x \quad \text{and} \quad \forall x > s(t), \quad u(t,x) > (K-x)_+. \quad (4.15)$$

Taking inequality (4.12) into account, we obtain $s(t) \geq x^*$, for all $t \in [0,T)$. The real number $s(t)$ is interpreted as the 'critical price' at time t: if the price of the underlying asset at time t is less than or equal to $s(t)$, the buyer of the option should exercise his or her option immediately; in the opposite case, he should keep it.

4.5 Implied volatility and local volatility models

One of the main features of the Black-Scholes model (and one of the reasons for its success) is the fact that the pricing formulae, as well as the hedging formulae, depend on only one non-observable parameter: the volatility σ (the drift parameter μ disappears by change of probability). In practice, two methods are used to evaluate σ:

1. The historical method: in the Black-Scholes model, $\sigma^2 T$ is the variance of $\log(S_T)$ and the variables

$$\log(S_T/S_0), \log(S_{2T}/S_T), \ldots, \log(S_{NT}/S_{(N-1)T})$$

   are independent and identically distributed. Therefore, σ can be estimated by statistical means using past observations of the asset prices (for example by calculating empirical variances; cf. Dacunha-Castelle and Duflo (1986b), Chapter 5).

2. The 'implied volatility' method: some options are quoted on organised markets; the price of options (calls and puts) being an increasing function of σ (cf. Exercise 24), we can associate an 'implied' volatility to each quoted option, by inversion of the Black-Scholes formula. Once the model is identified, it can be used to elaborate hedging schemes.

In those problems concerning volatility, one is soon confronted with the imperfections of the Black-Scholes model. Important differences between historical volatility and implied volatility are observed, the latter seeming to depend upon the strike price and the maturity. In spite of these inconsistencies, the model is used as a reference by practitioners, the implied volatility of an option being more meaningful to them than its price.

One way to construct a model that will be more consistent with market data is to replace the constant volatility σ by a stochastic process (σ_t), so that (4.2) becomes

$$dS_t = S_t(\mu dt + \sigma_t dB_t).$$

If the stochastic process σ_t is adapted to the natural filtration of $(B_t)_{0 \le t \le T}$ and bounded, as well as the process $(1/S_t)$, the approach of this chapter can be extended (see Problem 1 for the case of a deterministic function of time and Problem 5 for the general case). More precisely, European options can still be replicated and option prices can be computed using the probability $\mathbb{P}^*$, under which the discounted asset price is a martingale. In the so called *local volatility model*, we have

$$\sigma_t = \sigma(t, S_t),$$

where σ is now a deterministic function of time and the current price. It was observed in Dupire (1994) and Derman and Kani (1994) that, given market prices of call options, one can construct a local volatility model that will provide the same prices as the market. More precisely, if $C(T, K)$ is the market price, observed at time 0, of a call with strike price K and maturity T, the consistent local volatility is given by the following equation, known as Dupire's formula (see Problem 6 for a proof in the case $r = 0$):

$$\frac{\partial C}{\partial T}(T, K) = \frac{\sigma^2(T, K) K^2}{2} \frac{\partial^2 C}{\partial K^2}(T, K) - rK \frac{\partial C}{\partial K}(T, K). \qquad (4.1)$$

In practice, this formula is not easy to implement, as it involves partial derivatives of market prices with respect to strike and maturity, while only finitely many options are quoted. Local volatility models also have some unstability features, so that practitioners prefer to use more sophisticated models that combine stochastic volatility models and models with jumps.

Remark 4.5.1. The class of *stochastic volatility models* generally refers to models in which the volatility process (σ_t) satifies a stochastic differential equation governed by another Brownian motion (which may or may not be correlated with (B_t)). The volatility process can no longer be adapted to the natural filtration of (B_t). In fact, stochastic volatility models are *incomplete* models, in which replication of options by trading in the underlying asset may not be possible. See Chapter 7 for an introduction to incomplete markets.

4.6 The Black-Scholes model with dividends and call/put symmetry

In our discussion of the Black-Scholes approach to option pricing, we have implicitly assumed that the underlying asset does not distribute dividends. We will now show how the Black-Scholes methodology can be extended to options on a dividend paying stock, when dividends are paid in continuous-time, at a constant rate δ. This means that, in the infinitesimal time interval $[t, t + dt]$, the holder of one share receives $\delta S_t dt$. The interest of this rather unrealistic assumption is that it leads to closed-form formulae. The assumption becomes more realistic in the framework of foreign exchange options, where the dividend yield δ can be interpreted as the foreign interest rate (see Problem 2).

In the context of a dividend paying asset, the self-financing condition takes the form

$$dV_t = H_t^0 dS_t^0 + H_t(dS_t + \delta S_t dt),$$

and, in discounted terms,

$$d\tilde{V}_t = H_t(d\tilde{S}_t + \delta \tilde{S}_t dt) = H_t \sigma \tilde{S}_t dW_t^\delta,$$

where $W_t^\delta = B_t + (\mu + \delta - r)t/\sigma$. The pricing measure is now the probability $\mathbb{P}^\delta$ under which $(W_t^\delta)_{0 \leq t \leq T}$ is a standard Brownian motion. Note that, under $\mathbb{P}^\delta$, the process $(e^{(\delta-r)t} S_t)_{0 \leq t \leq T}$ is a martingale (see Problem 2 for details in the context of foreign exchange options).

We will establish an interesting symmetry relation between call and put prices. In order to clarify our statement, denote by $C_e(t, x; K, r, \delta)$ (resp. $P_e(t, x; K, r, \delta)$) the price at time t of a European call (resp. put) option with maturity T, and exercise price K, for a current stock price x, when the interest rate is r and the dividend yield is δ. Similarly, we denote by $C_a(t, x; K, r, \delta)$ (resp. $P_a(t, x; K, r, \delta)$) the price of the American call (resp. put). Note that

$$C_e(t, x; K, r, \delta) = \mathbb{E}^\delta e^{-r(T-t)} \left(x e^{(r-\delta-(\sigma^2/2))(T-t) + \sigma(W_T^\delta - W_t^\delta)} - K \right)_+.$$

Proposition 4.6.1. *We have*

$$C_e(t, x; K, r, \delta) = P_e(t, K; x, \delta, r) \quad and \quad C_a(t, x; K, r, \delta) = P_a(t, K; x, \delta, r).$$

Proof. We prove the result for American options. The case of European options is easier. We also assume that $t = 0$ for simplicity. We have

$$C_a(0, x; K, r, \delta) = \sup_{\tau \in \mathscr{T}_{0,T}} \mathbb{E}^\delta e^{-r\tau} \left(x e^{(r-\delta-\frac{\sigma^2}{2})\tau + \sigma W_\tau^\delta} - K \right)_+.$$

For $\tau \in \mathscr{T}_{0,T}$, we have, with the notation $\hat{W}_t^\delta$ for $W_t^\delta - \sigma t$ and $\hat{\mathbb{P}}^\delta$ for the probability measure with density given by $d\hat{\mathbb{P}}^\delta / d\mathbb{P}^\delta = e^{\sigma W_T^\delta - (\sigma^2/2)T}$,

$$\mathbb{E}^\delta \, e^{-r\tau} \left(x e^{(r-\delta-\frac{\sigma^2}{2})\tau + \sigma W_\tau^\delta} - K \right)^+ =$$

$$= \mathbb{E}^\delta e^{-\delta\tau} e^{\sigma W_\tau^\delta - (\sigma^2/2)\tau} \left(x - K e^{(\delta-r+\frac{\sigma^2}{2})\tau - \sigma W_\tau^\delta} \right)^+$$

$$= \mathbb{E}^\delta e^{-\delta\tau} e^{\sigma W_T^\delta - (\sigma^2/2)T} \left(x - K e^{(\delta-r-\frac{\sigma^2}{2})\tau - \sigma \hat{W}_\tau^\delta} \right)^+,$$

where the last equality comes from the fact that $(e^{\sigma W_t^\delta - (\sigma^2/2)t})_{t \geq 0}$ is a martingale. Therefore,

$$\mathbb{E}^\delta e^{-r\tau} \left(x e^{(r-\delta-\frac{\sigma^2}{2})\tau + \sigma W_\tau} - K \right)^+ = \hat{\mathbb{E}}^\delta e^{-\delta\tau} \left(x - K e^{(\delta-r-\frac{\sigma^2}{2})\tau - \sigma \hat{W}_\tau^\delta} \right)^+.$$

Now, under probability $\hat{\mathbb{P}}^\delta$, the process $(\hat{W}_t^\delta)_{0 \leq t \leq T}$ is a standard Brownian motion, as well as, by symmetry, the process $(-\hat{W}_t^\delta)_{0 \leq t \leq T}$. Hence, $C_a(t, x; K, r, \delta) = P_a(t, K; x, \delta, r)$. □

Notes: The presentation we have used, based on the Girsanov theorem, is inspired by Harrison and Pliska (1981) (also refer to Bensoussan (1984) and Section 5.8 in Karatzas and Shreve (1988)). The initial approach of Black and Scholes (1973) and Merton (1973) consisted in deriving a partial differential equation satisfied by the call price as a function of time and spot price. It is based on a no-arbitrage argument and the Itô formula. For more information on the statistical estimation of model parameters, the reader should refer to Dacunha-Castelle and Duflo (1986b) and Dacunha-Castelle and Duflo (1986a) and to the references in these books.

4.7 Exercises

Exercise 22 The objective of this exercise is to prove the Girsanov theorem 4.2.2 in the special case where the process (θ_t) is constant. Let $(B_t)_{0 \leq t \leq T}$ be a standard Brownian motion with respect to the filtration $(\mathscr{F}_t)_{0 \leq t \leq T}$ and let μ be a real number. We set, for $0 \leq t \leq T$, $L_t = \exp(-\mu B_t - (\mu^2/2)t)$.

1. Show that $(L_t)_{0 \leq t \leq T}$ is a martingale relative to the filtration $(\mathscr{F}_t)$ and that $\mathbb{E}(L_t) = 1$, for all $t \in [0, T]$.

2. Let $\mathbb{P}^{(L_t)}$ be the probability with density L_t with respect to the initial probability $\mathbb{P}$. Show that the probabilities $\mathbb{P}^{(L_T)}$ and $\mathbb{P}^{(L_t)}$ coincide on the σ-algebra $\mathscr{F}_t$.

3. Let Z be an $\mathscr{F}_T$-measurable, bounded random variable. Show that the conditional expectation of Z, under the probability $\mathbb{P}^{(L_T)}$, given $\mathscr{F}_t$, is

$$\mathbb{E}^{(L_T)}(Z \mid \mathscr{F}_t) = \frac{\mathbb{E}(ZL_T \mid \mathscr{F}_t)}{L_t}.$$

This is known as the *Bayes rule for conditional expectations.*

4. Let $W_t = \mu t + B_t$, for all $t \in [0, T]$. Show that for any real number u and for any s and t in $\in [0, T]$, with $s \leq t$, we have

$$\mathbb{E}^{(L_T)} \left(e^{iu(W_t - W_s)} \mid \mathscr{F}_s \right) = e^{-u^2(t-s)/2}.$$

Conclude using Proposition A.2.2 of the Appendix.

Exercise 23 Show that the portfolio replicating a European option in the Black-Scholes model is unique (in a sense to be specified).

Exercise 24 Consider an option described by $h = f(S_T)$ and denote by F the function of time and spot price corresponding to the option price (cf. equation (4.7)).

1. Show that if f is non-decreasing (resp. non-increasing), $F(t, x)$ is a non-decreasing (resp. non-increasing) function of x.

2. Assume that f is convex. Show that $F(t, x)$ is a convex function of x, a decreasing function of t if $r = 0$ and a non-decreasing function of σ in any case. (Hint: first consider equation (4.7) and make use of Jensen's inequality: $\Phi(\mathbb{E}(X)) \leq \mathbb{E}(\Phi(X))$, where Φ is a convex function and X is a random variable such that X and $\Phi(X)$ are integrable.)

3. Denote by F_c (resp. F_p) the function F obtained when $f(x) = (x - K)_+$ (resp. $f(x) = (K - x)_+$). Prove that $F_c(t, .)$ and $F_p(t, .)$ are non-negative for $t < T$. Study the functions $F_c(t, .)$ and $F_p(t, .)$ in the neighbourhood of 0 and $+\infty$.

Exercise 25 Calculate under the initial probability $\mathbb{P}$, the probability that a European call is exercised.

Exercise 26 Justify formulae (4.8) and (4.9) and calculate, for a call and a put, the delta, the gamma, the theta and the vega (cf. Remark 4.3.5).

Exercise 27 Let $(B_t)_{t\geq 0}$ be a standard Brownian motion. For any real numbers μ and b, let

$$T_b^\mu = \inf\{t \geq 0 \mid \mu t + B_t = b\},$$

with the convention: $\inf \emptyset = \infty$.

1. Use the Girsanov theorem to show the following equality:

$$\forall \alpha, t > 0, \quad \mathbb{E}(e^{-\alpha(T_b^\mu \wedge t)}) = \mathbb{E}\left(e^{-\alpha(T_b^\mu \wedge t)}e^{\mu B_{T_b^0 \wedge t} - \frac{\mu^2}{2}T_b^0 \wedge t}\right).$$

2. Prove the inequality

$$\forall \alpha, t > 0, \quad \mathbb{E}\left(e^{-\alpha(T_b^0 \wedge t)}e^{\mu B_{T_b^0 \wedge t} - \frac{\mu^2}{2}T_b^0 \wedge t}\mathbf{1}_{\{t < T_b^0\}}\right) \leq e^{-\alpha t}.$$

3. Deduce from above and Proposition 3.3.6 that

$$\forall \alpha, t > 0 \; \mathbb{E}\left(e^{-\alpha T_b^\mu}\mathbf{1}_{\{T_b^\mu < \infty\}}\right) = \exp\left(\mu b - |b|\sqrt{2\alpha + \mu^2}\right).$$

4. Calculate $\mathbb{P}(T_b^\mu < \infty)$.

Exercise 28

1. Let $\mathbb{P}$ and $\mathbb{Q}$ be two equivalent probabilities on a measurable space $(\Omega, \mathscr{A})$. Show that if a sequence (X_n) of random variables converges in probability under $\mathbb{P}$, it converges in probability under $\mathbb{Q}$ to the same limit.

2. Notations and hypothesis are those of Theorem 4.2.2. Let $(H_t)_{0\leq t\leq T}$ be an adapted process such that $\int_0^T H_s^2 ds < \infty$, $\mathbb{P}$ a.s. The stochastic integral of (H_t) relative to B_t is well defined under the probability $\mathbb{P}$. Set

$$X_t = \int_0^t H_s dB_s + \int_0^t H_s \theta_s ds.$$

Since $\mathbb{P}^{(L)}$ and $\mathbb{P}$ are equivalent, we have $\int_0^T H_s^2 ds < \infty$, $\mathbb{P}^{(L)}$ a.s. and we can define, under $\mathbb{P}^{(L)}$, the process

$$Y_t = \int_0^t H_s dW_s.$$

The question is to prove the equality of the two processes X and Y. To do so, consider first the case of simple processes; then use the fact that if $(H_t)_{0\leq t\leq T}$ is an adapted process satisfying $\int_0^T H_s^2 ds < \infty$ a.s., there is a sequence (H^n) of elementary processes such that $\int_0^T (H_s - H_s^n)^2 ds$ converges to 0 in probability.

Exercise 29 Let $(B_t)_{0\leq t\leq 1}$ be a standard Brownian motion defined on the time interval [0,1]. Denote by $(\mathscr{F}_t)_{0\leq t\leq 1}$ its natural filtration and consider τ an exponentially distributed random variable with parameter λ, independent of $\mathscr{F}_1$. For $t \in [0,1]$, let $\mathscr{G}_t$ be the σ-algebra generated by $\mathscr{F}_t$ and the random variable $\tau \wedge t$.

1. Show that $(\mathscr{G}_t)_{0\leq t\leq 1}$ is a filtration and that $(B_t)_{0\leq t\leq 1}$ is a Brownian motion with respect to $(\mathscr{G}_t)$.

2. For $t \in [0,1]$, let $M_t = \mathbb{E}\left(1_{\{\tau>1\}} \mid \mathscr{G}_t\right)$. Show that M_t is equal to $e^{-\lambda(1-t)}1_{\{\tau>t\}}$ a.s. The following property can be used: if $\mathscr{B}_1$ and $\mathscr{B}_2$ are two sub-σ-algebras and X a non-negative random variable such that the σ-algebra generated by $\mathscr{B}_2$ and X is independent of the σ-algebra $\mathscr{B}_1$, then $\mathbb{E}(X \mid \mathscr{B}_1 \vee \mathscr{B}_2) = \mathbb{E}(X \mid \mathscr{B}_2)$, where $\mathscr{B}_1 \vee \mathscr{B}_2$ is the σ-algebra generated by $\mathscr{B}_1$ and $\mathscr{B}_2$.

3. Show that there exists no continuous process (X_t) such that for all $t \in [0,1], \mathbb{P}(M_t = X_t) = 1$ (remark that we would necessarily have $\mathbb{P}(\forall t \in [0,1], \ M_t = X_t) = 1$). Deduce that the martingale (M_t) cannot be represented as a stochastic integral with respect to (B_t).

Exercise 30 The reader may use the results of Exercise 21 of Chapter 3. Let $(W_t)_{t\geq 0}$ be an $(\mathscr{F}_t)$-Brownian motion.

1. Prove that if $\mu \leq \lambda$ and $N(d) = \int_{-\infty}^{d} \exp(-x^2/2)dx/\sqrt{2\pi}$, we have

$$\mathbb{E}(e^{\alpha W_T}1_{\{W_T\leq\mu,\sup_{s\leq T} W_s\geq\lambda\}}) = \exp\left(\frac{\alpha^2 T}{2} + 2\alpha\lambda\right) N\left(\frac{\mu - 2\lambda - \alpha T}{\sqrt{T}}\right)$$

Deduce that if $\lambda \leq \mu$,

$$\mathbb{E}(e^{\alpha W_T}1_{\{W_T\geq\mu,\inf_{s\leq T} W_s\leq\lambda\}}) = \exp\left(\frac{\alpha^2 T}{2} + 2\alpha\lambda\right) N\left(\frac{2\lambda - \mu + \alpha T}{\sqrt{T}}\right).$$

2. Let $H \leq K$; we are looking for an analytic formula for

$$C = \mathbb{E}\left(e^{-rT}(X_T - K)_{+}1_{\left\{\inf_{s\leq T} X_s \geq H\right\}}\right),$$

where $X_t = x\exp((r - \sigma^2/2)t + \sigma W_t)$. Give a financial interpretation to this value and give an expression for the probability $\tilde{\mathbb{P}}$ that makes $\tilde{W}_t = (r/\sigma - \sigma/2)t + W_t$ a standard Brownian motion.

3. Write C as the expectation under $\tilde{\mathbb{P}}$ of a random variable function only of $\tilde{W}_T$ and $\sup_{0\leq s\leq T} \tilde{W}_s$.

4. Deduce an analytic formula for C.

4.8   Problems

Problem 1 Black-Scholes model with time-dependent parameters
Consider a variant of the Black-Scholes model, assuming that the asset prices are described by the following equations (we keep the same notations as in this chapter):

$$\begin{cases} dS_t^0 = r(t)S_t^0 dt \\ dS_t = S_t(\mu(t)dt + \sigma(t)dB_t), \end{cases}$$

where $r(t), \mu(t), \sigma(t)$ are *deterministic* continuous functions of time, on the interval $[0, T]$. We also assume that $\inf_{t \in [0,T]} \sigma(t) > 0$.

1. Prove that

$$S_t = S_0 \exp\left(\int_0^t \mu(s)ds + \int_0^t \sigma(s)dB_s - \frac{1}{2}\int_0^t \sigma^2(s)ds \right).$$

   Hint: consider the process

$$Z_t = S_t \exp\left[-\left(\int_0^t \mu(s)ds + \int_0^t \sigma(s)dB_s - \frac{1}{2}\int_0^t \sigma^2(s)ds \right) \right].$$

2. (a) Let (X_n) be a sequence of real-valued, zero-mean normal random variables converging to X in L_2. Show that X is a normal random variable.

   (b) By approximating σ by simple functions, show that $\int_0^t \sigma(s)dB_s$ is a normal random variable and calculate its variance.

3. Prove that there exists a probability $\mathbb{P}^*$ equivalent to $\mathbb{P}$, under which the discounted stock price is a martingale. Give its density with respect to $\mathbb{P}$.

4. In the remainder, we will tackle the problem of pricing and hedging a call with maturity T and strike price K.

   (a) Let $(H_t^0, H_t)_{0 \le t \le T}$ be a self-financing strategy, with value V_t at time t. Show that if (V_t/S_t^0) is a martingale under $\mathbb{P}^*$ and $V_T = (S_T - K)_+$, then

$$\forall t \in [0, T] \quad V_t = F(t, S_t),$$

   where F is the function defined by

$$F(t, x) = \mathbb{E}^* \left(x e^{\int_t^T \sigma(s)dW_s - \frac{1}{2}\int_t^T \sigma^2(s)ds} - K e^{-\int_t^T r(s)ds} \right)_+$$

   and (W_t) is a standard Brownian motion under $\mathbb{P}^*$.

   (b) Give an expression for the function F and compare it to the Black-Scholes formula.

(c) Construct a hedging strategy for the call (find H_t^0 and H_t; check the self-financing condition).

Problem 2 Garman-Kohlhagen model

The Garman-Kohlhagen model (1983) is the most commonly used model to price and hedge foreign-exchange options. It derives directly from the Black-Scholes model. For simplicity, we shall concentrate on 'dollar-euro' options. For example, a European *call* on the dollar, with maturity T and strike price K, is the right to buy, at time T, one dollar for K euros.

Let S_t be the price of the dollar at time t, i.e. the number of euros per dollar. The behavior of S_t through time is modelled by the following stochastic differential equation:

$$\frac{dS_t}{S_t} = \mu dt + \sigma dW_t,$$

where $(W_t)_{t \in [0,T]}$ is a standard Brownian motion on a probability space $(\Omega, \mathscr{F}, \mathbb{P})$, and μ and σ are real-valued, with $\sigma > 0$. We denote by $(\mathscr{F}_t)_{t \in [0,T]}$ the natural filtration of $(W_t)_{t \in [0,T]}$ and assume that $\mathscr{F}_t$ represents the accumulated information up to time t.

I

1. Express S_t as a function of S_0, t and W_t. Calculate the expectation of S_t.

2. Show that if $\mu > 0$, the process $(S_t)_{t \in [0,T]}$ is a submartingale.

3. Let $U_t = 1/S_t$ be the euro *vs* dollar exchange rate. Show that U_t satisfies the following stochastic differential equation:

$$\frac{dU_t}{U_t} = (\sigma^2 - \mu)dt - \sigma dW_t.$$

Deduce that if $0 < \mu < \sigma^2$, both processes $(S_t)_{t \in [0,T]}$ and $(U_t)_{t \in [0,T]}$ are submartingales. In what sense does this seem to be paradoxical?

II

We would like to price and hedge a European call on one dollar, with maturity T and strike price K, using a Black-Scholes-type method. From his premium, which represents his initial wealth, the writer of the option elaborates a strategy, defining at any time t a portfolio made of H_t^0 euros and H_t dollars, in order to create, at time T, a wealth equal to $(S_T - K)_+$ (in euros).

At time t, the value in euros of a portfolio made of H_t^0 euros and H_t dollars is obviously

$$V_t = H_t^0 + H_t S_t. \tag{4.16}$$

We suppose that euros are invested or borrowed at the *domestic* rate r_0 and US dollars are invested or borrowed at the *foreign* rate r_1. A self-financing strategy will thus be defined by an adapted process $(H_t^0, H_t)_{t \in [0,T]}$, such that

$$dV_t = r_0 H_t^0 dt + r_1 H_t S_t dt + H_t dS_t. \tag{4.17}$$

where V_t is defined by equation (4.16).

1. What integrability conditions should be imposed on the processes (H_t^0) and (H_t) in order for (4.17) to make sense?

2. Let $\tilde{V}_t = e^{-r_0 t} V_t$ be the discounted value of the (self-financing) portfolio (H_t^0, H_t). Prove the equality

$$d\tilde{V}_t = H_t e^{-r_0 t} S_t (\mu + r_1 - r_0) dt + H_t e^{-r_0 t} S_t \sigma dW_t.$$

3. (a) Show that there exists a probability $\tilde{\mathbb{P}}$, equivalent to $\mathbb{P}$, under which the process
$$\tilde{W}_t = \frac{\mu + r_1 - r_0}{\sigma} t + W_t$$
is a standard Brownian motion.

   (b) A self-financing strategy is said to be *admissible* if its discounted value $\tilde{V}_t$ is non-negative for all t and if $\sup_{t \in [0,T]} (\tilde{V}_t)$ is square-integrable under $\tilde{\mathbb{P}}$. Show that the discounted value of an admissible strategy is a martingale under $\tilde{\mathbb{P}}$.

4. Show that if an admissible strategy replicates the call, in other words it is worth $V_T = (S_T - K)_+$ at time T, then for any $t \leq T$ the value of the strategy at time t is given by

$$V_t = F(t, S_t),$$

where

$$F(t, x) = \tilde{\mathbb{E}} \left(x e - (r_1 + (\sigma^2/2))(T - t) + \sigma(\tilde{W}_T - \tilde{W}_t) - K e^{-r_0(T-t)} \right)_+.$$

(The symbol $\tilde{\mathbb{E}}$ stands for the expectation under the probability $\tilde{\mathbb{P}}$.)

5. Show (through detailed calculation) that

$$F(t, x) = e^{-r_1(T-t)} x N(d_1) - K e^{-r_0(T-t)} N(d_2),$$

where N is the distribution function of the standard normal law, and

$$d_1 = \frac{\log(x/K) + (r_0 - r_1 + (\sigma^2/2))(T - t)}{\sigma \sqrt{T - t}}, \quad d_2 = d_1 - \sigma \sqrt{T - t}.$$

6. The next step is to show that the option is effectively replicable.

   (a) Let $\tilde{S}_t = e^{(r_1 - r_0)t} S_t$. Derive the equality

$$d\tilde{S}_t = \sigma \tilde{S}_t d\tilde{W}_t.$$

(b) Let $\tilde{F}$ be the function defined by $\tilde{F}(t,x) = e^{-r_0 t}F(t, xe^{(r_0-r_1)t})$ (F is the function defined in Question 4). Let $C_t = F(t, S_t)$ and $\tilde{C}_t = e^{-r_0 t}C_t = \tilde{F}(t, \tilde{S}_t)$. Derive the equality

$$d\tilde{C}_t = \frac{\partial F}{\partial x}(t, S_t)\sigma e^{-r_0 t}S_t d\tilde{W}_t.$$

(c) Deduce that the call is replicable and give an explicit expression for the replicating portfolio (H_t^0, H_t).

7. Write down a put-call parity relationship, similar to the relationship we gave for stocks, and give an example of arbitrage opportunity when this relationship does not hold.

Problem 3 Option to exchange one asset for another

We consider a financial market in which there are two risky assets with respective prices S_t^1 and S_t^2 at time t and a riskless asset with price $S_t^0 = e^{rt}$ at time t. The dynamics of the prices S_t^1 and S_t^2 over time are modelled by the following stochastic differential equations:

$$\begin{cases} dS_t^1 = S_t^1(\mu_1 dt + \sigma_1 B_t^1) \\ dS_t^2 = S_t^2(\mu_2 dt + \sigma_2 dB_t^2), \end{cases}$$

where $(B_t^1)_{t\in[0,T]}$ and $(B_t^2)_{t\in[0,T]}$ are two independent Brownian motions defined on a probability space $(\Omega, \mathscr{F}, \mathbb{P})$; μ_1, μ_2, σ_1 and σ_2 are real numbers, with $\sigma_1 > 0$ and $\sigma_2 > 0$. Let $\mathscr{F}_t$ be the σ-algebra generated by the random variables B_s^1 and B_s^2 for $s \leq t$. Then the processes $(B_t^1)_{t\in[0,T]}$ and $(B_t^2)_{t\in[0,T]}$ are $(\mathscr{F}_t)$-Brownian motions and, for $t \geq s$, the vector $(B_t^1 - B_s^1, B_t^2 - B_s^2)$ is independent of $\mathscr{F}_s$.

I

We study the pricing and hedging of an option giving the right to exchange one of the risky assets for the other at time T.

1. Set $\theta = (\mu_1 - r)/\sigma_1$ and $\theta_2 = (\mu_2 - r)/\sigma_2$. Show that the process defined by

$$M_t = \exp\left(-\theta_1 B_t^1 - \theta_2 B_t^2 - \frac{1}{2}\left(\theta_1^2 + \theta_2^2\right)t\right)$$

is a martingale with respect to the filtration $(\mathscr{F}_t)_{t\in[0,T]}$.

2. Let $\tilde{\mathbb{P}}$ be the probability with density M_T with respect to $\mathbb{P}$. We introduce the processes W^1 and W^2 defined by $W_t^1 = B_t^1 - \theta_1 t$ and $W_t^2 = B_t^2 + \theta_2 t$. Derive, under the probability $\tilde{\mathbb{P}}$, the joint characteristic function of (W_t^1, W_t^2). Deduce that, for any $t \in [0, T]$, the random variables W_t^1 and W_t^2 are independent normal random variables with zero mean and variance t under $\tilde{\mathbb{P}}$.

In the remainder, we will admit that, under the probability $\tilde{\mathbb{P}}$, the processes $(W_t^1)_{0\leq t\leq T}$ and $(W_t^2)_{0\leq t\leq T}$ are $(\mathscr{F}_t)$-independent standard Brownian motions and that, for $t \geq s$, the vector $(W_t^1 - W_s^1, W_t^2 - W_s^2)$ is independent of $\mathscr{F}_s$.

3. Write $\tilde{S}_t^1$ and $\tilde{S}_t^2$ as functions of S_0^1, S_0^2, W_t^1 and W_t^2, and show that, under $\tilde{\mathbb{P}}$, the discounted prices $\tilde{S}_t^1 = e^{-rt} S_t^1$, and $\tilde{S}_t^2$, are martingales.

II

We want to price and hedge a European option, with maturity T, giving to the holder the right to exchange one unit of asset 2 for one unit of asset 1. To do so, we use the same method as in the Black-Scholes model. From his initial wealth, the premium, the writer of the option builds a strategy, defining at any time t a portfolio made of H_t^0 units of the riskless asset and H_t^1 and H_t^2 units of assets 1 and 2, respectively, in order to generate, at time T, a wealth equal to $(S_T^1 - S_T^2)_+$. A trading strategy will be defined by the three adapted processes H^0, H^1 and H^2.

1. Define precisely the self-financing strategies and prove that, if $\tilde{V}_t = e^{-rt} V_t$ is the discounted value of a self-financing strategy, we have

$$d\tilde{V}_t = H_t^1 e^{-rt} S_t^1 \sigma_1 dW_t^1 + H_t^2 e^{-rt} S_t^2 \sigma_2 dW_t^2.$$

2. Show that if the processes $(H_t^1)_{0 \le t \le T}$ and $(H_t^2)_{0 \le t \le T}$ of a self-financing strategy are uniformly bounded (which means that $\exists C > 0, \forall (t, \omega) \in [0, T] \times \Omega, |H_t^1(\omega)| \le C$, for $i = 1, 2$), then the discounted value of the strategy is a martingale under $\tilde{\mathbb{P}}$.

3. Prove that if a self-financing strategy satisfies the hypothesis of the previous question and has a terminal value equal to $V_T = (S_T^1 - S_T^2)_+$ then its value at any time $t < T$ is given by

$$V_t = F(t, S_t^1, S_t^2), \tag{4.18}$$

where the function F is defined by

$$F(t, x_1, x_2) = \tilde{\mathbb{E}} \left(x_1 e^{\sigma_1 (W_T^1 - W_t^1) - \frac{\sigma_1^2}{2}(T-t)} - x_2 e^{\sigma_2 (W_T^2 - W_t^2) - \frac{\sigma_2^2}{2}(T-t)} \right)_+, \tag{4.19}$$

the symbol $\tilde{\mathbb{E}}$ representing the expectation under $\tilde{\mathbb{P}}$. The existence of a strategy having this value will be proved later on. We will consider in the remainder that the value of the option $(S_T^1 - S_T^2)_+$ at time t is given by $F(t, S_t^1, S_t^2)$.

4. Find a parity relationship between the value of the option with payoff $(S_T^1 - S_T^2)_+$ and the symmetrical option with payoff $(S_T^2 - S_T^1)_+$, similar to the put-call parity relationship previously seen, and give an example of arbitrage opportunity when this relationship does not hold.

III

The objective of this section is to find an explicit expression for the function F defined by (4.19) and to establish a strategy replicating the option.

1. Let g_1 and g_2 be two independent standard normal random variables and let λ be a real number.

   (a) Show that under the probability $\mathbb{P}^{(\lambda)}$, with density with respect to $\mathbb{P}$ given by

   $$\frac{d\mathbb{P}^{(\lambda)}}{d\mathbb{P}} = e^{\lambda g_1 - \lambda^2/2},$$

   the random variables $g_1 - \lambda$ and g_2 are independent standard normal variables.

   (b) Deduce that for all real numbers y_1, y_2, λ_1 and λ_2, we have

   $$\mathbb{E}\left(e^{y_1 + \lambda_1 g_1} - e^{y_2 + \lambda_2 g_2}\right)_+ = e^{y_1 + \lambda_1^2/2} N\left(\frac{y_1 - y_2 + \lambda_1^2}{\sqrt{\lambda_1^2 + \lambda_2^2}}\right)$$
   $$- e^{y_2 + \lambda_2^2/2} N\left(\frac{y_1 - y_2 - \lambda_2^2}{\sqrt{\lambda_1^2 + \lambda_2^2}}\right),$$

   where N is the standard normal distribution function.

2. Deduce from the previous question an expression for F using the function N.

3. Let $\tilde{C}_t = e^{-rt} F(t, S_t^1, S_t^2)$. Observe that

   $$\tilde{C}_t = F(t, \tilde{S}_t^1, \tilde{S}_t^2) = \tilde{\mathbb{E}}\left(e^{-rT}(S_T^1 - S_T^2)_+ | \mathscr{F}_t\right),$$

   and prove the equality

   $$d\tilde{C}_t = \frac{\partial F}{\partial x_1}(t, \tilde{S}_t^1, \tilde{S}_t^2)\sigma_1 e^{-rt} S_t^1 dW_t^1 + \frac{\partial F}{\partial x_2}(t, \tilde{S}_t^1, \tilde{S}_t^2)\sigma_2 e^{-rt} S_t^2 dW_t^2.$$

   (Hint: use the fact that if (X_t) is an Itô process that can be written as $X_t = X_0 + \int_0^t J_s^1 dW_s^1 + \int_0^t J_s^2 dW_s^2 + \int_0^t K_s ds$, and if it is a martingale under $\tilde{\mathbb{P}}$, then $K_t = 0, dt d\tilde{\mathbb{P}}$ almost everywhere.)

4. Build a hedging scheme for the option.

Problem 4 A study of strategies with consumption

We consider a financial market in which there is one riskless asset, with price $S_t^0 = e^{rt}$ at time t (with $r \geq 0$) and one risky asset, with price S_t at time t. The model is studied on the time interval $[0, T](0 \leq T < \infty)$. In the following, $(S_t)_{0 \leq t \leq T}$ is a stochastic process defined on a probability space $(\Omega, \mathscr{F}, \mathbb{P})$, equipped with a filtration $(\mathscr{F}_t)_{0 \leq t \leq T}$. We assume that $(\mathscr{F}_t)_{0 \leq t \leq T}$ is the natural filtration of a standard Brownian motion $(B_t)_{0 \leq t \leq T}$ and that the process $(S_t)_{0 \leq t \leq T}$ is adapted to this filtration.

We want to study strategies in which consumption is allowed. The dynamics of $(S_t)_{0 \leq \leq T}$ is given by the Black-Scholes model,

$$dS_t = S_t(\mu dt + \sigma dB_t),$$

with $\mu \in \mathbb{R}$ and $\sigma > 0$. We denote by $\mathbb{P}^*$ the probability with density $\exp(-\theta B_T - \theta^2 T/2)$ with respect to $\mathbb{P}$, where $\theta = (\mu - r)/\sigma$. Under $\mathbb{P}^*$, the process $(W_t)_{0 \le t \le T}$, defined by $W_t = (\mu - r)t/\sigma + B_t$, is a standard Brownian motion.

A strategy with consumption is defined by three stochastic processes: $(H_t^0)_{0 \le t \le T}$, $(H_t)_{0 \le t \le T}$ and $(c(t))_{0 \le t \le T}$. The numbers H_t^0 and H_t respectively represent the quantities of riskless and risky asset held at time t, and $c(t)$ represents the consumption rate at time t. We say that such a strategy is *admissible* if the following conditions hold:

(i) The processes $(H_t^0)_{0 \le t \le T}$, $(H_t)_{0 \le t \le T}$ and $(c(t))_{0 \le t \le T}$ are adapted and satisfy

$$\int_0^T (|H_t^0| + H_t^2 + |c(t)|)dt < \infty, \quad \text{a.s.}$$

(ii) For all $t \in [0, T]$,

$$H_t^0 S_t^0 + H_t S_t = H_0^0 S_0^0 + H_0 S_0 + \int_0^t H_u^0 dS_u^0$$
$$+ \int_0^t H_u dS_u - \int_0^t c(u)du, \quad \text{a.s.}$$

(iii) For all $t \in [0, T], c(t) \ge 0$ a.s.

(iv) For $t \in [0, T]$, the random variable $H_t^0 S_t^0 + H_t S_t$ is non-negative and

$$\sup_{t \in [0,T]} \left(H_t^0 S_t^0 + H_t S_t + \int_0^t c(s)ds \right)$$

is square-integrable under the probability $\mathbb{P}^*$.

1. Let $(H_t^0)_{0 \le t \le T}$, $(H_t)_{0 \le t \le T}$ and $(c(t))_{0 \le t \le T}$ be three adapted processes satisfying condition (i) above. Let $V_t = H_t^0 S_t^0 + H_t S_t$ and $\tilde{V}_t = e^{-rt}V_t$. Show that condition (ii) is satisfied if and only if we have, for all $t \in [0, T]$,

$$\tilde{V}_t = V_0 + \int_0^t H_u d\tilde{S}_u - \int_0^t \tilde{c}(u)du, \quad \text{a.s.},$$

with $\tilde{S}_u = e^{-ru}S_u$ and $\tilde{c}(u) = e^{-ru}c(u)$.

2. Assume that conditions (i) to (iv) are satisfied and let $\tilde{V}_t = e^{-rt}V_t = e^{-rt} \left(H_t^0 S_t^0 + H_t S_t \right)$. Prove that the process $(\tilde{V}_t)_{0 \le t \le T}$ is a supermartingale under probability $\mathbb{P}^*$.

3. Let $(c(t))_{0 \le t \le T}$ be an adapted process with non-negative values such that $\mathbb{E}^* \left(\int_0^T c(t)dt \right)^2 < \infty$ and let $x > 0$. We say that $(c(t))_{0 \le t \le T}$ is a budget-feasible consumption process from the initial endowment x if there exist some processes $(H_t^0)_{0 \le t \le T}$ and $(H_t)_{0 \le t \le T}$ such that conditions (i) to (iv) are satisfied, and furthermore $V_0 = H_0^0 S_0^0 + H_0 S_0 = x$.

(a) Show that if the process $(c(t))_{0 \le t \le T}$ is budget-feasible from the initial endowment x, then we have $\mathbb{E}^* \left(\int_0^T e^{-rt} c(t) dt \right) \le x$.

(b) Let $(c(t))_{0 \le t \le T}$ be an adapted process, with non-negative values, such that

$$\mathbb{E}^* \left(\int_0^T c(t) dt \right)^2 < \infty \quad \text{and} \quad \mathbb{E}^* \left(\int_0^T e^{-rt} c(t) dt \right) \le x.$$

Prove that $(c(t))_{0 \le t \le T}$ is a budget-feasible consumption process with an initial endowment x. (Hint: introduce the martingale $(M(t))_{0 \le t \le T}$ defined by $M_t = \mathbb{E}^* \left(x + \int_0^T e^{-rs} c(s) ds \mid \mathcal{F}_t \right)$ and apply the Martingale Representation Theorem.)

(c) An investor with initial endowment x wants to consume a wealth corresponding to the sale of ρ risky assets by unit of time as long as S_t remains above some barrier K (this yields $c(t) = \rho S_t 1_{\{S_t > K\}}$). What conditions on ρ and x are necessary for this consumption process to be budget-feasible?

Problem 5 Monotonicity of option prices with respect to volatility
In this problem, we consider a financial market with one riskless asset (with price $S_t^0 = e^{rt}$ at time t) and one risky asset, with price S_t at time t. We assume that the process $(S_t)_{0 \le t \le T}$ is the solution of a stochastic differential equation of the following form:

$$dS_t = S_t(\mu dt + \sigma(t) dB_t), \tag{4.20}$$

where $\mu \in \mathbb{R}$ and $(\sigma(t))_{0 \le t \le T}$ is an adapted process with respect to the natural filtration of $(B_t)_{0 \le t \le T}$, satisfying

$$\forall t \in [0, T], \quad \sigma_1 \le \sigma(t) \le \sigma_2,$$

for some constants σ_1 and σ_2 such that $0 < \sigma_1 < \sigma_2$. We consider a European call option with maturity T and strike price K on one unit of the risky asset. We know (see Chapter 5) that if the process $(\sigma(t))_{0 \le t \le T}$ is constant (with $\sigma(t) = \sigma$ for any t), the price of the call at time t is $C(t, S_t)$, where the function $C(t, x)$ satisfies

$$\begin{cases} \dfrac{\partial C}{\partial t}(t, x) + \dfrac{\sigma^2 x^2}{2} \dfrac{\partial^2 C}{\partial x^2}(t, x) + rx \dfrac{\partial C}{\partial x}(t, x) - rC(t, x) = 0, \quad t \in [0, T), x > 0 \\ C(T, x) = (x - K)_+. \end{cases}$$

Denote by C_i the function C corresponding to the case $\sigma = \sigma_i$ $(i = 1, 2)$. We want to show that the price of the call at time 0 in the model with varying volatility belongs to the interval $[C_1(0, S_0), C_2(0, S_0)]$.

Recall that if $(\theta_t)_{0 \le t \le T}$ is a bounded adapted process, the process $(L_t)_{0 \le t \le T}$ defined by $L_t = \exp \left(\int_0^t \theta_s dB_s - \frac{1}{2} \int_0^t \theta_s^2 ds \right)$ is a martingale.

1. Prove (using the price formulae written as expectations) that the functions $x \mapsto C_i(t, x)$ $(i = 1, 2)$ are convex.

2. Show that the solution of equation (4.20) is given by

$$S_t = S_0 \exp \left(\mu t + \int_0^t \sigma(s) dB_s - \frac{1}{2} \int_0^t \sigma^2(s) ds \right).$$

3. Determine a probability $\mathbb{P}^*$ equivalent to $\mathbb{P}$, under which the process defined by $W_t = B_t + \int_0^t (\mu - r)/\sigma(s) ds$ is a standard Brownian motion.

4. Explain why the price of the call at time 0 is given by

$$C_0 = \mathbb{E}^* \left(e^{-rT} (S_T - K)_+ \right).$$

5. Let $\tilde{S}_t = e^{-rt} S_t$. Show that $\mathbb{E}^* \left(\tilde{S}_t^2 \right) \leq S_0^2 e^{\sigma \frac{2}{2} t}$.

6. Prove that the process defined by

$$M_t = \int_0^t e^{-ru} \frac{\partial C_1}{\partial x} (u, S_u) \sigma(u) S_u dW_u$$

is a martingale under probability $\mathbb{P}^*$.

7. Using Itô's formula and Questions 1 and 6, show that $(e^{-rt} C_1(t, S_t))$ is a submartingale under probability $\mathbb{P}^*$. Deduce that $C_1(0, S_0) \leq C_0$.

8. Derive the inequality $C_0 \leq C_2(0, S_0)$.

Problem 6 Dupire's formula

Consider a local volatility model, in which the risky asset price S_t satisfies

$$dS_t = \mu(t) S_t dt + \sigma(t, S_t) S_t dB_t, \tag{4.21}$$

where $B = (B_t)_{t \geq 0}$ is a standard Brownian motion, defined on $(\Omega, \mathscr{A}, \mathbb{P})$, $\mu : \mathbb{R}^+ \to \mathbb{R}$ is a (deterministic) continuous function and $\sigma : \mathbb{R}^+ \times \mathbb{R} \to \mathbb{R}$ is a continuous function such that

$$\forall t \geq 0, \quad \forall (x, y) \in \mathbb{R}^2, \quad |x \sigma(t, x) - y \sigma(t, y)| \leq M |x - y|$$

and

$$\forall (t, x) \in \mathbb{R}_+ \times \mathbb{R}, \quad \sigma(t, x) \geq m,$$

where m and M are positive constants. For simplicity, we assume that the interest rate is null. We denote by $\mathbf{F} = (\mathscr{F}_t)_{t \geq 0}$ the natural filtration of $(B_t)_{t \geq 0}$.

1. Prove that, for every $x \in \mathbb{R}$, equation (4.21) has a unique solution such that $S_0 = x$.

2. Prove that if S is a solution of (4.21), we have, for $t \geq 0$,

$$S_t = S_0 \exp\left(\int_0^t \mu(s)ds + \int_0^t \sigma(u, S_u)dB_u - \frac{1}{2}\int_0^t \sigma^2(u, S_u)du\right).$$

3. Throughout the problem, we assume that S solves (4.21) and that the initial price S_0 is a positive (deterministic) real number. Prove that the natural filtration of the process $(S_t)_{t\geq 0}$ is equal to $\mathbf{F}$. (Hint: write B_t as a stochastic integral with respect to the process $(S_t)_{t\geq 0}$.)

4. Let L be the martingale defined by $L_t = \exp\left(-\int_0^t \theta_u dB_u - \frac{1}{2}\int_0^t \theta_u^2 du\right)$, with $\theta_t = \mu(t)/\sigma(t, S_t)$. Fix the horizon $\bar{T}$ of the model $(0 < \bar{T} < +\infty)$ and let $\mathbb{P}^*$ be the probability given by $\dfrac{d\mathbb{P}^*}{d\mathbb{P}} = L_{\bar{T}}$. Given $T \in [0, \bar{T}]$, let $C(T, K)$ be the price of a call option with maturity T and strike price K.

   (a) Prove that, for $(t, x) \in \mathbb{R}^+ \times \mathbb{R}$, $\sigma(t, x) \leq M$. Deduce that, for $0 \leq t \leq \bar{T}$ and $p \geq 1$, $\mathbb{E}^*(S_t^p) \leq S_0^p \exp\left(\frac{p^2 - p}{2}M^2 t\right)$.

   (b) Briefly explain the equality $C(T, K) = \mathbb{E}^*\left(S_T - K\right)_+$, for $T \in [0, \bar{T}]$.

   (c) Use the martingale property of S under $\mathbb{P}^*$ to show that, for $K \geq 0$, $T \mapsto C(T, K)$ is nondecreasing on $[0, \bar{T}]$. Prove that $(T, K) \mapsto C(T, K)$ is continuous on $[0, \bar{T}] \times \mathbb{R}^+$.

   (d) Prove that

   $$\mathbb{E}^*\left[(S_T - K)_+\right]^2 = 2\int_K^{+\infty} C(T, y)dy.$$

5. Let $f_0(x) = (x_+)^2$ and, for $\varepsilon > 0$,

$$f_\varepsilon(x) = \begin{cases} 0 & \text{if } x < 0 \\ \frac{x^3}{3\varepsilon} & \text{if } x \in [0, \varepsilon] \\ x^2 - \varepsilon x + \frac{\varepsilon^2}{3} & \text{if } x > \varepsilon. \end{cases}$$

   (a) Prove that, for $\varepsilon > 0$, f_ε is of class C^2, that $\lim_{\varepsilon \to 0} f_\varepsilon(x) = f_0(x)$ for every $x \in \mathbb{R}$, and that

   $$\forall x \geq 0, \quad 0 \leq f_\varepsilon(x) \leq f_0(x), \quad 0 \leq f_\varepsilon'(x) \leq 2x, \quad 0 \leq f_\varepsilon''(x) \leq 2.$$

   (b) Prove that, for $K \geq 0$ and $T \in [0, \bar{T}]$,

   $$\mathbb{E}^* f_\varepsilon(S_T - K) = f_\varepsilon(S_0 - K)$$
   $$+ \frac{1}{2}\mathbb{E}^*\left(\int_0^T f_\varepsilon''(S_u - K)S_u^2\sigma^2(u, S_u)du\right).$$

6. Assume that, for every $t \in (0, \bar{T}]$, the random variable S_t has, under probability $\mathbb{P}^*$, a density $p(t, \cdot)$, where $(t, x) \mapsto p(t, x)$ is continuous on $(0, \bar{T}] \times \mathbb{R}^+$.

   (a) Prove that, for $0 < T \leq \bar{T}$ and $K > 0$, $p(T, K) = \dfrac{\partial^2 C}{\partial K^2}(T, K)$.

   (b) Prove, using Question 5, that

   $$\mathbb{E}^* \left[(S_T - K)_+ \right]^2 = (S_0 - K)_+^2$$
   $$+ \int_0^T \left(\int_K^{+\infty} y^2 \sigma^2(u, y) p(u, y) dy \right) du.$$

   (c) Deduce from above that $\dfrac{\partial C}{\partial T}(T, K) = \dfrac{K^2 \sigma^2(T, K)}{2} \dfrac{\partial^2 C}{\partial K^2}(T, K)$ (for $0 < T \leq \bar{T}$ and $K > 0$).

Problem 7 Compound option

We consider a financial market offering two investment opportunities. The first traded security is a riskless asset whose price is equal to $S_t^0 = e^{rt}$ at time t (with $r \geq 0$), and the second security is risky and its price is denoted by S_t at time $t \in [0, T]$. Let $(S_t)_{0 \leq t \leq T}$ be a stochastic process defined on a probability space $(\Omega, \mathscr{F}, \mathbb{P})$, equipped with a filtration $(\mathscr{F}_t)_{0 \leq t \leq T}$. We assume that $(\mathscr{F}_t)_{0 \leq t \leq T}$ is the natural filtration generated by a standard Brownian motion $(B_t)_{0 \leq t \leq T}$ and that $(S_t)_{0 \leq t \leq T}$ follows a Black-Scholes model,

$$dS_t = S_t(\mu dt + \sigma dB_t),$$

with $\mu \in \mathbb{R}$ and $\sigma > 0$.

We want to study an example of a compound option. We consider a call option with maturity $T_1 \in (0, T)$ and strike price K_1 on a call of maturity T and strike price K. The value of this option at time T_1 is equal to

$$h = (C(T_1, S_{T_1}) - K_1)_+,$$

where $C(t, x)$ is the price of the underlying call, given by the Black-Scholes formula.

1. (a) Graph the function $x \mapsto C(T_1, x)$. Show that the line $y = x - Ke^{-r(T-T_1)}$ is an asymptote. (Hint: use the put/call parity.)

   (b) Show that the equation $C(T_1, x) = K_1$ has a unique solution x_1.

2. Show that at time $t < T_1$, the compound option is worth $G(T_1 - t, S_t)$, where G is defined by

$$G(\theta, s) = \mathbb{E} \left[e^{-r\theta} \left(C \left(T_1, xe^{\left(r - \frac{\sigma^2}{2} \right) \theta + \sigma \sqrt{\theta} g} \right) - K_1 \right)^+ \right],$$

with g being a standard normal random variable.

3. (a) Show that $x \mapsto G(\theta, x)$ is an increasing convex function.

    (b) We now want to compute G explicitly. Denote by N the standard cumulative normal distribution function. Prove that, with the notation $\mu = r - (\sigma^2/2)$,

    $$G(\theta, x) = \mathbb{E}\left[e^{-r\theta} C\left(T_1, xe^{\mu\theta + \sigma\sqrt{\theta}g}\right) 1_{\{g > -d\}}\right] - K_1 e^{-r\theta} N(d),$$

    where

    $$d = \frac{\log(x/x_1) + (r - \sigma^2/2)\theta}{\sigma\sqrt{\theta}}.$$

    (c) Show that if g_1 is a standard normal variable independent of g, the function G satisfies

    $$G(\theta, x) + K_1 e^{-r\theta} N(d) = \mathbb{E}\left[\left(xe^{\sigma(\sqrt{\theta}g + \sqrt{\theta_1}g_1) - \frac{\sigma^2}{2}\bar{\theta}} - Ke^{-r\bar{\theta}}\right) 1_A\right],$$

    where $\theta_1 = T - T_1$, $\bar{\theta} = \theta + \theta_1$ and the event A is defined by

    $$A = \left\{ \sigma(\sqrt{\theta}g + \sqrt{\theta_1}g_1) > - \left(\log(x/K_1) + \mu\bar{\theta}\right) \text{ and } g > -d\right\}.$$

    (d) >From this, derive a formula for $G(\theta, x)$ in terms of N and N_2, the two-dimensional cumulative normal distribution function, defined by

    $$N_2(y, y_1, \rho) = \mathbb{P}(g < y, g + \rho g_1 < y_1), \quad y, y_1, \rho \in \mathbb{R}.$$

4. Show that we can replicate the compound option payoff by trading the underlying call and the riskless bond.

Problem 8 Behavior of the critical price close to maturity

We consider an American put maturing at T, with strike price K, on some risky asset with price S_t at time t. In the Black-Scholes model, the value of this option at time $t < T$ is equal to $P(t, S_t)$, where P is defined by

$$P(t, x) = \sup_{\tau \in \mathscr{T}_{0, T-t}} \mathbb{E}^* \left(Ke^{-r\tau} - xe^{\sigma W_\tau - \frac{\sigma^2}{2}\tau}\right)_+,$$

where $\mathscr{T}_{0, T-t}$ is the set of all stopping times with values in $[0, T - t]$ and $(W_t)_{0 \leq t \leq T}$ is a standard $\mathbb{P}^*$-Brownian motion. We also assume that $r > 0$. For $t \in [0, T]$, we denote by $s(t)$ the critical price, defined as

$$s(t) = \inf\{x > 0 \mid P(t, x) > K - x\}.$$

It can be proved that $\lim_{t \to T} s(t) = K$. We want to prove that $\lim_{t \to T} \dfrac{K - s(t)}{\sqrt{T - t}} = +\infty$.

1. Let P_e be the function pricing the European put with maturity T and strike price K:

$$P_e(t, x) = \mathbb{E}\left(e^{-r(T-t)}K - xe^{\sigma\sqrt{T-t}g - \frac{\sigma^2}{2}(T-t)}\right)_+,$$

where g is a standard normal variable. Show that if $t \in [0, T]$, the equation $P_e(t, x) = K - x$ has a unique solution in $[0, K]$. Call it $s_e(t)$.

2. Show that $s(t) \leq s_e(t)$, for any $t \in [0, T]$.

3. Show that

$$\liminf_{t \to T} \frac{K - s_e(t)}{\sqrt{T - t}} \geq \mathbb{E}\left(\liminf_{t \to T} \frac{K - s_e(t)}{\sqrt{T - t}} - \sigma K g\right)_+.$$

(Hint: use Fatou's lemma, which states that, for any sequence $(X_n)_{n \in N}$ of non-negative random variables, $\mathbb{E}(\liminf_{n \to \infty} X_n) \leq \liminf_{n \to \infty} \mathbb{E}(X_n).$)

4. (a) Show that for any real number η,

$$\mathbb{E}(\eta - K\sigma g)_+ > \eta.$$

(b) Deduce that

$$\lim_{t \to T} \frac{K - s_e(t)}{\sqrt{T - t}} = \lim_{t \to T} \frac{K - s(t)}{\sqrt{T - t}} = +\infty.$$

Problem 9 Asian option

Consider a financial market offering two investment opportunities. The first traded security is a riskless asset with price $S_t^0 = e^{rt}$ at time t (with $r \geq 0$), and the second security is risky with price S_t at time $t \in [0, T]$. The model is defined on a probability space $(\Omega, \mathscr{F}, \mathbb{P})$, equipped with a filtration $(\mathscr{F}_t)_{0 \leq t \leq T}$. We assume that $(\mathscr{F}_t)_{0 \leq t \leq T}$ is the natural filtration of a standard Brownian motion $(B_t)_{0 \leq t \leq T}$ and that $(S_t)_{0 \leq t \leq T}$ satisfies the Black-Scholes equation

$$dS_t = S_t(\mu dt + \sigma dB_t),$$

with $\mu \in \mathbb{R}$ and $\sigma > 0$. Let $\mathbb{P}^*$ be the probability measure with density $\exp(-\theta B_T - \theta^2 T/2)$ with respect to $\mathbb{P}$, where $\theta = (\mu - r)/\sigma$. Under $\mathbb{P}^*$, the process $(W_t)_{0 \leq t \leq T}$, defined by $W_t = (\mu - r)t/\sigma + B_t$, is a standard Brownian motion.

We are going to study the option with payoff

$$h = \left(\frac{1}{T}\int_0^T S_t dt - K\right)_+,$$

where K is a positive constant.

I

1. Explain briefly why the Asian option price at time t $(t \leq T)$ is given by

$$V_t = \mathbb{E}^* \left[e^{-r(T-t)} \left(\frac{1}{T} \int_0^T S_u du - K \right)_+ \bigg| \mathscr{F}_t \right].$$

2. Show that on the event $\left\{ \int_0^t S_u du \geq KT \right\}$, we have

$$V_t = \frac{e^{-r(T-t)}}{T} \int_0^t S_u du + \frac{1 - e^{-r(T-t)}}{rT} S_t - Ke^{-r(T-t)}.$$

3. Let $\tilde{S}_t = e^{-rt} S_t$, for $t \in [0, T]$.

   (a) Show that

   $$\mathbb{E}^*(\tilde{S}_t - Ke^{-rT})_+ \leq \mathbb{E}^*[e^{-rT}(S_T - K)_+].$$

   (Hint: use conditional expectations given $\mathscr{F}_t$.)

   (b) Deduce that
   $$V_0 \leq \mathbb{E}^*[e^{-rT}(S_T - K)_+],$$

   i.e. the Asian option price is smaller than its European counterpart.

   (c) For $t \leq u$, denote by $C_{t,u}$ the value at time t of a European call maturing at time u, with strike price K. Prove the following inequality:

   $$V_t \leq \frac{e^{-r(T-t)}t}{T} \left(\frac{1}{t} \int_0^t S_u du - K \right)_+ + \frac{1}{T} \int_t^T e^{-r(T-u)} C_{t,u} du.$$

II

Consider the process $(\xi_t)_{0 \leq t \leq T}$, defined by

$$\xi_t = \frac{1}{S_t} \left(\frac{1}{T} \int_0^t S_u du - K \right).$$

1. Show that $(\xi_t)_{0 \leq t \leq T}$ is the solution of the following stochastic differential equation:

$$d\xi_t = \left(\frac{1}{T} + (\sigma^2 - r)\xi_t \right) dt - \sigma \xi_t dW_t.$$

2. (a) Show that

$$V_t = e^{-r(T-t)} S_t \mathbb{E}^* \left[\left(\xi_t + \frac{1}{T} \int_t^T S_u^t du \right)_+ \bigg| \mathscr{F}_t \right],$$

with $S_u^t = \exp((r - \sigma^2/2)(u - t) + \sigma(W_u - W_t))$.

(b) Conclude that $V_t = e^{-r(T-t)} S_t F(t, \xi_t)$, with

$$F(t, \xi) = \mathbb{E}^* \left(\xi + \frac{1}{T} \int_t^T S_u^t du \right)_+ .$$

3. Find a replicating strategy to hedge the Asian option. For this, assume that the function F introduced earlier is of class C^2 on $[0, T) \times \mathbb{R}$ and use Itô's formula.

III

The purpose of this section is to suggest an approximation of V_0 obtained by considering the geometric average instead of the arithmetic one. Define

$$\hat{V}_0 = e^{-rT} \mathbb{E}^* \left(\exp \left(\frac{1}{T} \int_0^T \ln(S_t) dt \right) - K \right)_+ .$$

1. Show that $V_0 \geq \hat{V}_0$.

2. (a) Show that under $\mathbb{P}^*$, the random variable $\int_0^T W_t dt$ is normal with zero mean and a variance equal to $T^3/3$.

   (b) Deduce that

$$\hat{V}_0 = e^{-rT} \mathbb{E} \left(S_0 \exp \left((r - \sigma^2/2)(T/2) + \sigma \sqrt{T/3} g \right) - K \right)_+ ,$$

   where g is a standard normal variable. Give a closed-form formula for $\hat{V}_0$ in terms of the normal distribution function.

3. Prove the inequality

$$V_0 - \hat{V}_0 \leq S_0 e^{-rT} \left(\frac{e^{rT} - 1}{rT} - \exp((rT/2) - \sigma^2 T/12) \right) .$$

Chapter 5

Option pricing and partial differential equations

In the previous chapter, we saw how we could derive a closed-form formula for the price of a *vanilla* option (i.e. a call or a put) in the Black-Scholes environment. For more complex models, we are not able to find such explicit expressions. The same is true for American options, even in the Black-Scholes setting. That is why numerical methods are needed. The purpose of this chapter is to introduce the connection between diffusions and partial differential equations and to present some numerical methods based on this connection.

Firstly, we shall show how the problem of European option pricing is related to a parabolic partial differential equation(PDE). This link is based on the concept of the infinitesimal generator of a diffusion. We shall also address the problem of solving the PDE numerically.

The pricing of American options is rather difficult and we will not attempt to address it in its full generality. We will concentrate on the Black-Scholes model, and, in particular, we will underline the natural connection between the Snell envelope and a parabolic system of partial differential *inequalities*. We shall also explain how this kind of system can be solved numerically.

We shall only present classical numerical methods and the basics results that we need for option pricing. A more comprehensive introduction to numerical methods for solving parabolic PDEs can be found in Ciarlet and Lions (1990) or Raviart and Thomas (1983).

5.1 European option pricing and diffusions

In a Black-Scholes environment, the price of a European vanilla option is given by

$$V_t = \mathbb{E}(e^{-r(T-t)}f(S_T)|\mathscr{F}_t)$$

with $f(x) = (x - K)_+$ (for a call), $(K - x)_+$ (for a put) and

$$S_T = S_0 e^{(r - \sigma^2/2)T + \sigma W_T}.$$

In fact, we should point out that the pricing of a European option is only a special case of the following problem. Let $(X_t)_{t \geq 0}$ be a diffusion in $\mathbb{R}$, a solution of

$$dX_t = b(t, X_t)dt + \sigma(t, X_t)dW_t, \tag{5.1}$$

where b and σ are real-valued functions satisfying the assumptions of Theorem 3.5.3 in Chapter 3 and $r(t, x)$ is a bounded continuous function modelling the riskless interest rate. We generally want to compute

$$V_t = \mathbb{E}\left(e^{-\int_t^T r(s, X_s)ds} f(X_T) | \mathscr{F}_t\right).$$

In the same way as in the Black-Scholes model, V_t can be written as

$$V_t = F(t, X_t),$$

where

$$F(t, x) = \mathbb{E}\left(e^{-\int_t^T r\left(s, X_s^{t,x}\right)ds} f\left(X_T^{t,x}\right)\right),$$

and $X_s^{t,x}$ is the solution of (5.1) starting from x at time t. Intuitively

$$F(t, x) = \mathbb{E}\left(e^{-\int_t^T r(s, X_s)ds} f(X_T) | X_t = x\right).$$

Mathematically, this result is a consequence of Theorem 3.5.9 in Chapter 3. The computation of V_t is therefore equivalent to the computation of $F(t, x)$. Under some regularity assumptions that we shall specify, this function $F(t, x)$ is the unique solution of the following partial differential equation:

$$\begin{cases} \forall x \in \mathbb{R} \quad u(T, x) = f(x) \\ (\partial u/\partial t + A_t u - ru)(t, x) = 0 \quad \forall (t, x) \in [0, t) \times \mathbb{R}, \end{cases} \tag{5.2}$$

where

$$(A_t f)(x) = \frac{\sigma^2(t, x)}{2} f''(x) + b(t, x)f'(x).$$

Before proving this result, we will explain why the operator A_t appears naturally in connection with *stochastic differential equations*.

5.1.1   Infinitesimal generator of a diffusion

We assume that b and σ are time independent and we denote by $(X_t)_{t \geq 0}$ the solution of

$$dX_t = b(X_t)dt + \sigma(X_t)dW_t. \tag{5.3}$$

Proposition 5.1.1. *Let f be a C^2 function with bounded derivatives and A be the differential operator that maps a C^2 function f to Af such that*

$$(Af)(x) = \frac{\sigma^2(x)}{2} f''(x) + b(x)f'(x).$$

Then, the process $M_t = f(X_t) - \int_0^t Af(X_s)ds$ is an $(\mathscr{F}_t)$-martingale.

Proof. Itô's formula yields

$$f(X_t) = f(X_0) + \int_0^t f'(X_s)dX_s + \frac{1}{2}\int_0^t f''(X_s)\sigma^2(X_s)ds.$$

Hence

$$f(X_t) = f(X_0) + \int_0^t f'(X_s)\sigma(X_s)dW_s$$
$$+ \int_0^t \left[\frac{1}{2}\sigma^2(X_s)f''(X_s) + b(X_s)f'(X_s)\right] ds$$

and the result follows from the fact that the stochastic integral

$$\int_0^t f'(X_s)\sigma(X_s)dW_s$$

is a martingale. Indeed, according to Theorem 3.5.3 and since $|\sigma(x)|$ has sublinear growth, we obtain, for some positive constant K,

$$\mathbb{E}\left(\int_0^t |f'(X_s)|^2|\sigma(X_s)|^2 ds\right)$$
$$\leq KT \sup_{x\in\mathbb{R}} |f'(x)|^2 \left(1 + \mathbb{E}\left(\sup_{s\leq T} |X_s|^2\right)\right) < +\infty.$$

$\square$

Remark 5.1.2. If we denote by X_t^x the solution of (5.3) such that $X_0^x = x$, it follows from Proposition 5.1.1 that

$$\mathbb{E}(f(X_t^x)) = f(x) + \mathbb{E}\left(\int_0^t Af(X_s^x)ds\right).$$

Moreover, since the derivatives of f are bounded by a constant K_f, and since $|b(x)| + |\sigma(x)| \leq K(1 + |x|)$, we have

$$\mathbb{E}\left(\sup_{s\leq T} |Af(X_s^x)|\right) \leq K_f' \left(1 + \mathbb{E}\left(\sup_{s\leq T} |X_s^x|^2\right)\right) < +\infty.$$

Therefore, since $x \mapsto Af(x)$ and $s \mapsto X_s^x$ are continuous, the Lebesgue theorem is applicable and yields

$$\frac{d}{dt}\mathbb{E}\left(f\left(X_t^x\right)\right)|_{t=0} = \lim_{t\to 0}\mathbb{E}\left(\frac{1}{t}\int_0^t Af\left(X_s^x\right)ds\right) = Af(x).$$

The differential operator A is called the *infinitesimal generator* of the diffusion (X_t). The reader can refer to Bouleau (1988) or Revuz and Yor (1990) for more results on the infinitesimal generator of a diffusion.

Proposition 5.1.1 can also be extended to the time-dependent case. We assume that b and σ satisfy the assumptions of Theorem 3.5.3 in Chapter 3 that guarantee the existence and uniqueness of a solution of equation (5.1).

Proposition 5.1.3. *If $u(t, x)$ is a $C^{1,2}$ function with bounded derivatives with respect to x, and if X_t is a solution of (5.1), the process*

$$M_t = u(t, X_t) - \int_0^t \left(\frac{\partial u}{\partial t} + A_s u \right)(s, X_s) ds$$

is a martingale. Here, A_s is the operator defined by

$$(A_s u)(x) = \frac{\sigma^2(s, x)}{2} \frac{\partial^2 u}{\partial x^2} + b(s, x) \frac{\partial u}{\partial x}.$$

The proof is very similar to that of Proposition 5.1.1: the only difference is that we apply Itô's formula for a function of time and an Itô process (see Theorem 3.4.10).

In order to deal with discounted quantities, we state a slightly more general result in the following proposition.

Proposition 5.1.4. *Under the assumptions of Proposition 5.1.3, and if $r(t, x)$ is a bounded continuous function defined on $\mathbb{R}^+ \times \mathbb{R}$, the process*

$$M_t = e^{-\int_0^t r(s, X_s) ds} u(t, X_t) - \int_0^t e^{-\int_0^s r(v, X_v) dv} \left(\frac{\partial u}{\partial t} + A_s u - ru \right)(s, X_s) ds$$

is a martingale.

Proof. This proposition can be proved by using the integration by parts formula to differentiate the product (see Proposition 3.4.12 in Chapter 3)

$$e^{-\int_0^t r(s, X_s) ds} u(t, X_t),$$

and then applying Itô's formula to the process $u(t, X_t)$. □

This result is still true in a multidimensional model. Let us consider the stochastic differential equation

$$\begin{cases} dX_t^1 = b^1(t, X_t) dt + \sum_{j=1}^p \sigma_{1j}(t, X_t) dW_t^j \\ \quad\vdots \qquad\qquad\qquad \vdots \\ dX_t^n = b^n(t, X_t) dt + \sum_{j=1}^p \sigma_{nj}(t, X_t) dW_t^j. \end{cases} \tag{5.4}$$

We assume that the assumptions of Theorem 3.5.5 are still satisfied. For any time t we define the following differential operator A_t, which maps a C^2 function from $\mathbb{R}^n$ to $\mathbb{R}$ to a function characterized by

$$(A_t f)(x) = \frac{1}{2} \sum_{i,j=1}^{n} a_{i,j}(t,x) \frac{\partial^2 f}{\partial x_i \partial x_j}(x) + \sum_{j=1}^{n} b_j(t,x) \frac{\partial f}{\partial x_j}(x),$$

where $(a_{ij}(t,x))$ is the matrix with entries

$$a_{ij}(t,x) = \sum_{k=1}^{p} \sigma_{ik}(t,x) \sigma_{jk}(t,x).$$

In other words, $a(t,x) = \sigma(t,x)\sigma^*(t,x)$, where σ^* is the transpose of $\sigma(t,x) = (\sigma_{ij}(t,x))$.

Proposition 5.1.5. *If (X_t) is a solution of system (5.4) and $u(t,x)$ is a real-valued function of class $C^{1,2}$ defined on $\mathbb{R}^+ \times \mathbb{R}^n$ with bounded derivatives with respect to x and, also, $r(t,x)$ is a continuous bounded function defined on $\mathbb{R}^+ \times \mathbb{R}^n$, then the process*

$$M_t = e^{-\int_0^t r(s,X_s)ds} u(t,X_t) - \int_0^t e^{-\int_0^s r(v,X_v)dv} \left(\frac{\partial u}{\partial t} + A_s u - ru \right)(s,X_s)ds$$

is a martingale.

The proof is based on the multidimensional Itô formula stated in Proposition 3.4.18.

Remark 5.1.6. The differential operator $\partial/\partial t + A_t$ is sometimes called the *Dynkin operator of the diffusion.*

5.1.2 Conditional expectations and partial differential equations

In this section, we want to emphasize the link between pricing a European option and solving a parabolic partial differential equation. Let us, consider $(X_t)_{t \geq 0}$, a solution of system (5.4); $f(x)$ a function from $\mathbb{R}^n$ to $\mathbb{R}$; and $r(t,x)$, a bounded continuous function. We want to compute

$$V_t = \mathbb{E}\left(e^{-\int_t^T r(s,X_s)ds} f(X_T) | \mathscr{F}_t \right).$$

In a similar way as in the scalar case, we can prove that

$$V_t = F(t, X_t),$$

where

$$F(t,x) = \mathbb{E}\left(e^{-\int_t^T r(s,X_s^{t,x})ds} f\left(X_T^{t,x}\right) \right),$$

when we denote by $X^{t,x}$ the unique solution of (5.4) starting from x at time t.

The following result characterizes the function F as a solution of a partial differential equation.

Theorem 5.1.7. *Let u be a $C^{1,2}$ function with a bounded derivative with respect to x, defined on $[0, T) \times \mathbb{R}^n$. If u satisfies*

$$\forall x \in \mathbb{R}^n \qquad u(T, x) = f(x)$$

and

$$\left(\frac{\partial u}{\partial t} + A_t u - ru \right)(t, x) = 0 \quad \forall (t, x) \in [0, T) \times \mathbb{R}^n,$$

then

$$\forall (t, x) \in [0, T) \times \mathbb{R}^n \quad u(t, x) = F(t, x) = \mathbb{E}\left(e^{-\int_t^T r(s, X_s^{t,x}) ds} f\left(X_T^{t,x} \right) \right).$$

Proof. Let us prove the equality $u(t, x) = F(t, x)$ at time $t = 0$. From Proposition 5.1.5, we know that the process

$$M_t = e^{-\int_0^t r(s, X_s^{0,x}) ds} u(t, X_t^{0,x})$$

is a martingale. Therefore the relation $\mathbb{E}(M_0) = \mathbb{E}(M_T)$ yields

$$u(0, x) = \mathbb{E}\left(e^{-\int_0^T r(s, X_s^{0,x}) ds} u\left(T, X_T^{0,x} \right) \right)$$

$$= \mathbb{E}\left(e^{-\int_0^T r(s, X_s^{0,x}) ds} f\left(X_T^{0,x} \right) \right),$$

since $u(T, x) = f(x)$. The proof runs similarly for $t > 0$. $\qquad \square$

Remark 5.1.8. Obviously, Theorem 5.1.7 suggests the following method to price the option. In order to compute

$$F(t, x) = \mathbb{E}\left(e^{-\int_t^T r(s, X_s^{t,x}) ds} f\left(X_t^{t,x} \right) \right)$$

for a given f, we just need to find u such that

$$\begin{cases} \dfrac{\partial u}{\partial t} + A_t u - ru = 0 & \text{in } [0, T) \times \mathbb{R}^n \\ u(T, x) = f(x), & x \in \mathbb{R}^n. \end{cases} \qquad (5.5)$$

Problem (5.5) is a parabolic equation with a *final* condition (as soon as the function $u(T, .)$ is given). If we have a solution u to this problem, and if it satisfies the regularity assumptions of Proposition 5.1.4, we can conclude that $F = u$. We have to rely on existence and uniqueness theorems and regularity results for parabolic equations. These results generally require that the operator A_t be elliptic, which means that, for some positive constant C, we have, for all $(t, x) \in [0, T] \times \mathbb{R}^n$,

$$\forall (\xi_1, \ldots, \xi_n) \in \mathbb{R}^n, \quad \sum_{ij} a_{ij}(t, x)\xi_i\xi_j \geq C \left(\sum_{i=1}^n \xi_i^2 \right). \qquad (5.6)$$

In the absence of ellipticity and for irregular payoff functions f, a notion of weak solution can be used to characterise F (see Fleming and Soner (2006)).

5.1.3   Application to the Black-Scholes model

We are working under the risk neutral probability $\mathbb{P}^*$. The process $(W_t)_{t \geq 0}$ is a standard Brownian motion and the asset price S_t satisfies

$$dS_t = S_t(rdt + \sigma dW_t).$$

The operator A_t is now time independent and is equal to

$$A_t = A^{bs} = \frac{\sigma^2}{2} x^2 \frac{\partial^2}{\partial x^2} + rx \frac{\partial}{\partial x}.$$

It is straightforward to check that the call price, given by $F(t, x) = xN(d_1) - Ke^{-r(T-t)}N(d_1 - \sigma\sqrt{T-t})$ with

$$d_1 = \frac{\log(x/K) + (r + \sigma^2/2)(T-t)}{\sigma\sqrt{T-t}}$$

$$N(d) = \frac{1}{\sqrt{2\pi}} \int_{-\infty}^d e^{-x^2/2}dx,$$

is solution of the equation

$$\begin{cases} \dfrac{\partial u}{\partial t} + A_u^{bs} - ru = 0 & \text{in } [0, T) \times [0, +\infty) \\ u(T, x) = (x - K)_+, & \forall x \in [0, +\infty). \end{cases}$$

The same type of result holds for the put.

Note that the operator A^{bs} does not satisfy the ellipticity condition (5.6). However, the trick is to consider the diffusion $X_t = \log(S_t)$, which is a solution of

$$dX_t = \left(r - \frac{\sigma^2}{2} \right) dt + \sigma dW_t,$$

since $S_t = S_0 e^{(r-\sigma^2/2)t + \sigma W_t}$. Its infinitesimal generator can be written as

$$A^{bs-\log} = \frac{\sigma^2}{2} \frac{\partial^2}{\partial x^2} + \left(r - \frac{\sigma^2}{2} \right) \frac{\partial}{\partial x}.$$

It is clearly elliptic because $\sigma^2 > 0$ and, moreover, it has constant coefficients. We write

$$\tilde{A}^{bs-\log} = \frac{\sigma^2}{2}\frac{\partial^2}{\partial x^2} + \left(r - \frac{\sigma^2}{2}\right)\frac{\partial}{\partial x} - r. \qquad (5.7)$$

The connection between the parabolic problem associated with $\tilde{A}^{bs-\log}$ and the computation of the price of an option in the Black-Scholes model can be highlighted as follows: if we want to compute the price $F(t,x)$ at time t and for a spot price x of an option paying off $f(S_T)$ at time T, we need to find a regular solution v of

$$\begin{cases} \dfrac{\partial v}{\partial t}(t,x) + \tilde{A}^{bs-\log}v(t,x) = 0 & \text{in } [0,t] \times \mathbb{R} \\ v(T,x) = f(e^x), & x \in \mathbb{R}, \end{cases} \qquad (5.8)$$

so that $F(t,x) = v(t, \log(x))$.

5.1.4 Partial differential equations on a bounded open set and computation of expectations

Throughout the rest of this section, we shall assume that there is only one asset and that $b(x), \sigma(x)$ and $r(x)$ are all time independent. The number $r(x)$ can be interpreted as the instantaneous interest rate. Consider the differential operator A, defined by

$$(Af)(x) = \frac{1}{2}\sigma(x)^2\frac{\partial^2 f(x)}{\partial x^2} + b(x)\frac{\partial f(x)}{\partial x}.$$

We denote by $\tilde{A}$ the *discount* operator such that $\tilde{A}f(x) = Af(x) - r(x)f(x)$. Equation (5.5) becomes

$$\begin{cases} \dfrac{\partial u}{\partial t}(t,x) + \tilde{A}u(t,x) = 0 & \text{on } [0,T] \times \mathbb{R} \\ u(T,x) = f(x), & \forall x \in \mathbb{R}. \end{cases} \qquad (5.9)$$

If we want to solve problem (5.9) on the open interval $\mathscr{O} = (a,b)$ instead of $\mathbb{R}$, we need to introduce boundary conditions at the endpoints a and b. We will concentrate on the case when the function takes the value zero on the boundaries. These are the so-called Dirichlet boundary conditions. The problem to be solved is then

$$\begin{cases} \frac{\partial u}{\partial t}(t,x) + \tilde{A}u(t,x) = 0 & \text{on } (0,T) \times \mathscr{O} \\ u(t,a) = u(t,b) = 0 & \forall t \in (0,T) \\ u(T,x) = f(x) & \forall x \in \mathscr{O}. \end{cases} \qquad (5.10)$$

The following theorem shows that a regular solution of (5.10) can also be expressed in terms of the diffusion $X^{t,x}$, which is the solution of (5.3) starting from x at time t.

Theorem 5.1.9. *Let u be a continuous function on $[0,T] \times [a,b]$. Assume that u is $C^{1,2}$ on $(0,T) \times \mathcal{O}$ and that $\partial u/\partial x$ is bounded on $(0,T) \times \mathcal{O}$. Then, if u satisfies (5.10), we have*

$$\forall (t,x) \in [0,T] \times \mathcal{O}, \quad u(t,x) = \mathbb{E}\left(1_{\{\forall s \in [t,T], X_s^{t,x} \in \mathcal{O}\}} e^{-\int_t^T r(X_s^{t,x})ds} f\left(X_T^{t,x}\right)\right).$$

Proof. We prove the result for $t=0$ since the argument is similar at other times. In order to avoid technicalities, we assume that there exists an extension of the function u from $[0,T] \times \mathcal{O}$ to $[0,T] \times \mathbb{R}$ that is still of class $C^{1,2}$. We also denote by u such an extension. From Proposition 5.1.4, we know that

$$M_t = e^{-\int_0^t r(X_s^{0,x})ds} u(t, X_t^{0,x})$$
$$- \int_0^t e^{-\int_0^s r(X_v^{0,x})dv} \left(\frac{\partial u}{\partial t} + Au - ru\right)(s, X_s^{0,x})\, ds$$

is a martingale. Now, let

$$\tau^x \inf\{0 \leq s \leq T, X_s^{0,x} \notin \mathcal{O}\} \wedge T \text{ with the convention } \inf \emptyset = +\infty.$$

Note that τ^x is a bounded stopping time, because $\tau^x = T_a^x \wedge T_b^x \wedge T$, where

$$T_l^x = \inf\{0 \leq s \leq T, X_s^{t,x} = l\}$$

and indeed T_l^x is a stopping time according to Proposition 3.3.6. By applying the optional sampling theorem with the stopping times 0 and τ^x, we get $\mathbb{E}(M_0) = \mathbb{E}(M_{\tau^x})$. Thus, by noticing that if $s \in [0, \tau^x]$, $Af(X_s^{0,x}) = 0$, it follows that

$$u(0,x) = \mathbb{E}\left(e^{-\int_0^{\tau^x} r(s, X_s^{0,x})ds} u\left(\tau^x, X_{\tau^x}^{0,x}\right)\right)$$
$$= \mathbb{E}\left(1_{\{\forall s \in [t,T], X_s^{t,x} \in \mathcal{O}\}} e^{-\int_0^T r(s, X_s^{0,x})ds} u\left(T, X_T^{0,x}\right)\right)$$
$$+ \mathbb{E}\left(1_{\{\exists s \in [t,T], X_s^{t,x} \notin \mathcal{O}\}} e^{-\int_0^{\tau^x} r(s, X_s^{0,x})ds} u\left(\tau^x, X_{\tau^x}^{0,x}\right)\right).$$

Furthermore, $f(x) = u(T,x)$ and $u\left(\tau^x, X_{\tau^x}^{0,x}\right) = 0$ on the event

$$\{\exists s \in [t,T], X_s^{t,x} \notin \mathcal{O}\};$$

consequently

$$u(0,x) = \mathbb{E}\left(1_{\{\forall s \in [t,T], X_s^{t,x} \in \mathcal{O}\}} e^{-\int_0^T r(s, X_s^{0,x})ds} f\left(X_T^{0,x}\right)\right),$$

which completes the proof for $t=0$. $\qquad\square$

Remark 5.1.10. An option defined by the $\mathscr{F}_T$-measurable random variable

$$1_{\{\forall s \in [t,T], X_s^{t,x} \in \mathcal{O}\}} e^{-\int_t^T r(X_s^{t,x})ds} f\left(X_T^{t,x}\right)$$

is called *extinguishable*. Indeed, as soon as the asset price exits in the open set $\mathscr{O}$, the option becomes worthless. In the Black-Scholes model, if $\mathscr{O}$ is of the form $(0, l)$ or $(l, +\infty)$, we are able to compute explicit formulae for the option price (see Exercise 30 for the pricing of *Down and Out* options and Cox and Rubinstein (1985) and Shreve (2004) for more on barrier options).

5.2   Solving parabolic equations numerically

We saw under which conditions the option price coincided with the solution of the partial differential equation (5.9). We now want to address the problem of solving a PDE such as (5.9) numerically, and we shall see how we can approximate its solution using the so-called *finite difference* method. This method is obviously useless in the Black-Scholes model since we are able to derive a closed-form solution, but it proves to be useful when we are dealing with more general diffusion models. We shall only state the most important results, but the reader can refer to Glowinsky et al. (1976) or Raviart and Thomas (1983) for a detailed analysis.

5.2.1   Localization

Problem (5.9) is set on $\mathbb{R}$. In order to discretize, we will have to work on a bounded open set $\mathscr{O}_l = (-l, l)$, where l is a constant to be chosen carefully in order to optimize the algorithm. We also need to specify the boundary conditions (i.e. at l and $-l$). Typically, we shall impose Dirichlet conditions (i.e. $u(l) = u(-l) = 0$ or some more relevant constants) or Neumann conditions (involving $(\partial u/\partial x)(l), (\partial u/\partial x)(-l)$). If we specify Dirichlet boundary conditions, the PDE becomes

$$\begin{cases} \dfrac{\partial u(t, x)}{\partial t} + \tilde{A}u(t, x) = 0 \text{ on } (0, T) \times \mathscr{O}_l \\ u(t, l) = u(t, -l) = 0 \text{ if } t \in (0, T) \\ u(T, x) = f(x) \text{ if } x \in \mathscr{O}_l. \end{cases}$$

We are going to show how we can estimate the error that we make by restricting the state space to $\mathscr{O}_l$. We shall work in a Black-Scholes environment and, thus, the logarithm of the asset price solves the following stochastic differential equation:

$$dX_t = (r - \sigma^2/2)dt + \sigma dW_t.$$

We want to compute the price of an option whose payoff can be written as $f(S_T) = f(S_0 e^{X_T})$. We write $\bar{f}(x) = f(e^x)$. For simplicity, we adopt Dirichlet boundary conditions. We also assume that the solution u of (5.9) and the solutions u_l of (5.10) are smooth enough, so that

$$u(t, x) = \mathbb{E}\left(e^{-r(T-t)} \bar{f}\left(X_T^{t,x}\right)\right)$$

and

$$u_l(t,x) = \mathbb{E}\left(1_{\{\forall s \in [t,T], |X_s^{t,x}| < l\}} e^{-r(T-t)} \bar{f}\left(X_T^{t,x}\right)\right),$$

where $X_s^{t,x} = x \exp((r - \sigma^2/2)(s-t) + \sigma(W_s - W_t))$. We assume that the function f (hence $\bar{f}$) is bounded by a constant M and that $r \geq 0$. Then, it is easy to show that

$$|u(t,x) - u_l(t,x)| \leq M\mathbb{P}\left(\exists s \in [t,T], |X_s^{t,x}| \geq l\right).$$

Let $r' = r - \sigma^2/2$. We have

$$\{\exists s \in [t,T], |X_s^{t,x}| \geq l\} \ \subset \left\{\sup_{t \leq s \leq T} |x + r'(s-t) + \sigma(W_s - W_t)| \geq l\right\}$$

$$\subset \left\{\sup_{t \leq s \leq T} |x + \sigma(W_s - W_t)| \geq l - |r'T|\right\}.$$

Thus

$$|u(t,x) - u_l(t,x)| \leq M\mathbb{P}\left(\sup_{t \leq s \leq T} |x + \sigma(W_s - W_t)| \geq l - |r'T|\right)$$

$$= M\mathbb{P}\left(\sup_{0 \leq s \leq T-t} |x + \sigma W_s| \geq l - |r'T|\right)$$

$$\leq M\mathbb{P}\left(\sup_{0 \leq s \leq T} |x + \sigma W_s| \geq l - |r'T|\right).$$

From Proposition 3.3.6, we know that if we define $T_a = \inf\{s > 0, W_s = a\}$, then $\mathbb{E}(\exp(-\lambda T_a)) = \exp(-\sqrt{2\lambda}|a|)$. It infers that for any $a > 0$, and for any $\lambda \geq 0$,

$$\mathbb{P}\left(\sup_{s \leq T} W_s \geq a\right) = \mathbb{P}(T_a \leq T) \leq e^{\lambda T}\mathbb{E}(e^{-\lambda T_a}) \leq e^{\lambda T} e^{-a\sqrt{2\lambda}}.$$

Minimizing with respect to λ yields

$$\mathbb{P}\left(\sup_{s \leq T} W_s \geq a\right) \leq \exp\left(-\frac{a^2}{T}\right),$$

and therefore

$$\mathbb{P}\left(\sup_{s \leq T}(x + \sigma W_s) \geq a\right) \leq \exp\left(-\frac{|a-x|^2}{\sigma^2 T}\right).$$

Since $(-W_s)_{s \geq 0}$ is also a standard Brownian motion, we also have

$$\mathbb{P}\left(\inf_{s \leq T}(x + \sigma W_s) \leq -a\right) = \mathbb{P}\left(\sup_{s \leq T}(-x - \sigma W_s) \geq a\right) \leq \exp\left(-\frac{|a+x|^2}{\sigma^2 T}\right).$$

These two results imply that

$$\mathbb{P}\left(\sup_{s\leq T}|x+\sigma W_s|\geq a\right)\leq\exp\left(-\frac{|a-x|^2}{\sigma^2 T}\right)+\exp\left(-\frac{|a+x|^2}{\sigma^2 T}\right).$$

Hence

$$|u(t,x)-u_l(t,x)|\leq M\left(\exp\left(-\frac{|l-|r'T|-x|^2}{\sigma^2 T}\right)\right.$$
$$\left.+\exp\left(-\frac{|l-|r'T|+x|^2}{\sigma^2 T}\right)\right).$$

This proves that for fixed t and x, $\lim_{l\to+\infty}u_l(t,x)=u(t,x)$. The convergence is even uniform in t and x as long as x remains in a compact set of $\mathbb{R}$.

Remark 5.2.1.

- It can be proved that $\mathbb{P}(\sup_{s\leq T}W_s\geq a)=2\mathbb{P}(W_T\geq a)$ (see Exercise 21 in Chapter 3). This would lead to a slightly better estimate than the one above.

- The fundamental advantage of the localization method is that it can be used for pricing American options, and in that case the numerical approximation is compulsory. The estimate of the error will give us a hint to choose the domain of integration of the PDE. It is quite a crucial choice that determines how efficient our numerical procedure will be.

5.2.2   The finite difference method

Once the problem has been localised, we obtain the following system with Dirichlet boundary conditions:

$$(E)\begin{cases}\frac{\partial u(t,x)}{\partial t}+\tilde{A}u(t,x)=0 \text{ on } [0,T)\times\mathcal{O}_l\\ u(t,l)=u(t,-l)=0 \text{ if } t\in[0,T]\\ u(T,x)=f(x) \text{ if } x\in\mathcal{O}_l.\end{cases}$$

The finite difference method is basically a discretization in time and space of equation (E).

We shall start by discretizing the differential operator $\tilde{A}$ on $\mathcal{O}_l$. In order to do this, we need to associate with the function $(f(x))_{x\in\mathcal{O}_l}$, which is an element of an infinite-dimensional vector space, a vector $(f_i)_{1\leq i\leq N}$ in $\mathbb{R}^N$. We proceed as follows: for $i=0,1,\ldots,N$, let $x_i=-l+2il/(N+1)$. The number f_i is supposed to be an approximation of $f(x_i)$. The boundary conditions are stated as $f_0=0$, $f_{N+1}=0$ in the Dirichlet case, and $f_0=f_1$, $f_N=f_{N+1}$ in the Neumann case.

Let $h=2l/(N+1)$. The discretized version of the operator $\tilde{A}$ is the operator $\tilde{A}_h$ on $\mathbb{R}^N$, defined as follows. Think of a vector $u_h=(u_h^i)_{1\leq i\leq N}$ in $\mathbb{R}^N$

as the discrete approximation of a function u (i.e. $u_h^i \approx u(x_i)$) and replace the first derivative $\frac{\partial u(x_i)}{\partial x}$ with

$$\partial_h u_h^i = \frac{u_h^{i+1} - u_h^{i-1}}{2h}.$$

Similarly, replace the second derivative $\frac{\partial^2 u(x_i)}{\partial x^2}$ with

$$\partial_h^2 u_h^i = \frac{\frac{u_h^{i+1} - u_h^i}{h} - \frac{u_h^i - u_h^{i-1}}{h}}{h} = \frac{u_h^{i+1} - 2u_h^i + u_h^{i-1}}{h^2}.$$

The vector $\tilde{A}_h u_h$ is defined by

$$(\tilde{A}_h u_h)_i = \frac{\sigma^2(x_i)}{2} \partial_h^2 u_h^i + b(x_i) \partial_h u_h^i - r u_h^i, \quad i = 1, \dots, N.$$

Remark 5.2.2. In the Black-Scholes case (after the usual logarithmic change of variables),

$$\tilde{A}^{bs-\log} u(x) = \frac{\sigma^2}{2} \frac{\partial^2 u(x)}{\partial x^2} + \left(r - \frac{\sigma^2}{2} \right) \frac{\partial u(x)}{\partial x} - r u(x)$$

is associated with

$$(\tilde{A}_h u_h)_i = \frac{\sigma^2}{2h^2} \left(u_h^{i+1} - 2u_h^i + u_h^{i-1} \right) + \left(r - \frac{\sigma^2}{2} \right) \frac{1}{2h} \left(u_h^{i+1} - u_h^{i-1} \right) - r u_h^i.$$

If we specify null Dirichlet boundary conditions, $\tilde{A}_h$ is then represented by the following matrix:

$$((\tilde{A}_h)_{ij})_{1 \leq i,j \leq N} = \begin{pmatrix} \beta & \gamma & 0 & \cdots & 0 & 0 \\ \alpha & \beta & \gamma & 0 & \cdots & 0 \\ 0 & \alpha & \beta & \gamma & \cdots & 0 \\ 0 & \vdots & \ddots & \ddots & \ddots & \vdots \\ 0 & 0 & \cdots & \alpha & \beta & \gamma \\ 0 & 0 & 0 & \cdots & \alpha & \beta \end{pmatrix},$$

where

$$\begin{cases} \alpha = \frac{\sigma^2}{2h^2} - \frac{1}{2h} \left(r - \frac{\sigma^2}{2} \right) \\ \beta = -\frac{\sigma^2}{h^2} - r \\ \gamma = \frac{\sigma^2}{2h^2} + \frac{1}{2h} \left(r - \frac{\sigma^2}{2} \right). \end{cases}$$

If we specify null Neumann conditions, the matrix has the following form:

$$\begin{pmatrix} \beta + \alpha & \gamma & 0 & \cdots & 0 & 0 \\ \alpha & \beta & \gamma & 0 & \cdots & 0 \\ 0 & \alpha & \beta & \gamma & \cdots & 0 \\ 0 & \vdots & \ddots & \ddots & \ddots & \vdots \\ 0 & 0 & \cdots & \alpha & \beta & \gamma \\ 0 & 0 & 0 & \cdots & \alpha & \beta + \gamma \end{pmatrix}. \tag{5.1}$$

The space discretization transforms (E) into an ordinary differential equation (E_h):

$$(E_h) \begin{cases} \frac{du_h(t)}{dt} + \tilde{A}_h u_h(t) = 0 & \text{if } 0 \leq t \leq T \\ u_h(T) = f_h \end{cases},$$

where $f_h = \left(f_j^i\right)_{1 \leq i \leq N}$ is the vector $f_h^i = f(x_i)$.

A time discretization of this equation can now be defined, using the so-called θ-schemes. Consider $\theta \in [0, 1]$, and let k a be a time-step such that $T = Mk$ (where M is a (large) positive integer). We approximate the solution u_h of (E_h) at time nk by $u_{h,k}^n$, where the sequence $(u_{h,k}^n)_{n=0,\ldots,M}$ solves the following recursive equations:

$$(E_{h,k}) \begin{cases} u_{h,k}^M = f_h \\ \dfrac{u_{h,k}^{n+1} - u_{h,k}^n}{k} + \theta \tilde{A}_h u_{h,k}^n + (1-\theta) \tilde{A}_h u_{h,k}^{n+1} = 0, & \text{if } 0 \leq n \leq M-1. \end{cases}$$

Note that the system $(E_{h,k})$ is solved by backward induction, starting from $n = M$.

Remark 5.2.3.

- When $\theta = 0$, the scheme is explicit because $u_{h,k}^n$ is computed directly from $u_{h,k}^{n+1}$. But when $\theta > 0$, we have to solve at each step a system of the form $R u_{h,k}^n = v$, with

$$\begin{cases} R = (I - \theta k \tilde{A}_h) \\ v = (I + (1-\theta)k\tilde{A}_h)u_{h,k}^{n+1}, \end{cases}$$

  where R is a tridiagonal matrix. This is obviously more complex and more time consuming. However, these schemes are often used in practice because of their good convergence properties, as we shall see shortly.

- When $\theta = 1/2$, the algorithm is called the *Crank and Nicholson* scheme. It is often used to solve systems of type (E) when $b = 0$ and σ is constant.

- When $\theta = 1$, the scheme is said to be completely implicit.

We shall now state convergence results for the solution $u_{h,k}$ of $(E_{h,k})$ towards the solution $u(t, x)$ of (E), assuming the ellipticity condition. The reader ought to refer to Raviart and Thomas (1983) for proofs. Define the function u_h^k on $[0, T] \times \mathcal{O}_l$ by

$$u_h^k(t, x) = \left(u_{h,k}^n\right)_i \quad \text{for } nk - k \leq t < nk, \; x_i - h/2 \leq x < x_i + h/2.$$

We also denote by $\delta\phi$ the approximate derivative of a function ϕ, defined by

$$(\delta\phi)(x) = \frac{1}{h}(\phi(x + h/2) - \phi(x - h/2)), \quad x \in \mathcal{O}_l.$$

Recall that $\tilde{A}f(x)$ is equal to $1/2\sigma(x)^2(\partial^2 f(x)/\partial x^2) + b(x)(\partial f(x)/\partial x) - r(x)f(x)$. In the following statement, the usual inner product on the Hilbert space $L^2(\mathcal{O}_l)$ is denoted by $(.,.)_{L^2(\mathcal{O}_l)}$, and the Hilbert norm by $|.|_{L^2(\mathcal{O}_l)}$.

Theorem 5.2.4. *Assume that b and σ are Lipschitz and that r is a non-negative continuous function. We assume that the operator $\tilde{A}$ is elliptic, i.e. there exists a constant $\varepsilon > 0$ such that for every C^2 function u with compact support in $\mathcal{O}_l$,*

$$(-\tilde{A}u, u)_{L^2(\mathcal{O}_l)} \geq \varepsilon(|u|_{L^2(\mathcal{O}_l)} + |u'|_{L^2(\mathcal{O}_l)}).$$

Then:

- *If $1/2 \leq \theta \leq 1$, we have, as h, k tend to 0,*

$$\lim u_h^k = u \quad in \ L^2([0,T] \times \mathcal{O}_l)$$
$$\lim \delta u_h^k = \partial u/\partial x \quad in \ L^2([0,T] \times \mathcal{O}_l).$$

- *If $0 \leq \theta < 1/2$, we have, as h, k tend to 0, with $\lim k/h^2 = 0$,*

$$\lim u_h^k = u \quad in \quad L^2([0,T] \times \mathcal{O}_l)$$
$$\lim \delta u_h^k = \partial u/\partial x \quad in \ L^2([0,T] \times \mathcal{O}_l).$$

Remark 5.2.5.

- In the case $0 \leq \theta < 1/2$, we say that the scheme is *conditionally convergent* because the algorithm converges only if h, k and k/h^2 tend to 0. These algorithms are rather tricky to implement numerically and therefore they are rarely used except when $\theta = 0$.

- In the case $1/2 \leq \theta \leq 1$, we say that the scheme is *unconditionally convergent* because it converges as soon as h and k tend to 0.

Let us now examine in detail how problem $(E_{h,k})$ can be solved in practice. At each time-step n we are looking for a solution of $RX = G$, where

$$\begin{cases} X &= u_{h,k}^n \\ G &= (I + (1-\theta)k\tilde{A}_h)u_{h,k}^{n+1} \\ R &= I - k\theta\tilde{A}_h. \end{cases}$$

The matrix R is tridiagonal. The following algorithm, known as the Gauss method, solves the system with a number of multiplications proportional to N. Let $X = (x_i)_{1 \leq i \leq N}, G = (g_i)_{1 \leq i \leq N}$ and

$$R = \begin{pmatrix} b_1 & c_1 & 0 & \cdots & 0 & 0 \\ a_2 & b_2 & c_2 & 0 & \cdots & 0 \\ 0 & a_3 & b_3 & c_3 & \cdots & 0 \\ 0 & \ddots & \ddots & \ddots & \ddots & \vdots \\ 0 & 0 & \cdots & a_{N-1} & b_{N-1} & c_{N-1} \\ 0 & 0 & 0 & \cdots & a_N & b_N \end{pmatrix}.$$

The algorithm runs as follows: first, we transform R into a lower triangular matrix using the Gauss method from bottom to top:

$$
\begin{vmatrix}
\textbf{Upward:} \\
b'_N = b_N \\
g'_N = g_N \\
\text{For } 1 \le i \le N-1, \ i \text{ decreasing:} \\
\qquad b'_i = b_i - c_i a_{i+1}/b'_{i+1} \\
\qquad g'_i = g_i - c_i g'_{i+1}/b'_{i+1}
\end{vmatrix}
$$

We have obtained an equivalent system $R'X = G'$, where

$$
R' = \begin{pmatrix}
b'_1 & 0 & 0 & \cdots & 0 & 0 \\
a_2 & b'_2 & 0 & 0 & \cdots & 0 \\
0 & a_3 & b'_3 & 0 & \cdots & 0 \\
0 & \vdots & \ddots & \ddots & \ddots & \vdots \\
0 & 0 & \cdots & a_{N-1} & b'_{N-1} & 0 \\
0 & 0 & 0 & \cdots & a_N & b'_N
\end{pmatrix}.
$$

To conclude, we just have to compute X starting from the top of the matrix:

$$
\begin{vmatrix}
\textbf{Downward:} \\
x_1 = g'_1/b'_1 \\
\text{For } 2 \le i \le N, \quad i \text{ increasing} \\
\qquad x_i = (g'_i - a_i x_{i-1})/b'_i.
\end{vmatrix}
$$

Remark 5.2.6. If the matrix R is not invertible, the above algorithm will not work (as it will imply dividing by 0 at some point). It can be proved that the matrix R is invertible if, for any i, we have $|a_i| + |c_i| \le |b_i|$. In the Black-Scholes case, it is easy to check that this condition is satisfied as soon as $|r - \sigma^2/2| \le \sigma^2/h$, i.e. for sufficiently small h.

5.3 American options

5.3.1 Statement of the problem

The analysis of American options in continuous-time is not straightforward. In the Black-Scholes model, we obtained the following formula for the price of an American call $(f(x) = (x - K)_+)$ or an American put $(f(x) = (K - x)_+)$:

$$
V_t = \Phi(t, S_t),
$$

where

$$
\Phi(t, x) = \sup_{\tau \in \mathscr{T}_{t,T}} \mathbb{E}^* \left(e^{-r(\tau - t)} f \left(x e^{(r - \sigma^2/2)(\tau - t) + \sigma(W_\tau - W_t)} \right) \right)
$$

and, under $\mathbb{P}^*$, $(W_t)_{t\geq 0}$ is a standard Brownian motion and $\mathscr{T}_{t,T}$ is the set of all stopping times with values in $[t, T]$. We showed that the American call price (on a stock offering no dividend) is equal to the European call price. For the American put price, there is no explicit formula and numerical methods are needed.

The problem to be solved is a particular case of the following general problem: given a *good* function f and a diffusion $(X_t)_{t\geq 0}$ in $\mathbb{R}^n$, the solution of (5.4), compute the function

$$\Phi(t, x) = \sup_{\tau \in \mathscr{T}_{t,T}} \mathbb{E}\left(e^{-\int_t^T r(s,X_s^{t,x})ds} f\left(X_\tau^{t,x}\right)\right).$$

By considering the stopping time $\tau = t$, we get $\Phi(t, x) \geq f(x)$. Also note that, for $t = T$, we clearly have $\Phi(T, x) = f(x)$.

Remark 5.3.1. It can be proved (see Chapter 2 for the analogy with discrete time models and Chapter 4 for the Black-Scholes case) that the process

$$e^{-\int_0^t r(s,X_s)ds} \Phi(t, X_t)$$

is the smallest martingale that dominates the process $f(X_t)$ at all times.

We just stressed the fact that the European option price is the solution of a parabolic partial differential equation. As far as American options are concerned, we obtain a similar result in terms of a *parabolic system of differential inequalities*. The following theorem, stated in rather loose terms (see Remark 5.3.3), tries to explain that.

Theorem 5.3.2. *Assume that u is a regular solution of the following system of partial differential inequalities:*

$$\begin{cases} \dfrac{\partial u}{\partial t} + A_t u - ru \leq 0, \quad u \geq f & in \quad [0,T) \times \mathbb{R}^n \\[2mm] \left(\dfrac{\partial u}{\partial t} + A_t u - ru\right)(f - u) = 0 & in \quad [0,T) \times \mathbb{R}^n \\[2mm] u(T, x) = f(x) & in \quad \mathbb{R}^n. \end{cases} \quad (5.13)$$

Then

$$u(t, x) = \Phi(t, x) = \sup_{\tau \in \mathscr{T}_{t,T}} \mathbb{E}\left(e^{-\int_t^T r(s,X_s^{t,x})ds} f\left(X_\tau^{t,x}\right)\right).$$

Proof. We shall only sketch the proof of this result. For a complete proof, the reader ought to refer to Bensoussan and Lions (1982) (Chapter 3, Section 2) and Jaillet, Lamberton, and Lapeyre (1990) (Section 3). We only consider the case $t = 0$ since the proof is very similar for arbitrary t. Denote by X_t^x the solution of (5.4) starting from x at time 0. Proposition 5.1.3 shows that the process

$$M_t = e^{-\int_0^t r(s, X_s^x)ds} u(t, X_t^x)$$
$$- \int_0^t e^{-\int_0^s r(v, X_v^x)dv} \left(\frac{\partial u}{\partial t} + A_s u - ru\right)(s, X_s^x)\, ds$$

is a martingale. By applying the Optional Sampling Theorem to this martingale with the stopping times 0 and τ, we get $\mathbb{E}(M_\tau) = \mathbb{E}(M_0)$, and since $\partial u/\partial t + A_s u - ru \leq 0$,

$$u(0, x) \geq \mathbb{E}\left(e^{-\int_0^\tau r(s, X_s^x)ds} u(\tau, X_\tau^x)\right).$$

Recall that $u(t, x) \geq f(x)$; thus

$$u(0, x) \geq \mathbb{E}\left(e^{-\int_0^\tau r(s, X_s^x)ds} f(X_\tau^x)\right).$$

This proves that

$$u(0, x) \geq \sup_{\tau \in \mathscr{T}_{0,T}} \mathbb{E}\left(e^{-\int_0^\tau r(s, X_s^x)ds} f(X_\tau^x)\right) = \Phi(0, x).$$

Now, let $\tau_{\mathrm{opt}} = \inf\{0 \leq s \leq T,\ u(s, X_s^x) = f(X_s^x)\}$. It can be proved that τ_{opt} is a stopping time. Also, for s between 0 and τ_{opt}, we have $(\partial u/\partial t + A_s u - ru)(s, X_s^x) = 0$. The Optional Sampling Theorem yields

$$u(0, x) = \mathbb{E}\left(e^{-\int_0^{\tau_{\mathrm{opt}}} r(s, X_s^x)ds} u\left(\tau_{\mathrm{opt}}, X_{\tau_{\mathrm{opt}}}^x\right)\right).$$

But, at time τ_{opt}, $u\left(\tau_{\mathrm{opt}}, X_{\tau_{\mathrm{opt}}}^x\right) = f\left(X_{\tau_{\mathrm{opt}}}^x\right)$, so that

$$u(0, x) = \mathbb{E}\left(e^{-\int_0^{\tau_{\mathrm{opt}}} r(s, X_s^x)ds} f\left(X_{\tau_{\mathrm{opt}}}^x\right)\right).$$

This proves that $u(0, x) \leq \Phi(0, x)$, hence $u(0, x) = \Phi(0, x)$. We even proved that τ_{opt} is an optimal stopping time (i.e. the supremum is attained for $\tau = \tau_{\mathrm{opt}}$). □

Remark 5.3.3. The precise meaning of system (5.13) is awkward because, even for a regular function f, the solution u is generally not C^2. A precise definition can be given through the variational formulation of the problem (see Bensoussan and Lions (1982)), which is called a variational inequality. The difficulty to make the proof that we have just sketched rigorous comes from the fact that we cannot directly apply Itô's formula to a solution of the variational inequality.

5.3.2 The American put in the Black-Scholes model

We are now leaving the general framework to concentrate on the *pricing of the American put in the Black-Scholes model.*

We are working under the risk neutral probability measure $\mathbb{P}^*$, under which the process $(W_t)_{t\geq 0}$ is a standard Brownian motion and the stock price S_t satisfies

$$dS_t = S_t(rdt + \sigma dW_t).$$

We saw in Section 5.1.3 how we can get an elliptic operator by introducing the process

$$X_t = \log(S_t) = \log(S_0) + \left(r - \frac{\sigma^2}{2}\right)t + \sigma W_t.$$

Its infinitesimal generator A is actually time independent and

$$\tilde{A}^{bs-\log} = A^{bs-\log} - r = \frac{\sigma^2}{2}\frac{\partial^2}{\partial x^2} + \left(r - \frac{\sigma^2}{2}\right)\frac{\partial}{\partial x} - r.$$

If we consider $\phi(x) = (K - e^x)_+$, the partial differential inequality corresponding to the price of the American put is

$$
\begin{cases}
\frac{\partial v}{\partial t}(t,x) + \tilde{A}^{bs-\log}v(t,x) \leq 0 & \text{a.e. in} \quad [0,T) \times \mathbb{R} \\
v(t,x) \geq \phi(x) & \text{a.e. in} \quad [0,T] \times \mathbb{R} \\
(v(t,x) - \phi(x))\left(\frac{\partial v}{\partial t}(t,x) + \tilde{A}^{bs-\log}v(t,x)\right) = 0 \text{ a.e.in} & [0,T) \times \mathbb{R} \\
v(T,x) = \phi(x).
\end{cases}
\tag{5.14}
$$

The following theorem states existence and uniqueness resuts for a solution of (5.14) and establishes the connection with the American put price.

Theorem 5.3.4. *The variational inequality (5.14) has a unique continuous bounded solution $v(t,x)$ such that its partial derivatives in the distribution sense $\partial v/\partial x, \partial v/\partial t, \partial^2 v/\partial x^2$ are locally bounded. Moreover, this solution satisfies*

$$v(t,\log(x)) = \Phi(t,x) = \sup_{\tau \in \mathcal{T}_{t,T}} \mathbb{E}^*\left(e^{-r(\tau - t)}f\left(xe^{(r-\sigma^2/2)(\tau - t + \sigma(W_\tau - W_t))}\right)\right).$$

The proof of this theorem can be found in Jaillet, Lamberton, and Lapeyre (1990).

Numerical solution to this inequality

We are going to show how we can numerically solve equation (5.14). Essentially, the method is similar to the one used in the European case. First, we localize the problem to work in the interval $\mathcal{O}_l = (-l, l)$. Then, we must impose boundary conditions at $\pm l$. Here is the inequality with Neumann boundary conditions:

$$(A) \begin{cases} \dfrac{\partial v}{\partial t}(t,x) + \tilde{A}^{bs-\log} v(t,x) \leq 0 \quad \text{a.e. in} \quad [0,T] \times \mathcal{O}_l \\[2mm] v(t,x) \geq \phi(x) \quad \text{a.e. in} \quad [0,T] \times \mathcal{O}_l \\[2mm] (v-\phi)\left(\dfrac{\partial v}{\partial t}(t,x) + \tilde{A}^{bs-\log} v(t,x)\right) = 0 \quad \text{a.e. in} \quad [0,T] \times \mathcal{O}_l \\[2mm] v(T,x) = \phi(x) \\[2mm] \dfrac{\partial v}{\partial x}(t,\pm l) = 0. \end{cases}$$

We can now discretize (A) using the finite difference method. The notations are the same as in Section 5.2.2. In particular, the integer M is such that $Mk = T$, the vector f_h is given by $f_h^i = \phi(x_i)$, where $x_i = -l + 2il/(N+1)$, and $\tilde{A}_h$ is represented by the matrix (5.1). If u and v are two vectors in $\mathbb{R}^n$, we write $u \leq v$ if $u_i \leq v_i$ for $i = 1, \ldots, N$. Formally, the method is the same as in the European case: the discretization in time leads to the finite-dimensional inequality $(A_{h,k})$:

$$(A_{h,k}) \begin{cases} u_{h,k}^M = f_h \\[2mm] \text{and if } 0 \leq n \leq M-1 \\[2mm] u_{h,k}^n \geq f_h \\[2mm] u_{h,k}^{n+1} - u_{h,k}^n + k\left(\theta \tilde{A}_h u_{h,k}^n + (1-\theta)\tilde{A}_h u_{h,k}^{n+1}\right) \leq 0 \\[2mm] \left(u_{h,k}^{n+1} - u_{h,k}^n + k\left(\theta \tilde{A}_h u_{h,k}^n + (1-\theta)\tilde{A}_h u_{h,k}^{n+1}\right), u_{h,k}^n - f_h\right) = 0, \end{cases}$$

where (x,y) is the scalar product of the vectors x and y in $\mathbb{R}^N$ and $\tilde{A}_h$ is given by (5.1). Introduce the following notations:

$$\begin{cases} R = I - k\theta \tilde{A}_h \\[2mm] X = u_{h,k}^n \\[2mm] G = \left(I + k(1-\theta)\tilde{A}_h\right) u_{h,k}^{n+1} \\[2mm] F = f_h. \end{cases}$$

We have to solve, at each time n, the system of inequalities

$$(AD) \begin{cases} RX \geq G \\[2mm] X \geq F \\[2mm] (RX - G, X - F) = 0, \end{cases}$$

where R is the tridiagonal matrix

$$R = \begin{pmatrix} a+b & c & 0 & \cdots & 0 & 0 \\ a & b & c & 0 & \cdots & 0 \\ 0 & a & b & c & \cdots & 0 \\ 0 & \vdots & \ddots & \ddots & \ddots & \vdots \\ 0 & 0 & \cdots & a & b & c \\ 0 & 0 & 0 & \cdots & a & b+c \end{pmatrix}$$

with

$$\begin{cases} a = \theta k \left(-\frac{\sigma^2}{2h^2} + \frac{1}{2h}\left(r - \frac{\sigma^2}{2} \right) \right) \\ b = 1 + \theta k \left(\frac{\sigma^2}{h^2} + r \right) \\ c = -\theta k \left(\frac{\sigma^2}{2h^2} + \frac{1}{2h}\left(r - \frac{\sigma^2}{2} \right) \right). \end{cases}$$

The problem (AD) is known as a *linear complementarity problem*. It can be solved both theoretically and numerically if the matrix R is coercive (i.e. $x.Rx \geq \alpha x.x$, with $\alpha > 0$). In our case, R will satisfy this assumption if $|r - \sigma^2/2| \leq \sigma^2/h$ and if $|r - \sigma^2/2|k/2h < 1$. Indeed, this condition implies that a and c are negative, and, therefore, by using the fact that $(a+b)^2 \leq 2(a^2+b^2)$ we have

$$x.Rx = \sum_{i=2}^{n} ax_{i-1}x_i + \sum_{i=1}^{n} bx_i^2 + \sum_{i=1}^{n-1} cx_i x_{i+1} + ax_1^2 + cx_n^2$$

$$\geq (a/2) \sum_{i=2}^{n} (x_{i-1}^2 + x_i^2)$$

$$+ \sum_{i=1}^{n} bx_i^2 + (c/2) \sum_{i=1}^{n-1} (x_i^2 + x_{i+1}^2) + ax_1^2 + cx_n^2$$

$$= \frac{a-c}{2}(x_1^2 - x_n^2) + (a+b+c) \sum_{i=1}^{n} x_i^2$$

$$\geq \left(a+b+c - \frac{1}{2}|a-c| \right) \sum_{i=1}^{n} x_i^2 \geq \left(1 - \frac{k}{2h}\left| r - \frac{\sigma^2}{2} \right| \right) \sum_{i=1}^{n} x_i^2.$$

Under the coercitivity assumption, it can be can proved that there exists a unique solution to the problem $(A_{h,k})$ (see Exercise 31).

The following theorem analyses explicitly the nature of the convergence of a solution of $(A_{h,k})$ to the solution of (A). We note

$$u_h^k(t,x) = \sum_{n=1}^{M} \sum_{i=1}^{N} (u_{h,k}^n)_i \, \mathbf{1}_{(x_i - h/2, x_i + i/2]} \times \mathbf{1}_{((n-1)k, nk]}.$$

Theorem 5.3.5. Let u be the solution of (A).

1. *If $\theta < 1$, the convergence is conditional: if h and k converge to 0 and if k/h^2 converges to 0, then*

$$\lim u_h^k = u \quad in \quad L^2([0,T]) \times \mathscr{O}_l$$
$$\lim \delta u_h^k = \frac{\partial u}{\partial x} \quad in \quad L^2([0,T] \times \mathscr{O}_l).$$

2. *If $\theta = 1$, the convergence is* unconditional, *i.e. the previous convergence is true when h and k converge to 0 without restriction.*

The reader will find the proof of this result in Glowinsky et al. (1976). See also Zhang (1997).

Remark 5.3.6. In practice, we normally use $\theta = 1$ in order to have unconditional convergence.

Numerical solution of a finite-dimensional linear complementarity problem

In the *American put* case, when *the step h is sufficiently small*, we can solve the system (AD) very efficiently by a slight modification of the algorithm used to solve tridiagonal systems of equations. We shall proceed as follows (we denote by $b_1, \ldots, b_N$ the diagonal entries of the matrix R):

Upward:
$b_N' = b_N$
$g_N' = g_N$
For $1 \leq i \leq N - 1$, decreasing i
$\qquad b_i' = b_i - ca/b_{i+1}'$
$\qquad g_i' = g_i - cg_{i+1}'/b_{i+1}'$

'American' downward:
$x_1 = g_1'/b_1'$
For $2 \leq i \leq N$, increasing i
$\qquad \tilde{x}_i = (g_i' - ax_{k-1})/b_i'$
$\qquad x_i = \sup(\tilde{x}_i, f_i).$

Jaillet, Lamberton, and Lapeyre (1990) prove that, under the previous assumptions, this algorithm does compute a solution of (AD).

Remark 5.3.7. The algorithm is exactly the same as in the European case, apart from the step $x_i = \sup(\tilde{x}_i, f_i)$. That makes it very effective.

There exist other algorithms to solve inequalities in finite dimensions. Some exact methods are presented in Jaillet, Lamberton, and Lapeyre (1990) and some iterative methods are discussed in Glowinsky et al. (1976).

Remark 5.3.8. When we plug in $\theta = 1$ in $(A_{h,k})$, and we impose Neumann boundary conditions, the previous algorithm is due to Brennan and Schwartz (1977).

We must emphasize the fact that the previous algorithm only computes the exact solution of system (AD) if the assumptions stated above are satisfied. In particular, it works specifically for the American put. There exist some cases where the result computed by the previous algorithm is not the solution of (AD). Here is an example:

$$R = \begin{pmatrix} 1 & -1 & 0 \\ -\varepsilon & 1 & 0 \\ 0 & 0 & 1 \end{pmatrix}, \quad F = \begin{pmatrix} 1 \\ 2 \\ 0 \end{pmatrix}, \quad G = \begin{pmatrix} 0 \\ 0 \\ 1 \end{pmatrix}.$$

The algorithm gives

$$X = \begin{pmatrix} 1 \\ 2 \\ 0 \end{pmatrix},$$

which is not a solution of (AD).

5.3.3 American put pricing by a binomial method

We shall now explain another numerical method that is widely used to price the American put in the Black-Scholes model. Let r, a, b be three real numbers such that $0 < d < 1 + r < u$. Let $(S_n)_{n \geq 0}$ be the binomial model defined by $S_0 = x$ and $S_{n+1} = S_n T_n$, where $(T_n)_{n \geq 0}$ is a sequence of IID random variables such that $\mathbb{P}(T_n = d) = p = (u - 1 - r)/(u - d)$ and $\mathbb{P}(T_n = u) = 1 - p$. We saw in Chapter 2, Exercise 7, that the American put price in this model could be written as

$$\mathscr{P}_n = P_{am}(n, S_n),$$

and that the function $P_{am}(n, x)$ could be computed by induction according to the equation

$$P_{am}(n, x) = \max \left((K - x)^+, \frac{p P_{am}(n + 1, xd) + (1 - p) P_{am}(n + 1, xu)}{1 + r} \right)$$
$$(5.15)$$

with the final condition $P_{am}(N, x) = (K - x)^+$. On the other hand, we proved in Chapter 1, Section 1.4, that if the parameters are chosen as follows:

$$\begin{cases} r & = & RT/N \\ d & = & \exp(-\sigma\sqrt{T/N}) \\ u & = & \exp(+\sigma\sqrt{T/N}) \\ p & = & (u - 1 - r)/(u - d), \end{cases} \quad (5.16)$$

then the European option price in this model approximates the Black-Scholes price computed for a riskless rate equal to R and a volatility equal to σ. This suggests that in order to price the American put, we can proceed as follows.

Given some discretization parameter N, we fix the values r, d, u, p according to (5.16) and we compute the price $P^N_{am}(n, .)$ at the nodes $xd^{n-i}u^i$, $0 \leq i \leq n$ by induction from (5.15). It seems quite natural to take $P^N_{am}(0, x)$ as an approximation of the American Black-Scholes price $P(0, x)$. Indeed, it can be proved that $\lim_{n \to +\infty} P^N_{am}(0, x) = P(0, x)$. For a proof, see Kushner (1977), Lamberton and Pagès (1990) or Amin and Khanna (1994).

The above method is the so-called Cox-Ross-Rubinstein method and it is presented in detail in Cox and Rubinstein (1985).

5.4   Exercises

Exercise 31   We denote by (X, Y) the scalar product of two vectors $X = (x_i)_{1 \leq i \leq n}$ and $Y = (y_i)_{1 \leq i \leq n}$. The notation $X \geq Y$ means that for all i between 1 and n, $x_i \geq y_i$. We assume that for all X in $\mathbb{R}^n$, R satisfies $(X, RX) \geq \alpha(X, X)$ with $\alpha > 0$. We want to to study the system

$$\begin{cases} RX \geq G \\ X \geq F \\ (RX - G, X - F) = 0. \end{cases}$$

1. Show that this is equivalent to finding $X \geq F$ such that

$$\forall V \geq F \quad (RX - G, V - X) \geq 0. \tag{5.17}$$

2. Prove the uniqueness of a solution of (5.17).

3. Show that if R is the identity matrix, there exists a unique solution to (5.17).

4. Let ρ be positive; we denote by $S_\rho(X)$ the unique vector $Y \geq F$ such that
$$\forall V \geq F \quad (Y - X + \rho(RX - G), V - Y) \geq 0.$$

   Show that for sufficiently small ρ, S_ρ is a contraction.

5. Derive the existence of a solution to (5.17).

Exercise 32   The purpose of this exercise is to propose an approximation for the Black-Scholes American put price $u(t, x)$. Recall that u is a solution of the following problem:

$$\begin{cases} \dfrac{\partial u}{\partial t}(t, x) + \tilde{A}^{bs} u(t, x) \leq 0 \quad \text{a.e. in} \quad [0, T] \times [0, +\infty) \\[2mm] u(t, x) \geq (K - x)_+ \quad \text{a.e. in} \quad [0, T] \times [0, +\infty) \\[2mm] (u - (K - x)_+) \left(\dfrac{\partial u}{\partial t}(t, x) + \tilde{A}^{bs} u(t, x) \right) = 0 \quad \text{a.e. in} \quad [0, T] \times [0, +\infty) \\[2mm] u(T, x) = (K - x)_+, \end{cases}$$

where

$$\tilde{A}^{bs} = \frac{\sigma^2 x^2}{2} \frac{\partial^2}{\partial x^2} + rx \frac{\partial}{\partial x} - r.$$

1. Denote by $u_e(t, x)$ the price of the European put with the same strike and maturity in the Black-Scholes model. Derive the system of inequalities satisfied by $v = u - u_e$.

2. We want to approximate the solution $v = u - u_e$ of this inequality by discretizing it in time, *using one time-step only*. Show that, with a totally implicit scheme, the one-step approximation $\tilde{v}(x)$ of $v(0, x)$ satisfies

$$\begin{cases} -\tilde{v}(x) + T\tilde{A}^{bs}\tilde{v}(x) \leq 0 & a.e. \ in \ \ [0, +\infty) \\ \tilde{v}(t, x) \geq \tilde{\psi}(x) = (K - x)_+ - u_e(0, x) & a.e. \ in \ \ [0, +\infty) \quad\quad (5.18) \\ (\tilde{v}(x) - \tilde{\psi}(x))(-\tilde{v}(x) + T\tilde{A}^{bs}\tilde{v}(x)) = 0 & a.e. \ in \ \ [0, +\infty), \end{cases}$$

where $\tilde{\psi}(x) = (K - x)_+ - u_e(0, x)$.

3. Find the unique negative value for α such that $v(x) = x^\alpha$ is a solution of $-v(x) + T\tilde{A}^{bs}v(x) = 0$.

4. We look for a continuously differentiable solution of (5.18) of the following form:

$$\tilde{v}(x) = \begin{cases} \lambda x^\alpha \ if \ x \geq x^* \\ \tilde{\psi}(x) \ otherwise. \end{cases} \quad\quad (5.19)$$

Write down the equations satisfied by λ and α so that $\tilde{v}$ is continuous with continuous derivative at x^*. Deduce that if $\tilde{v}$ is continuously differentiable, then x^* is a solution of $f(x) = x$ where

$$f(x) = |\alpha| \frac{K - u_e(0, x)}{u'_e(0, x) + 1 + |\alpha|}$$

and $u'_e(t, x) = (\partial u_e(t, x)/\partial x)$.

5. Using the closed-form formula for $u_e(0, x)$ (see Chapter 4, equation 4.9), prove that $f(0) > 0$, that $f(K) < K$ (hint: use the convexity of the function u_e) and that $f(x) - x$ is non-increasing. Conclude that there exists a unique solution to the equation $f(x) = x$.

6. Prove that $\tilde{v}(x)$ defined by (5.19) where x^* is the solution of $f(x) = x$ is a solution of (5.18).

7. Suggest an iterative algorithm (using a dichotomy argument) to compute x^* with an arbitrary accuracy.

8. From the previous results, write a computer program to compute the American put price.

The above algorithm is a marginally different version of the MacMillan algorithm (see MacMillan (1986), Barone-Adesi and Whaley (1987) and Chevance (1990)).

Chapter 6

Interest rate models

Interest rate models are mainly used to price and hedge bonds and interest rate options. Hitherto, there has not been any reference model equivalent to the Black-Scholes model for stock options. Over the last fifteen years, the research in this area has been very active and an exhaustive presentation of currently used models and techniques would be beyond the scope of this book. In this chapter, we will present the main features of interest rate modelling (following essentially, Artzner and Delbaen (1989)), discuss the concept of forward measures and change of numéraire, and review some of the most widely used models.

6.1 Modelling principles

6.1.1 The yield curve

In most of the models that we have already studied, the interest rate was assumed to be constant. In the real world, it is observed that the loan interest rate depends both on the date t of the loan emission and on the date T of the end or *maturity* of the loan.

Someone borrowing one dollar at time t, until maturity T, will have to pay back an amount $F(t, T)$ at time T, which is equivalent to an average interest rate $R(t, T)$ given by the equality

$$F(t, T) = e^{(T-t)R(t,T)}.$$

If we consider the future as certain, i.e. if we assume that all interest rates $(R(t, T))_{t \leq T}$ are known, then, in an arbitrage-free world, the function F must satisfy

$$\forall t < u < s, \quad F(t, s) = F(t, u)F(u, s).$$

Indeed, it is easy to derive arbitrage schemes when this equality does not hold. From this relationship and the equality $F(t, t) = 1$, it follows that, if F

is smooth, there exists a function $r(t)$ such that

$$F(t,T) = \exp\left(\int_t^T r(s)ds\right), \quad 0 \leq t < T,$$

and consequently

$$R(t,T) = \frac{1}{T-t}\int_t^T r(s)ds.$$

The function $r(s)$ is interpreted as the instantaneous interest rate.

In an uncertain world, this rationale does not hold any more. At time t, the future interest rates $R(u,T)$ for $T > u > t$ are not known. Nevertheless, intuitively, it makes sense to believe that there should be some relationships between the different rates. The aim of the modelling is to determine them.

The first issue is to price bonds. We call a *zero-coupon bond* a security paying 1 dollar at a maturity date T and denote by $P(t,T)$ the value of this security at time t. Obviously we have $P(T,T) = 1$ and, in a world where the future is certain,

$$P(t,T) = e^{-\int_t^T r(s)ds}. \tag{6.1}$$

6.1.2 Yield curve for an uncertain future

For an uncertain future, one must think of the instantaneous rate in terms of a random process: between times t and $t + dt$, it is possible to borrow at the rate $r(t)$ (in practice it corresponds to a short rate, for example the overnight rate). We specify the mathematical setup by introducing a filtered probability space $(\Omega, \mathscr{F}, \mathbb{P}, (\mathscr{F}_t)_{0 \leq t \leq \bar{T}})$. Note that, here, the finite horizon is denoted by $\bar{T}$, because we want to consider various maturities T, with $0 \leq T \leq \bar{T}$. We will assume that the filtration $(\mathscr{F}_t)_{0 \leq t \leq \bar{T}}$ is the natural filtration of a standard Brownian motion $(W_t)_{0 \leq t \leq \bar{T}}$ and that $\mathscr{F}_{\bar{T}} = \mathscr{F}$. As in the models we previously studied, we introduce a *riskless* asset, whose price at time t is given by

$$S_t^0 = e^{\int_0^t r(s)ds},$$

where $(r(t))_{0 \leq t \leq \bar{T}}$ is an adapted process satisfying $\int_0^{\bar{T}} |r(t)|dt < \infty$, almost surely. It might seem strange that we should call such an asset riskless even though its price is random; we will see later why this asset is less risky than the others. The risky assets here are the zero-coupon bonds with maturity less than or equal to the horizon $\bar{T}$. For a given maturity $T \in [0, \bar{T}]$, we denote by $P(t,T)$ the price at time t $(0 \leq t \leq T)$ of the zero-coupon bond with maturity T. Note that $P(T,T) = 1$. We assume that the price process $(P(t,T))_{0 \leq t \leq T}$, is adapted.

In Chapter 1, we characterized the absence of arbitrage opportunities by the existence of an equivalent probability under which discounted asset prices are martingales. The extension of this result to continuous-time models is rather technical (cf. Harrison and Kreps (1979), Harrison and Pliska (1981),

Stricker (1990), Delbaen and Schachermayer (1994, 2006), Artzner and Delbaen (1989)), but we were able to check in Chapter 4 that such a probability exists in the Black-Scholes model. In the light of these considerations, the starting point of the modelling will be based upon the following hypothesis:

(H) There is a probability $\mathbb{P}^*$ equivalent to $\mathbb{P}$, under which, for all $T \in [0, \bar{T}]$, the process $(\tilde{P}(t, T))_{0 \le t \le T}$ defined by

$$\tilde{P}(t, T) = e^{-\int_0^t r(s)ds} P(t, T)$$

is a martingale.

This hypothesis has some interesting consequences. Indeed, the martingale property under $\mathbb{P}^*$ and the equality $P(T, T) = 1$ yield

$$\tilde{P}(t, T) = \mathbb{E}^*(\tilde{P}(T, T)|\mathscr{F}_t) = \mathbb{E}^*\left(e^{-\int_0^T r(s)ds}\Big|\mathscr{F}_t\right)$$

and, eliminating the discounting,

$$P(t, T) = \mathbb{E}^*\left(e^{-\int_t^T r(s)ds}\Big|\mathscr{F}_t\right). \tag{6.2}$$

This equality, which should be compared to (6.1), shows that the prices $P(t, T)$ only depend on the behavior of the process $(r(s))_{0 \le s \le \bar{T}}$ under the probability $\mathbb{P}^*$. The hypothesis we made on the filtration $(\mathscr{F}_t)_{0 \le t \le \bar{T}}$ allows us to express the density of the probability $\mathbb{P}^*$ with respect to $\mathbb{P}$. We denote by $L_{\bar{T}}$ this density. For any non-negative random variable X, we have $\mathbb{E}^*(X) = \mathbb{E}(X L_T)$, and, if X is $\mathscr{F}_t$-measurable, $\mathbb{E}^*(X) = \mathbb{E}(X L_t)$, where $L_t = \mathbb{E}(L_T|\mathscr{F}_t)$. Thus the random variable L_t is the density of $\mathbb{P}^*$ restricted to $\mathscr{F}_t$ with respect to $\mathbb{P}$.

Proposition 6.1.1. *There is an adapted process $(q(t))_{0 \le t \le \bar{T}}$ such that, for all $t \in [0, \bar{T}]$,*

$$L_t = \exp\left(\int_0^t q(s)dW_s - \frac{1}{2}\int_0^t q(s)^2 ds\right) \quad a.s. \tag{6.3}$$

Proof. The process $(L_t)_{0 \le t \le \bar{T}}$ is a martingale relative to $(\mathscr{F}_t)$, which is the natural filtration of the Brownian motion (W_t). It follows (cf. Chapter 4, Section 4.2.3) that there exists an adapted process $(H_t)_{0 \le t \le \bar{T}}$ satisfying $\int_0^{\bar{T}} H_t^2 dt < \infty$ a.s. and, for all $t \in [0, \bar{T}]$,

$$L_t = L_0 + \int_0^t H_s dW_s \quad a.s.$$

Since $L_{\bar{T}}$ is a probability density, we have $\mathbb{E}(L_{\bar{T}}) = 1 = L_0$ and, $\mathbb{P}^*$ being equivalent to $\mathbb{P}$, we have $L_{\bar{T}} > 0$ a.s. and more generally $\mathbb{P}(L_t > 0) = 1$ for any t. To obtain the formula (6.3), we apply Itô's formula with the logarithmic function.

To do so, we need to check that $\mathbb{P}\left(\forall t \in [0, \bar{T}], L_0 + \int_0^t H_s dW_s > 0\right) = 1$. The proof of this fact relies in a crucial way on the martingale property and it is the purpose of Exercise 33. Itô's formula then yields

$$\log(L_t) = \int_0^t \frac{1}{L_s} H_s dW_s - \frac{1}{2} \int_0^t \frac{1}{L_s^2} H_s^2 ds \quad \text{a.s.,}$$

which leads to (6.3) with $q(t) = H_t / L_t$. $\qquad\square$

Corollary 6.1.2. *The price at time t of the zero-coupon bond with maturity $T \geq t$ can be expressed as*

$$P(t, T) = \mathbb{E}\left(\exp\left(-\int_t^T r(s)ds + \int_t^T q(s)dW_s - \frac{1}{2}\int_t^T q(s)^2 ds\right) | \mathscr{F}_t\right).$$
$$(6.4)$$

Proof. This follows immediately from Proposition 6.1.1 and from the following formula, which is easy to derive for any non-negative random variable X (see Chapter 4, Exercise 22, Bayes rule for conditional expectations):

$$\mathbb{E}^*(X|\mathscr{F}_t) = \frac{\mathbb{E}(XL_{\bar{T}}|\mathscr{F}_t)}{L_t}. \qquad (6.5)$$

$\qquad\square$

The following proposition suggests an economic interpretation for the process $(q(t))$ (see Remark 6.1.4 below).

Proposition 6.1.3. *For each maturity T, there is an adapted process $(\sigma_t^T)_{0 \leq t \leq T}$ such that*

$$\frac{dP(t, T)}{P(t, T)} = (r(t) - \sigma_t^T q(t))dt + \sigma_t^T dW_t, \quad 0 \leq t \leq T. \qquad (6.6)$$

Proof. The process $(\tilde{P}(t, T))_{0 \leq t \leq T}$ is a martingale under $\mathbb{P}^*$, so, using Exercise 34, $(\tilde{P}(t, T)L_t)_{0 \leq t \leq T}$ is a martingale under $\mathbb{P}$. Moreover, we have $\tilde{P}(t, T)L_t > 0$ a.s., for all $t \in [0, T]$. Then, using the same rationale as in the proof of Proposition 6.1.1, we see that there exists an adapted process $(\theta_t^T)_{0 \leq t \leq T}$ such that $\int_0^T (\theta_t^T)^2 dt < \infty$ and

$$\tilde{P}(t, T)L_t = \tilde{P}(0, T)e^{\int_0^t \theta_s^T dW_s - \frac{1}{2}\int_0^t (\theta_s^T)^2 ds}.$$

Hence, using the explicit expression for L_t and getting rid of the discounting factor,

$$P(t, T) = P(0, T)\exp\left(\int_0^t r(s)ds + \int_0^t (\theta_s^T - q(s))\, dW_s\right.$$
$$\left. - \frac{1}{2}\int_0^t \left((\theta_s^T)^2 - q(s)^2\right) ds\right).$$

Applying Itô's formula with the exponential function, we get

$$\frac{dP(t,T)}{P(t,T)} = r(t)dt + \left(\theta_t^T - q(t)\right)dW_t - \frac{1}{2}\left(\left(\theta_t^T\right)^2 - q(t)^2\right)dt$$

$$+ \frac{1}{2}\left(\theta_t^T - q(t)\right)^2 dt$$

$$= (r(t) + q(t)^2 - \theta_t^T(t))dt + \left(\theta_t^T - q(t)\right)dW_t,$$

which gives (6.6) with $\sigma_t^u = \theta_t^u - q(t)$. □

Remark 6.1.4. The formula (6.6) is to be related with the equality $dS_t^0 = r(t)S_t^0 dt$, satisfied by the so-called riskless asset. It is the term in dW_t which makes the bond riskier. Furthermore, the term $r(t) - \sigma_t^T q(t)$ corresponds intuitively to the average yield (i.e. in expectation) and the term $-\sigma_t^T q(t)$ is the difference between the average yield of the bond and the riskless rate, hence the interpretation of $-q(t)$ as a *risk premium*. Under probability $\mathbb{P}^*$, the process $(\tilde{W}_t)$ defined by $\tilde{W}_t = W_t - \int_0^t q(s)ds$ is a standard Brownian motion (Girsanov theorem), and we have

$$\frac{dP(t,T)}{P(t,T)} = r(t)dt + \sigma_t^T d\tilde{W}_t. \tag{6.7}$$

Therefore, under $\mathbb{P}^*$, the mean rate of return of the bonds is equal to the riskless interest rate. For this reason, the probability $\mathbb{P}^*$ is called the *risk neutral* probability. Note that, by solving equation (6.7), we get

$$P(t,T) = P(0,T)e^{\int_0^t r(s)ds}\exp\left(\int_0^t \sigma_s^T d\tilde{W}_s - \frac{1}{2}\int_0^t (\sigma_s^T)^2 ds\right). \tag{6.8}$$

6.1.3 Bond options

Let us first consider a European option with maturity θ on the zero-coupon bond with maturity T, where $0 \leq \theta \leq T \leq \bar{T}$. If it is a call with strike price K, the value of the option at time θ is obviously $(P(\theta,T) - K)_+$ and it seems reasonable to hedge this call with a portfolio involving the riskless asset and the zero-coupon bond with maturity T. A strategy (on the period $[0,\theta]$) is then defined by an adapted process $(H_t^0, H_t)_{0 \leq t \leq \theta}$ with values in $\mathbb{R}^2$, with H_t^0 representing the quantity of riskless asset and H_t the number of bonds with maturity T held in the portfolio at time t. The value of the portfolio at time t is given by

$$V_t = H_t^0 S_t^0 + H_t P(t,T) = H_t^0 e^{\int_0^t r(s)ds} + H_t P(t,T)$$

and the self-financing condition is written, as in Chapter 4, as

$$dV_t = H_t^0 dS_t^0 + H_t dP(t,T).$$

Taking into account Proposition 6.1.3, we impose the following integrability conditions: $\int_0^\theta |H_t^0 r(t)| dt < \infty$ and $\int_0^\theta (H_t \sigma_t^T)^2 dt < \infty$ a.s.. As in Chapter 4, we define admissible strategies in the following way.

Definition 6.1.5. A strategy $\phi = ((H_t^0, H_t))_{0 \le t \le T}$ is admissible if it is self-financing and if the discounted value $\tilde{V}_t(\phi) = H_t^0 + H_t \tilde{P}(t, T)$ of the corresponding portfolio is, for all t, non-negative and such that $\sup_{t \in [0,\theta]} \tilde{V}_t$ is square-integrable under $\mathbb{P}^*$.

The following proposition shows that under some assumptions, it is possible to hedge all European options with maturity $\theta < T$.

Proposition 6.1.6. *Assume* $\sup_{0 \le t \le T} |r(t)| < \infty$ *a.s. and* $\sigma_t^T \ne 0$ *a.s. for all* $t \in [0, \theta]$. *Let* $\theta < T$ *and let* h *be an* $\mathscr{F}_\theta$-*measurable random variable such that* $h e^{-\int_0^\theta r(s)ds}$ *is square-integrable under* $\mathbb{P}^*$. *Then there exists an admissible strategy whose value at time* θ *is equal to* h. *The value at time* $t \le \theta$ *of such a strategy is given by*

$$V_t = \mathbb{E}^* \left(e^{-\int_t^\theta r(s)ds} h \,\middle|\, \mathscr{F}_t \right).$$

Proof. The method is the same as in Chapter 4. We first observe that if $\tilde{V}_t$ is the discounted value at time t of an admissible strategy $(H_t^0, H_t)_{0 \le t \le T}$, we obtain, using the self-financing condition, the integration by parts formula and Remark 6.1.4 (cf. equation (6.7)),

$$\begin{aligned} d\tilde{V}_t &= H_t d\tilde{P}(t, T) \\ &= H_t \tilde{P}(t, T) \sigma_t^T d\tilde{W}_t. \end{aligned}$$

We deduce, bearing in mind that $\sup_{t \in [0,T]} \tilde{V}_t$ is square-integrable under $\mathbb{P}^*$, that $(\tilde{V}_t)$ is a martingale under $\mathbb{P}^*$. Thus we have

$$\tilde{V}_t = \mathbb{E}^* (\tilde{V}_\theta | \mathscr{F}_t), 0 \le \theta,$$

and, if we impose the condition $V_\theta = h$, we get

$$V_t = e^{\int_0^t r(s)ds} \mathbb{E}^* \left(e^{-\int_0^\theta ds} h | \mathscr{F}_t \right).$$

To complete the proof, it is sufficient to find an admissible strategy having the same value at any time. To do so, one proves that there exists a process $(J_t)_{0 \le t \le \theta}$ such that $\int_0^\theta J_t^2 dt < \infty$ a.s., and

$$h e^{-\int_0^\theta r(s)ds} = \mathbb{E}^* \left(h e^{-\int_0^\theta r(s)ds} \right) + \int_0^\theta J_s d\tilde{W}_s.$$

Note that this property is not a trivial consequence of the Martingale Representation Theorem because we do not know whether the martingale

$\mathbb{E}^* \left(h e^{- \int_0^\theta r(s)ds} \mid \mathscr{F}_t \right)$ is adapted to the natural filtration of $\tilde{W}$ (see Exercise 35 for this particular point). Once this property is proved, it is sufficient to set

$$H_t = \frac{J_t}{\tilde{P}(t,T)\sigma_t^T} \quad \text{and} \quad H_t^0 = \mathbb{E}^* \left(h e^{- \int_0^\theta r(s)ds} \bigg| \mathscr{F}_t \right) - \frac{J_t}{\sigma_t^T}$$

for $t \leq \theta$. It is easy to check that $(H_t^0, H_t)_{0 \leq t \leq \theta}$ defines an admissible strategy (the hypothesis $\sup_{0 \leq t \leq \bar{T}} |r(t)| < \infty$ a.s. guarantees that the condition $\int_0^\theta |r(s) H_s^0| ds < \infty$ holds) whose value at time θ is indeed equal to h.   □

In light of Proposition 6.1.6, it is natural to define the fair price of the option h at time t as the quantity

$$\mathbb{E}^* \left(e^{- \int_t^\theta r(s)ds} h \mid \mathscr{F}_t \right), \quad 0 \leq t \leq \theta.$$

Remark 6.1.7. We have not investigated the uniqueness of the probability $\mathbb{P}^*$ and it is not clear that the risk process $(q(t))$ is defined without ambiguity. Actually, it can be shown (cf. Artzner and Delbaen (1989)) that $\mathbb{P}^*$ is the unique probability equivalent to $\mathbb{P}$ under which $(\tilde{P}(t,T))_{0 \leq t \leq T}$ is a martingale if and only if the process (σ_t^T) satisfies $\sigma_t^T \neq 0, dtd\mathbb{P}$ almost everywhere. This condition, slightly weaker than the hypothesis of Proposition 6.1.6, is exactly what is needed to hedge options with bonds of maturity T, which is not surprising when one keeps in mind the characterization of complete markets we gave in Chapter 1.

Remark 6.1.8. The implementation of admissible strategies as defined in Definition 6.1.5 is not quite clear because trading in the so-called riskless asset may not be possible or appropriate. In practice, the hedging strategy of an option with maturity θ on a zero-coupon bond with maturity T will rather involve zero-coupon bonds of both maturities θ and T. See Remark 6.1.14 and Exercise 36.

Remark 6.1.9. For the pricing of options on bonds with coupons, the reader is referred to Jamshidian (1989) and El Karoui and Rochet (1989).

6.1.4   Forward measures and change of numéraire

Definition 6.1.10. *For a fixed maturity date $T \in [0, \bar{T}]$, the T-forward measure is the probability measure $\mathbb{P}^T$ defined by*

$$\frac{d\mathbb{P}^T}{d\mathbb{P}^*} = \frac{e^{- \int_0^T r(s)ds}}{P(0,T)}.$$

Note that $\mathbb{P}^T$ is a probability measure, because

$$P(0,T) = \mathbb{E}^* \left(e^{-\int_0^T r(s)ds} \right).$$

Since the density $d\mathbb{P}^T/d\mathbb{P}^*$ is positive, the probability $\mathbb{P}^T$ is equivalent to $\mathbb{P}^*$ and $\mathbb{P}$. We can also rewrite (6.2) as follows:

$$\mathbb{E}^* \left(\frac{d\mathbb{P}^T}{d\mathbb{P}^*} \mid \mathscr{F}_t \right) = \frac{\tilde{P}(t,T)}{P(0,T)}, \quad 0 \leq t \leq T.$$

The following proposition can be used to compute the value of an option with maturity θ by using the θ-forward measure (see Remark 6.1.12 below for more comments).

Proposition 6.1.11. *Let θ be a maturity date ($0 \leq \theta \leq \bar{T}$).*

1. *If h is a non-negative random variable, we have, for $t \in [0,\theta]$,*

$$\mathbb{E}^* \left(e^{-\int_t^\theta r(s)ds} h \mid \mathscr{F}_t \right) = P(t,\theta) \mathbb{E}^\theta \left(h \mid \mathscr{F}_t \right).$$

2. *If $(X_t)_{0 \leq t \leq \theta}$ is an adapted stochastic process, the process $(X_t/S_t^0)_{0 \leq t \leq \theta}$ is a $\mathbb{P}^*$-martingale if and only if $(X_t/P(t,\theta))_{0 \leq t \leq \theta}$ is a $\mathbb{P}^\theta$-martingale.*

Proof. The first assertion follows from the Bayes rule for conditional expectations:

$$\mathbb{E}^\theta \left(h \mid \mathscr{F}_t \right) = \frac{\mathbb{E}^* \left(h L_\theta^\theta \mid \mathscr{F}_t \right)}{L_t^\theta} \quad \text{with} \quad L_t^\theta = \tilde{P}(t,\theta)/P(0,\theta).$$

For the second assertion, observe that $(M_t)_{0 \leq t \leq \theta}$ is a $\mathbb{P}^\theta$-martingale if and only if $(M_t L_t^\theta)_{0 \leq t \leq \theta}$ is a $\mathbb{P}^*$-martingale (see Exercise 34). □

Remark 6.1.12. Proposition 6.1.11 has a nice interpretation in terms of *change of numéraire*. A *numéraire* is an asset that is used as a price unit. If (S_t) is the price process of an asset, the discounted price (S_t/S_t^0) can be viewed as the price of the asset when the riskless asset is taken as a numéraire. If we choose as a numéraire the zero-coupon bond with maturity θ, the price of the asset at time t ($t \leq \theta$) becomes $F_S(t,\theta) = S_t/P(t,\theta)$. This is called the θ-forward price of the asset. We know that the discounted price process of a risky asset is a martingale under the risk neutral probability. It follows from the second assertion of Proposition 6.1.11 that the θ-forward price is a martingale under the θ-forward measure. The first assertion means that the θ-forward value of a European option with maturity θ is the conditional expectation under the θ-forward measure of the payoff attached to the option. Note that the forward price $F_S(t,\theta)$ is the value at time t of an option with maturity θ and payoff S_θ, i.e. an option that delivers one unit of the asset at time θ.

Suppose we want to price a call option with maturity θ and strike price K on the zero-coupon bond with maturity T. Then we have $h = (P(\theta, T) - K)_+$. In order to compute the θ-forward price of the option at time t, we need the conditional distribution of $P(\theta, T)$ given $\mathscr{F}_t$ under the θ-forward measure. Let $P^\theta(t, T)$ be the θ-forward price of the zero-coupon bond with maturity T. Note that $P(\theta, T) = P^\theta(\theta, T)$. The following result specifies the dynamics of the process $(P^\theta(t, T))$ under the θ-forward measure.

Proposition 6.1.13. *Given two maturity dates θ and T, the θ-forward price $P^\theta(t, T) = P(t, T)/P(t, \theta)$ of the zero-coupon bond with maturity T satisfies*

$$\frac{dP^\theta(t, T)}{P^\theta(t, T)} = \left(\sigma_t^T - \sigma_t^\theta\right) dW_t^\theta, \quad 0 \le t \le \theta \wedge T, \tag{6.9}$$

where $W_t^\theta = \tilde{W}_t - \int_0^t \sigma_s^\theta ds$, and the process $(W_t^\theta)_{0 \le t \le \theta}$ is a standard $(\mathscr{F}_t)$-Brownian motion under the θ-forward measure $\mathbb{P}^\theta$.

Proof. From (6.8), we have

$$P^\theta(t, T) = \frac{P(0, T)}{P(0, \theta)} \exp\left(\int_0^t \left(\sigma_s^T - \sigma_s^\theta\right) d\tilde{W}_s - \frac{1}{2} \int_0^t \left((\sigma_s^T)^2 - (\sigma_s^\theta)^2\right) ds\right)$$

$$= P^\theta(0, T) \exp\left(\int_0^t \left(\sigma_s^T - \sigma_s^\theta\right) dW_s^\theta - \frac{1}{2} \int_0^t \left(\sigma_s^T - \sigma_s^\theta\right)^2 ds\right),$$

where $W_t^\theta = \tilde{W}_t - \int_0^t \sigma_s^\theta ds$. A straightforward application of Itô's formula yields (6.9). It remains to prove that, under $\mathbb{P}^\theta$, $(W_t^\theta)_{0 \le t \le \theta}$ is a standard Brownian motion. Let $L_t^\theta = \exp\left(\int_0^t \sigma_s^\theta dW_s - \frac{1}{2} \int_0^t (\sigma_s^\theta)^2 ds\right) = \tilde{P}(t, \theta)/P(0, \theta)$. We have, using $P(\theta, \theta) = 1$, $d\mathbb{P}^\theta/d\mathbb{P}^* = L_\theta^\theta$, and we know from (H) that $(L_t^\theta)_{0 \le t \le \theta}$ is a martingale. So, the result follows from Girsanov's theorem. □

Remark 6.1.14. A consequence of Proposition 6.1.13 is that, if the volatilities of the zero-coupon bonds σ_t^T, σ_t^θ are *deterministic*, the θ-forward price of a European call with maturity θ and strike price K on the zero-coupon bond with maturity T, is given by

$$C_t^\theta = \mathbb{E}^\theta\left((P^\theta(\theta, T) - K)_+ \mid \mathscr{F}_t\right)$$

$$= \mathbb{E}^\theta\left((P^\theta(t, T)e^{Z(t, \theta)} - K)_+ \mid \mathscr{F}_t\right), \quad 0 \le t \le \theta,$$

where

$$Z(t, \theta) = \int_t^\theta \left(\sigma_s^T - \sigma_s^\theta\right) dW_s^\theta - \frac{1}{2} \int_t^\theta \left(\sigma_s^T - \sigma_s^\theta\right)^2 ds.$$

Note that, under $\mathbb{P}^\theta$, $Z(t, \theta)$ is independent of $\mathscr{F}_t$ and Gaussian with mean $-\frac{1}{2}\Sigma^2(t, \theta)$ and variance $\Sigma^2(t, \theta)$, where

$$\Sigma^2(t, \theta) = \int_t^\theta \left(\sigma_s^T - \sigma_s^\theta\right)^2 ds.$$

Therefore, using Proposition A.2.5 of the Appendix, we have

$$C_t^\theta = B(t, P^\theta(t, T)), \tag{6.10}$$

with

$$\begin{aligned} B(t, x) &= \mathbb{E}^*(xe^{Z(t,\theta)} - K)_+ \\ &= xN(d_1(t, x)) - KN(d_2(t, x)), \end{aligned}$$

where N is the normal distribution function,

$$d_1(t, x) = \frac{\log(x/K) + (\Sigma^2(t, \theta)/2)}{\Sigma(t, \theta)} \quad \text{and} \quad d_2(t, x) = d_1(t, x) - \Sigma(t, \theta).$$

The above formula (which corresponds to the Black-Scholes formula with no interest rate) is known as Black's formula. Going back to (6.10), we can write the value of the call at time t as

$$\begin{aligned} C_t &= P(t, \theta)C_t^\theta \\ &= P(t, \theta)B(t, P^\theta(t, T)) \\ &= P(t, T)H_t^T + P(t, \theta)H_t^\theta, \end{aligned}$$

where $H_t^T = N(d_1(t, P^\theta(t, T)))$ and $H_t^\theta = -KN(d_2(t, P^\theta(t, T)))$. In this framework, the option can be hedged by holding H_t^T zero-coupon bonds with maturity T and H_t^θ zero-coupon bonds with maturity θ at time t (see Exercise 36).

6.2   Some classical models

Equations (6.2) and (6.4) show that in order to calculate the price of bonds, we need to know either the dynamics of the process $(r(t))$ under $\mathbb{P}^*$, or the dynamics of the pair $(r(t), q(t))$ under $\mathbb{P}$. The first models we are about to examine describe the dynamics of $r(t)$ under $\mathbb{P}$ by a diffusion equation and determine the form that $q(t)$ should have in order to get a similar equation under $\mathbb{P}^*$. Then the prices of bonds and options depend explicitly on *risk parameters*, which are difficult to estimate. One advantage of the Heath-Jarrow-Morton model, which we will explain briefly in Section 6.2.3, is to provide formulae that only depend on the parameters of the dynamics of interest rates under $\mathbb{P}$.

6.2.1   The Vasicek model

In this model, we assume that the process $r(t)$ satisfies

$$dr(t) = a(b - r(t))dt + \sigma dW_t, \tag{6.8}$$

where a, b, σ are non-negative constants. We also assume that the process $q(t)$ is a constant $q(t) = -\lambda$, with $\lambda \in \mathbb{R}$. Then

$$dr(t) = a(b^* - r(t))dt + \sigma d\tilde{W}_t, \tag{6.9}$$

where $b^* = b - \lambda\sigma/a$ and $\tilde{W}_t = W_t + \lambda t$. Before calculating the price of bonds according to this model, let us give some consequences of equation (6.8). If we set

$$X_t = r(t) - b,$$

we see that (X_t) is a solution of the stochastic differential equation

$$dX_t = -aX_t dt + \sigma dW_t,$$

which means that (X_t) is an Ornstein-Uhlenbeck process (cf. Chapter 3, Section 3.5.2). We deduce that $r(t)$ can be written as

$$r(t) = r(0)e^{-at} + b(1 - e^{-at}) + \sigma e^{-at} \int_0^t e^{as} dW_s \qquad (6.10)$$

and that $r(t)$ is normally distributed with mean and variance given by

$$\mathbb{E}(r(t)) = r(0)e^{-at} + b(1 - e^{-at}), \quad \mathrm{Var}(r(t)) = \sigma^2(1 - e^{-2at})/2a.$$

It follows that $r(t)$ can be negative with positive probability, which is not very satisfactory from a practical point of view (unless this probability is always very small). Note that, when t tends to infinity, $r(t)$ converges in law to a Gaussian random variable with mean b and variance $\sigma^2/2a$.

To calculate the price of zero-coupon bonds, we proceed under probability $\mathbb{P}^*$ and use equation (6.9). From (6.2), we have

$$P(t, T) = \mathbb{E}^* \left(e^{-\int_t^T r(s)ds} \Big| \mathscr{F}_t \right)$$
$$= e^{-b^*(T-t)} \mathbb{E}^* \left(e^{-\int_t^T X_s^* ds} \Big| \mathscr{F}_t \right), \qquad (6.11)$$

where $X_t^* = r(t) - b^*$. Since (X_t^*) is a solution of the diffusion equation with time-independent coefficients

$$dX_t = -aX_t dt + \sigma d\tilde{W}_t, \qquad (6.12)$$

we can write

$$\mathbb{E}^* \left(e^{-\int_t^T X_s^* ds} \Big| \mathscr{F}_t \right) = F(T - t, X_t^*) = F(T - t, r(t) - b^*), \qquad (6.13)$$

where F is the function defined by $F(\theta, x) = \mathbb{E}^* \left(e^{-\int_0^\theta X_s^x ds} \right)$, and (X_t^x) is the unique solution of equation (6.12) that satisfies $X_0^x = x$ (cf. Chapter 3, Remark 3.5.11).

It is possible to compute $F(\theta, x)$ explicitly. We know (cf. Chapter 3) that the process (X_t^x) is Gaussian with continuous paths. It follows that $\int_0^\theta X_s^x ds$ is a normal random variable, since the integral is the limit of Riemann sums, which are Gaussian. Thus, from the expression of the Laplace transform of a Gaussian,

$$\mathbb{E}^*\left(e^{-\int_0^\theta X_s^x ds}\right) = \exp\left(-\mathbb{E}^*\left(\int_0^\theta X_s^x ds\right) + \frac{1}{2}\text{Var}\left(\int_0^\theta X_s^x ds\right)\right).$$

From equality $\mathbb{E}^*(X_s^x) = xe^{-as}$, we deduce

$$\mathbb{E}^*\left(\int_0^\theta X_s^x ds\right) = x\frac{1-e^{-a\theta}}{a}.$$

For the computation of the variance, we write

$$\text{Var}\left(\int_0^\theta X_s^x ds\right) = \text{Cov}\left(\int_0^\theta X_s^x ds, \int_0^\theta X_s^x ds\right)$$

$$= \int_0^\theta \int_0^\theta \text{Cov}\left(X_t^x, X_u^x\right) du\, dt. \qquad (6.14)$$

Since $X_t^x = xe^{-at} + \sigma e^{-at}\int_0^t e^{as}d\tilde{W}_s$, we have

$$\text{Cov}\left(X_t^x, X_u^x\right) = \sigma^2 e^{-a(t+u)}\mathbb{E}^*\left(\int_0^t e^{as}d\tilde{W}_s \int_0^u e^{as}d\tilde{W}_s\right)$$

$$= \sigma^2 e^{-a(t+u)}\int_0^{t\wedge u} e^{2as}ds$$

$$= \sigma^2 e^{-a(t+u)}\frac{(e^{2a(t\wedge u)}-1)}{2a},$$

so that, going back to (6.14), we get

$$\text{Var}\left(\int_0^\theta X_s^x ds\right) = \frac{\sigma^2\theta}{a^2} - \frac{\sigma^2}{a^3}(1-e^{-a\theta}) - \frac{\sigma^2}{2a^3}(1-e^{-a\theta})^2.$$

From (6.11) and (6.13), we derive the following formula:

$$P(t,T) = \exp[-(T-t)R(T-t,r(t))],$$

where $R(T-t, r(t))$, which can be seen as the average interest rate on the period $[t, T]$, is given by the formula

$$R(\theta, r) = R_\infty - \frac{1}{a\theta}\left[(R_\infty - r)(1 - e^{-a\theta}) - \frac{\sigma^2}{4a^2}(1 - e^{-a\theta})^2\right]$$

with $R_\infty = \lim_{\theta\to\infty} R(\theta, r) = b^* - \sigma^2/(2a^2)$. The yield R_∞ can be interpreted as a long-term rate. Note that it does not depend on the 'instantaneous spot rate' r. This last property is considered as a drawback of the model by practitioners.

In the Vasicek model, the volatilities of the zero-coupon bonds are deterministic, so that closed-form formulae for pricing and hedging bond options can be derived (see Exercise 36).

Remark 6.2.1. In practice, parameters must be estimated and a value for r must be chosen. For the value of r, one may choose a short rate (for example, the overnight rate); then the parameters b, a, σ can, in principle, be estimated by statistical methods from historical data on the instantaneous rate. Finally, λ can be determined from market data by inverting the Vasicek formula. What practitioners really do is determine the parameters, including r, by fitting the Vasicek formula on market data.

6.2.2   The Cox-Ingersoll-Ross model

Cox, Ingersoll and Ross (1985) suggest modelling the behavior of the instantaneous rate by the following equation:

$$dr(t) = (a - br(t))dt + \sigma\sqrt{r(t)}dW_t \tag{6.15}$$

with σ and a non-negative, $b \in \mathbb{R}$, and the process $(q(t))$ being equal to $q(t) = -\alpha\sqrt{r(t)}$, with $\alpha \in \mathbb{R}$. Note that we cannot apply the existence and uniqueness theorem that we gave in Chapter 3 because the square root function is only defined on $\mathbb{R}^+$ and is not Lipschitz. However, from the Hölder property of the square root function, one can show the following result.

Theorem 6.2.2. *Suppose that (W_t) is a standard Brownian motion defined on $[0, \infty)$. For any real number $x \geq 0$, there is a unique continuous, adapted process (X_t), taking values in $\mathbb{R}^+$, satisfying $X_0 = x$ and*

$$dX_t = (a - bX_t)dt + \sigma\sqrt{X_t}dW_t \quad on\ [0, \infty). \tag{6.16}$$

For a proof of this result, the reader is referred to Ikeda and Watanabe (1981), p. 221. Before investigating the Cox-Ingersol-Ross model, we give some properties of this equation. Denote by (X_t^x) the solution of (6.16) starting at x and τ_0^x the stopping time defined by

$$\tau_0^x = \inf\{t \geq 0 \mid X_t^x = 0\}$$

with, as usual, $\inf \emptyset = \infty$.

Proposition 6.2.3.

1. *If $a \geq \sigma^2/2$, we have $\mathbb{P}(\tau_0^x = \infty) = 1$, for all $x > 0$.*

2. *If $0 \leq a < \sigma^2/2$ and $b \geq 0$, we have $\mathbb{P}(\tau_0^x < \infty) = 1$, for all $x > 0$.*

3. *If $0 \leq a < \sigma^2/2$ and $b < 0$, we have $0 < \mathbb{P}(\tau_0^x < \infty) < 1$, for all $x > 0$.*

This proposition is proved in Exercise 37.

The following proposition, which enables us to characterize the joint law of $\left(X_t^x, \int_0^t X_s^x ds\right)$, is the key to any pricing within the Cox-Ingersoll-Ross model.

Proposition 6.2.4. *For any non-negative λ and μ, we have*

$$\mathbb{E}\left(e^{-\lambda X_t^x} e^{-\mu \int_0^t X_s^x ds}\right) = \exp(-a\phi_{\lambda,\mu}(t)) \exp(-x\psi_{\lambda,\mu}(t)),$$

where the functions $\phi_{\lambda,\mu}$ and $\psi_{\lambda,\mu}$ are given by

$$\phi_{\lambda,\mu}(t) = -\frac{2}{\sigma^2} \log \left(\frac{2\gamma e^{\frac{t(\gamma+b)}{2}}}{\sigma^2 \lambda(e^{\gamma t} - 1) + \gamma - b + e^{\gamma t}(\gamma + b)} \right)$$

and

$$\psi_{\gamma,\mu}(t) = \frac{\lambda(\gamma + b + e^{\gamma t}(\gamma - b)) + 2\mu(e^{\gamma t} - 1)}{\sigma^2 \lambda(e^{\gamma t} - 1) + \gamma - b + e^{\gamma t}(\gamma + b)}$$

with $\gamma = \sqrt{b^2 + 2\sigma^2 \mu}$.

Proof.   The fact that this expectation can be written as $e^{-a\phi(t)-x\psi(t)}$ is due to the additivity property of the process (X_t^x) relative to the parameter a and the initial condition x (cf. Ikeda and Watanabe (1981), p. 225, Revuz and Yor (1990)). If, for λ and μ fixed, we consider the function $F(t,x)$ defined by

$$F(t,x) = \mathbb{E}\left(e^{-\lambda X_t^x} e^{-\mu \int_0^t X_s^x ds}\right), \qquad (6.17)$$

it is natural to look for F as a solution of the problem

$$\begin{cases} \dfrac{\partial F}{\partial t} = \dfrac{\sigma^2}{2} x \dfrac{\partial^2 F}{\partial x^2} + (a - bx)\dfrac{\partial F}{\partial x} - \mu x F \\ F(0,x) = e^{-\lambda x}. \end{cases}$$

Indeed, if F satisfies these equations and has bounded derivatives, we deduce from Itô's formula that, for any T, the process $(M_t)_{0 \le t \le T}$, defined by

$$M_t = e^{-\mu \int_0^t X_s^x ds} F\left(T - t, X_t^x\right),$$

is a martingale and the equality $\mathbb{E}(M_T) = M_0$ leads to (6.17). If F can be written as $F(t,x) = e^{-a\phi(t)-x\psi(t)}$, the equations above become $\phi(0) = 0, \psi(0) = \lambda$ and

$$\begin{cases} -\psi'(t) = \frac{\sigma^2}{2}\psi^2(t) + b\psi(t) - \mu \\ \phi'(t) = \psi(t). \end{cases}$$

Solving these two differential equations gives the desired expressions for ϕ and ψ.   □

When applying Proposition 6.2.4 with $\mu = 0$, we obtain the Laplace transform of X_t^x,

$$\mathbb{E}(e^{-\lambda X_t^x}) = \left(\frac{b}{(\sigma^2/2)\lambda(1 - e^{-bt}) + b}\right)^{2a/\sigma^2} \exp\left(-x\frac{\lambda b e^{-bt}}{(\sigma^2/2)\lambda(1 - e^{-bt}) + b}\right)$$

$$= \frac{1}{(2\lambda L + 1)^{2a/\sigma^2}} \exp\left(-\frac{\lambda L \zeta}{2\lambda L + 1}\right)$$

with $L = (\sigma^2/4b)(1 - e^{-bt})$ and $\zeta = 4xb/(\sigma^2(e^{bt} - 1))$. With these notations, the Laplace transform of X_t^x/L is given by the function $g_{4a/\sigma^2, \zeta}$, where $g_{\delta, \zeta}$ is defined by

$$g_{\delta, \zeta}(\lambda) = \frac{1}{(2\lambda + 1)^{\delta/2}} \exp\left(-\frac{\lambda \zeta}{2\lambda + 1}\right).$$

This function is the Laplace transform of the non-central chi-square distribution with δ degrees of freedom and parameter ζ (see Exercise 38 for this matter). The *density* of this law is given by the function $f_{\delta, \zeta}$, defined by

$$f_{\delta, \zeta}(x) = \frac{e^{-\zeta/2}}{2\zeta^{\delta/4 - 1/2}} e^{-x/2} x^{\delta/4 - 1/2} I_{\delta/2 - 1}(\sqrt{x\zeta}) \quad \text{for} \quad x > 0,$$

where I_ν is the first-order modified Bessel function with index ν, defined by

$$I_\nu(x) = \left(\frac{x}{2}\right)^\nu \sum_{n=0}^\infty \frac{(x/2)^{2n}}{n!\Gamma(\nu + n + 1)}.$$

The reader can find many properties of Bessel functions and some approximations of distribution functions of non-central chi-squared laws in Abramowitz and Stegun (1992), Chapters 9 and 26.

Let us go back to the Cox-Ingersoll-Ross model. From the hypothesis on the processes $r(t))$ and $(q(t))$, we get

$$dr(t) = (a - (b + \sigma\alpha)r(t))dt + \sigma\sqrt{r(t)}d\tilde{W}_t,$$

where, under probability $\mathbb{P}^*$, the process $(\tilde{W}_t)_{0 \le t \le T}$ is a standard Brownian motion. The price of a zero-coupon bond with maturity T is then given, at time 0, by

$$P(0, T) = \mathbb{E}^*\left(e^{-\int_0^T r(s)ds}\right)$$

$$= e^{-a\phi(T) - r(0)\psi(T)}, \tag{6.18}$$

where the functions ϕ and ψ are given by the following formulae:

$$\phi(t) = -\frac{2}{\sigma^2} \log\left(\frac{2\gamma^* e^{\frac{t(\gamma^* + b^*)}{2}}}{\gamma^* - b^* + e^{\gamma^* t}(\gamma^* + b^*)}\right)$$

and

$$\psi(t) = \frac{2(e^{\gamma^* t} - 1)}{\gamma^* - b^* + e^{\gamma^* t}(\gamma^* + b^*)}$$

with $b^* = b + \sigma\alpha$ and $\gamma^* = \sqrt{(b^*)^2 + 2\sigma^2}$. The price at time t is given by

$$P(t, T) = \exp(-a\phi(T - t) - r(t)\psi(T - t)).$$

Let us now price a European call with maturity θ and exercise price K, on a zero-coupon bond with maturity T. The call price at time 0 is given by

$$C_0 = P(0, \theta)\mathbb{E}^\theta \left[(P(\theta, T) - K)_+\right],$$

where $\mathbb{P}^\theta$ is the θ-forward measure (see Proposition 6.1.11). Note that

$$\{P(\theta, T) > K\} = \left\{e^{-a\phi(T-\theta) - r(\theta)\psi(T-\theta)} > K\right\} = \{r(\theta) < r^*\},$$

where r^* is defined by

$$r^* = -\frac{a\phi(T - \theta) + \log(K)}{\psi(T - \theta)}.$$

Hence

$$C_0 = P(0, \theta)\mathbb{E}^\theta \left(P(\theta, T)\mathbf{1}_{\{r(\theta) < r^*\}}\right) - KP(0, \theta)\mathbb{P}^\theta \left(r(\theta) < r^*\right)$$
$$= P(0, \theta)\mathbb{E}^\theta \left(P(\theta, T)\right)\mathbb{P}^{\theta,T}(r(\theta) < r^*) - KP(0, \theta)\mathbb{P}^\theta \left(r(\theta) < r^*\right),$$

where we have introduced the probability measure $\mathbb{P}^{\theta,T}$, defined by

$$\frac{d\mathbb{P}^{\theta,T}}{d\mathbb{P}^\theta} = \frac{P(\theta, T)}{\mathbb{E}^\theta \left(P(\theta, T)\right)}.$$

Note that the process $(P(t, T)/P(t, \theta))_{0 \le t \le \theta}$ is a $\mathbb{P}^\theta$-martingale, because $(\tilde{P}(t, T))_{0 \le t \le \theta}$ is a $\mathbb{P}^*$-martingale (see Proposition 6.1.11). Therefore,

$$\mathbb{E}^\theta \left(P(\theta, T)\right) = \frac{P(0, T)}{P(0, \theta)}.$$

We can now write the price of the option as

$$C_0 = P(0, T)\mathbb{P}^{\theta,T}(r(\theta) < r^*) - KP(0, \theta)\mathbb{P}^\theta(r(\theta) < r^*).$$

It can be proved (see Exercise 39) that, if we set

$$L^\theta = \frac{\sigma^2}{2}\frac{(e^{\gamma^*\theta} - 1)}{\gamma^*(e^{\gamma^*\theta} + 1) + b^*(e^{\gamma^*\theta} - 1)}$$

and

$$L^{\theta,T} = \frac{\sigma^2}{2}\frac{(e^{\gamma^*\theta} - 1)}{\gamma^*(e^{\gamma^*\theta} + 1) + (\sigma^2\psi(T - \theta) + b^*)(e^{\gamma^*\theta} - 1)},$$

the law of $r(\theta)/L^\theta$ under $\mathbb{P}^\theta$ (resp. $r(\theta)/L^{\theta,T}$ under $\mathbb{P}^{\theta,T}$) is a non-central chi-squared distribution with $4a/\sigma^2$ degrees of freedom and parameter equal to ζ_θ (resp. $\zeta_{\theta,T}$), with

$$\zeta_\theta = \frac{8r(0)\gamma^{*2}e^{\gamma^*\theta}}{\sigma^2(e^{\gamma^*\theta} - 1)(\gamma^*(e^{\gamma^*\theta} + 1) + b^*)(e^{\gamma^*\theta} - 1))}$$

and

$$\zeta_{\theta,T} = \frac{8r(0)\gamma^{*2}e^{\gamma^*\theta}}{\sigma^2(e^{\gamma^*\theta}-1)(\gamma^*(e^{\gamma^*\theta}+1)+(\sigma^2\psi(T-\theta)+b^*)(e^{\gamma^*\theta}-1))}.$$

With these notations, introducing the distribution function $F_{\delta,\zeta}$ of the non-cenral chi-squared law with δ degrees of freedom and parameter ζ, we have

$$C_0 = P(0,T)F_{4a/\sigma^2,\zeta_1}\left(\frac{r^*}{L_1}\right) - KP(0,\theta)F_{4a/\sigma^2,\zeta_2}\left(\frac{r^*}{L_2}\right).$$

Remark 6.2.5. Observe that in the CIR model, the price of zero-coupon bonds is the exponential of an affine function of the interest rate (cf. equation (6.18)). This property is shared by the Vasicek model, and is closely related to the fact that the coefficients of the infinitesimal generator of the underlying diffusions (6.9) and (6.16) are affine functions of the state (see Duffie, Filipović, and Schachermayer (2003)).

6.2.3 The Heath-Jarrow-Morton methodology

The main drawback to the Vasicek model and the Cox-Ingersoll-Ross model lies in the fact that prices are explicit functions of the instantaneous 'spot' interest rate, so that these models are unable to take the whole yield curve observed on the market into account in the price structure.

Some authors have resorted to a two-dimensional analysis to improve the models in terms of discrepancies between short and long rates; cf.Brennan and Schwartz (1979), Schaefer and Schwartz (1984) and Courtadon (1982). These more complex models do not lead to explicit formulae and require the solution of partial differential equations. More recently, Ho and Lee (1986) have proposed a discrete-time model describing the behavior of the whole yield curve. The continuous-time model we present now is based on the same idea and has been introduced by Heath, Jarrow, and Morton (1992) (see also Morton (1989)). First of all, we define the *forward* instantaneous interest rate $f(t,s)$, for $t \geq s$, characterized by the following equality:

$$P(t,T) = \exp\left(-\int_t^T f(t,s)ds\right), \quad 0 \leq t \leq T, \tag{6.19}$$

for any maturity T. So $f(t,s)$ represents the instantaneous interest rate at time s as 'anticipated' by the market at time t. For each T, the process $(f(t,T))_{0\leq t\leq T}$ must then be an adapted process and it is natural to set $f(t,t) = r(t)$. Moreover, we constrain the map $(t,s) \mapsto f(t,s)$, defined for $t \leq s$, to be continuous. Then the next step of the modelling consists in assuming that, for each maturity T, the process $(f(t,T))_{0\leq t\leq T}$ has the following form:

$$f(t,T) = f(0,T) + \int_0^t \alpha(v,T)dv + \int_0^t \beta(v,T)dW_v, \tag{6.20}$$

where $(\alpha(t,T))_{0\leq t\leq T}$ and $(\beta(t,T))_{0\leq t\leq T}$ are continuous adapted processes. We also assume joint continuity with respect to (t,T).

We then have to make sure that this model is compatible with hypothesis (H). This will imply some conditions on the processes α and β. To find them, we derive the differential $dP(t,T)/P(t,T)$ and we compare it to (6.6). Let $X_t = -\int_t^T f(t,s)ds$. We have $P(t,T) = e^{X_t}$ and, from (6.20),

$$
\begin{aligned}
X_t &= \int_t^T (-f(s,s) + f(s,s) - f(t,s))ds \\
&= -\int_t^T f(s,s)ds + \int_t^T \left(\int_t^s \alpha(v,s)dv \right) ds \\
&\quad + \int_t^T \left(\int_t^s \beta(v,s)dW_v \right) ds \\
&= -\int_t^T f(s,s)ds + \int_t^T \left(\int_v^T \alpha(v,s)ds \right) dv \\
&\quad + \int_t^T \left(\int_v^T \beta(v,s)ds \right) dW_v \\
&= X_0 + \int_0^t f(s,s)ds - \int_0^t \left(\int_v^T \alpha(v,s)ds \right) dv \\
&\quad - \int_0^t \left(\int_v^T \beta(v,s)ds \right) dW_v.
\end{aligned}
$$
(6.21)

The fact that the integrals commute in (6.21) is justified in Exercise 40. We then have

$$
dX_t = \left(f(t,t) - \int_t^T \alpha(t,s)ds \right) dt - \left(\int_t^T \beta(t,s)ds \right) dW_t
$$

and, by Itô's formula,

$$
\begin{aligned}
\frac{dP(t,T)}{P(t,T)} &= dX_t + \frac{1}{2}d\langle X,X\rangle_t \\
&= \left(f(t,t) - \left(\int_t^T \alpha(t,s)ds \right) + \frac{1}{2} \left(\int_t^T \beta(t,s)ds \right)^2 \right) dt \\
&\quad - \left(\int_t^T \beta(t,s)ds \right) dW_t.
\end{aligned}
$$

If (H) holds, we must have, from Proposition 6.1.3 and equality $f(t,t) = r(t)$,

$$
\sigma_t^T = -\int_t^T \beta(t,s)ds \quad \text{and} \quad \sigma_t^T q(t) = \left(\int_t^T \alpha(t,s)ds \right) - \frac{1}{2} \left(\int_t^T \beta(t,s)ds \right)^2.
$$

Whence

$$\int_t^T \alpha(t,s)ds = \frac{1}{2}\left(\int_t^T \beta(t,s)ds\right)^2 - q(t)\int_t^T \beta(t,s)ds$$

and, differentiating with respect to T,

$$\alpha(t,T) = \beta(t,T)\left(\int_t^T \beta(t,s)ds - q(t)\right).$$

Equation (6.20) becomes, if written in differential form,

$$df(t,T) = \beta(t,T)\left(\int_t^T \beta(t,s)ds\right)dt + \beta(t,T)d\tilde{W}_t.$$

The point is that the dynamics of the instantaneous forward rates under $\mathbb{P}^*$ will not depend on the process α. The model can be specified by imposing the following form for β:

$$\beta(t,T) = \sigma(f(t,T)),$$

where $\sigma : \mathbb{R} \to \mathbb{R}$ is a continuous function (which could also depend on time; cf. Morton (1989)). In this case, the two-parameter process $(f(t,T))_{0 \le t \le T}$ must satisfy the following stochastic differential equation:

$$df(t,T) = \sigma(f(t,T))\left(\int_t^T \sigma(f(t,s))ds\right)dt + \sigma(f(t,T))d\tilde{W}_t. \qquad (6.22)$$

The following theorem, due to Heath, Jarrow, and Morton (1992), gives sufficient conditions under which equation (6.22) has a unique solution.

Theorem 6.2.6. *If the function σ is Lipschitz and bounded, for any continuous function ϕ from $[0,\tilde{T}]$ into $\mathbb{R}^+$ there exists a unique continuous process with two indices $(f(t,T))_{0 \le t \le T \le \tilde{T}}$ such that, for all T, the process $(f(t,T))_{0 \le t \le T}$ is adapted and satisfies (6.22), with $f(0,T) = \phi(T)$.*

We see that, for any continous process $(q(t))$, it is then possible to build a model of the form (6.20): take a solution of (6.22) and set

$$\alpha(t,T) = \sigma(f(t,T))\left(\int_t^T \sigma(f(t,s))ds - q(t)\right).$$

The striking feature of this model is that the law of forward rates under $\mathbb{P}^*$ only depends on the function σ. This is a consequence of equation (6.22), in which only σ and $(\tilde{W}_t)$ appear. It follows that the option prices will only depend on the function σ. This situation is similar to Black-Scholes. Observe that if σ is a constant, the volatilities of zero-coupon bonds $(\sigma_t^T), 0 \le t \le T$) are deterministic, so that Remark 6.1.14 applies (see Exercise 41). Note that the boundedness condition on σ is essential since, for $\sigma(x) = x$, there is no solution (see Heath, Jarrow, and Morton (1992); Morton (1989)).

6.2.4   Forward Libor model

The Heath-Jarrow-Morton framework focuses on the dynamics of forward *instantaneous* rates. In practice, these rates are not directly observable. In fact, typical interest rate derivatives involve forward rates over a finite time period. In this section, we describe the so-called forward LIBOR (London Inter Bank Offered Rate) and introduce the forward LIBOR model, also known as the BGM model, as it was introduced by Brace, Gatarek, and Musiela (1977).

The forward LIBOR for the expiry T can be seen as a simply compounded interest rate over the time interval $[T, T + \delta]$, where δ is called the *tenor*. Its value $L(t, T)$ at time t ($0 \leq t \leq T$) is related to the forward *instantaneous* rate or to zero-coupon bond prices as follows:

$$1 + \delta L(t, T) = \exp\left(\int_T^{T+\delta} f(t, s)ds\right) = \frac{P(t, T)}{P(t, T + \delta)}.$$

This is clearly equivalent to

$$L(t, T) = \frac{P(t, T) - P(t, T + \delta)}{\delta P(t, T + \delta)}.$$

A typical interest rate option is a *caplet*, which pays at time $T + \delta$ the difference between the LIBOR and a fixed rate K. More precisely, the payoff at time $T + \delta$ is given by

$$\delta \left(L(T, T) - K\right)_+.$$

An agent who borrows one dollar at time T will have to pay $1 + \delta L(T, T)$ at time $T + \delta$. If he/she holds a caplet with strike price K, his/her net payment at time $T + \delta$ will be

$$1 + \delta L(T, T) - \delta \left(L(T, T) - K\right)_+ = 1 + \delta \min(L(T, T), K).$$

In words, the interest rate is reduced to $\min(L(T, T), K)$. In case of several payment dates, $T_1 = T_0 + \delta, \ldots, T_n = T_0 + n\delta$, the interest rate can be locked at the level K, by holding a *cap*, i.e. a collection of caplets. The holder of a cap receives $\delta \left(L(T_{i-1}, T_{i-1}) - K\right)_+$ at times T_i, $i = 1, \ldots, n$.

Denote by C_t the price at time t of a caplet that pays $\delta \left(L(T, T) - K\right)_+$ at time $T + \delta$. We have, using Proposition 6.1.11,

$$C_t = P(t, T + \delta)\mathbb{E}^{T+\delta}\left(\delta \left(L(T, T) - K\right)_+ \mid \mathscr{F}_t\right), \quad 0 \leq t \leq T + \delta.$$

Observe that the process $(L(t, T))_{0 \leq t \leq T}$ is a martingale under the $(T + \delta)$-forward measure, because $1 + \delta L(t, T)$ is the $(T + \delta)$-forward price of the zero-coupon bond with maturity T (cf. Proposition 6.1.11 and Remark 6.1.14).

If we assume that the dynamics of $(L(t, T))_{0 \leq t \leq T}$ is given by

$$\frac{dL(t, T)}{L(t, T)} = \gamma(t, T)dW_t^{T+\delta}, \quad 0 \leq t \leq T, \tag{6.23}$$

where $\gamma(t, T)$ is a *deterministic* bounded function, and $W^{T+\delta}$ a standard Brownian motion under $\mathbb{P}^{T+\delta}$, then we have a closed-form expression for C_t, known as Black's caplet formula (see Remark 6.1.14 and Exercise 42). The main result of Brace, Gatarek, and Musiela (1977) (see also Musiela and Rutkowski (2005)) is that it is possible to construct a consistent model in which the dynamics of each process $(L(t, T_i))_{0 \leq t \leq T_i + \delta}$ $(T_i = T_0 + i\delta, \, i = 1, \ldots, n\delta)$ has the form (6.23), with a deterministic volatility $\gamma(t, T_i)$. In this model, cap prices are easily derived, as sums of caplet prices. For more complex options, such as swaptions, which involve the joint distribution of LIBORs with different maturities, say $L(t, T_i)$, with $T_i = i\delta, \, i = 1, \ldots n$, one may need the dynamics of $L(t, T_i)$ under another forward measure (see Exercice 42).

Notes: We have restricted our presentation to models driven by a single Brownian motion. The main results of Section 6.1 can be extended to *multifactor* models, where the underlying Brownian motion is multidimensional. We refer to Brigo and Mercurio (2006) for an exhaustive account of interest rate modelling and a thorough discussion of practical issues. As mentioned above, the BGM model does not provide simple formulae for swaptions. Jamshidian (1997) proposed a model in which swaptions can be priced using Black's formula. This model is not compatible with the BGM model. The first systematic presentation of change of numéraire techniques is due to Geman, El Karoui, and Rochet (1995). These techniques can be applied in many situations apart from interest rate modelling. In fact, some of the problems in Chapter 4 can be addressed in this way (especially, the problem on the Garman-Kohlagen model and the one on Asian options: see Shreve (2004), Chapter 9).

6.3 Exercises

Exercise 33 Let $(M_t)_{0 \leq t \leq T}$ be a continuous martingale such that, for any $t \in [0, T], \mathbb{P}(M_t > 0) = 1$. Set

$$\tau = (\inf\{t \in [0, T] | M_t = 0\}) \wedge T.$$

1. Show that τ is a stopping time.

2. Using the optional sampling theorem, show that

$$\mathbb{E}(M_T) = \mathbb{E}(M_T 1_{\{\tau = T\}}).$$

Deduce that $\mathbb{P}(\{\forall t \in [0, T], M_t > 0\}) = 1$.

Exercise 34 Let $(\Omega, \mathscr{F}, (\mathscr{F}_t)_{0 \leq t \leq T}, \mathbb{P})$ be a filtered probability space and let $\mathbb{Q}$ be a probability measure absolutely continuous with respect to $\mathbb{P}$. Denote by L_t the density of the restriction of $\mathbb{Q}$ to $\mathscr{F}_t$. Let $(M_t)_{0 \leq t \leq T}$ be an adapted process. Show that $(M_t)_{0 \leq t \leq T}$ is a martingale under $\mathbb{Q}$ if and only if the process $(L_t M_t)_{0 \leq t \leq T}$ is a martingale under $\mathbb{P}$.

Exercise 35 The notations are those of Section 6.1.3. Let $(M_t)_{0 \le t \le T}$ be a process adapted to the filtration $(\mathcal{F}_t)$. Suppose that (M_t) is a martingale under $\mathbb{P}^*$. Using Exercise 34, show that there exists an adapted process $(H_t)_{0 \le t \le T}$ such that $\int_0^T H_t^2 \, dt < \infty$ a.s. and

$$M_t = M_0 + \int_0^t H_s d\tilde{W}_s \quad \text{a.s.,} \quad 0 \le t \le T.$$

Exercise 36 Assume that the volatilities of the zero-coupon bonds $(\sigma_t^T, 0 \le t \le T \le \bar{T})$ are deterministic. Denote by C_t the value at time t of a call with maturity θ and strike price K on a zero-coupon bond with maturity $T > \theta$. We use the same notation as in Remark 6.1.14.

1. Prove that

$$C_t^\theta = C_0^\theta + \int_0^t \frac{\partial B}{\partial x}(t, P^\theta(t, T)) dP^\theta(t, T), \quad 0 \le t \le \theta.$$

   (Hint: apply Itô's formula and recall that $(C_t^\theta)_{0 \le t \le \theta}$ is a $\mathbb{P}^\theta$-martingale.)

2. Show that $(\partial B/\partial x)(t, x) = N(d_1(t, x))$.

3. Let $H_t^T = N(d_1(t, P^\theta(t, T)))$ and $H_t^\theta = -KN(d_2(t, P^\theta(t, T))$. Prove that $dC_t = H_t^T dP(t, T) + H_t^\theta dP(\theta, T)$. Explain why the option can be hedged by holding H_t^T (resp. H_t^θ) zero-coupon bonds with maturity T (resp. θ) at time t.

4. In the framework of the Vasicek model, prove that $\sigma_t^T = -\sigma(1 - e^{-a(T-t)})/a$, where a is the mean reversion factor (cf. (6.8)). Compute the value of the call option and the hedge ratios.

Exercise 37 The aim of this exercise is to prove Proposition 6.2.3. For $x, M > 0$, define the stopping time defined by $\tau_M^x = \inf\{t \ge 0 | X_t^x = M\}$.

1. Let s be the function defined on $(0, \infty)$ by

$$s(x) = \int_1^x e^{2by/\sigma^2} y^{-2a/\sigma^2} dy.$$

   Prove that s satisfies

$$\frac{\sigma^2}{2} x \frac{d^2 s}{dx^2} + (a - bx) \frac{ds}{dx} = 0.$$

2. For $0 < \varepsilon < x < M$, let $\tau_{\varepsilon,M}^x = \tau_\varepsilon^x \wedge \tau_M^x$. Show that, for any $t > 0$, we have

$$s\left(X_{t \wedge \tau_{\varepsilon,M}^x}^x\right) = s(x) + \int_0^{t \wedge \tau_{\varepsilon,M}^x} s'(X_s^x) \sigma \sqrt{X_s^x} dW_s.$$

   Deduce, taking the variance on both sides and using the fact that s' is bounded from below on the interval $[\varepsilon, M]$, that $\mathbb{E}\left(\tau_{\varepsilon,M}^x\right) < \infty$, which implies that $\tau_{\varepsilon,M}^x$ is infinite a.s.

3. Show that if $\varepsilon < x < M$, $s(x) = s(\varepsilon)\mathbb{P}\left(\tau_\varepsilon^x < \tau_M^x\right) + s(M)\mathbb{P}\left(\tau_\varepsilon^x > \tau_M^x\right)$.

4. We assume $a \geq \sigma^2/2$. Prove that $\lim_{x \to 0} s(x) = -\infty$. Deduce that

$$\mathbb{P}\left(\tau_0^x < \tau_M^x\right) = 0$$

for all $M > 0$, so that $\mathbb{P}\left(\tau_0^x < \infty\right) = 0$.

5. We now assume that $0 \leq a < \sigma^2/2$ and we set $s(0) = \lim_{x \to 0} s(x)$. Show that, for all $M > x$, we have

$$s(x) = s(0)\mathbb{P}\left(\tau_0^x < \tau_M^x\right) + s(M)\mathbb{P}\left(\tau_0^x > \tau_M^x\right),$$

and complete the proof of Proposition 6.2.4.

Exercise 38 Let d be an integer and let $X_1, \ldots, X_d$, be d independent Gaussian random variables with unit variance and respective means m_1, m_2, $\ldots$, m_d. Show that the random variable $X = \sum_{i=1}^{d} X_i^2$ follows a non-central chi-squared law with d degrees of freedom and parameter $\zeta = \sum_{i=1}^{d} m_i^2$.

Exercise 39 Use Proposition 6.2.4 to derive the distribution of $r(\theta)$ under the probablities $\mathbb{P}^\theta$ and $\mathbb{P}^{\theta,T}$, introduced at the end of Section 6.2.2.

Exercise 40 Let $(\Omega, \mathscr{F}, (\mathscr{F}_t)_{0 \leq t \leq T}, \mathbb{P})$ be a filtered probability space and let $(W_t)_{0 \leq t \leq T}$ be a standard Brownian motion with respect to $(\mathscr{F}_t)$. We consider a process with two indices $(H(t, s))_{0 \leq t, s \leq T}$ satisfying the following properties: for any ω, the map $(t, s) \mapsto H(t, s)(\omega)$ is continuous and for any $s \in [0, T]$, the process $(H(t, s))_{0 \leq t \leq T}$ is adapted. We would like to justify the equality

$$\int_0^T \left(\int_0^T H(t, s) dW_t\right) ds = \int_0^T \left(\int_0^T H(t, s) ds\right) dW_t.$$

For simplicity, we assume that $\int_0^T \mathbb{E}\left(\int_0^T H^2(t, s) dt\right) ds < \infty$ (which is sufficient to justify (6.21)).

1. Prove that

$$\int_0^T \mathbb{E}\left(\left|\int_0^T H(t, s) dW_t\right|\right) ds \leq \int_0^T \left[\mathbb{E}\left(\int_0^T H^2(t, s) dt\right)\right]^{1/2} ds.$$

Deduce that the integral $\int_0^T \left(\int_0^T H(t, s) dW_t\right) ds$ exists.

2. Let $0 = t_0 < t_1 < \cdots < t_N = T$ be a partition of the interval $[0, T]$. Observe that

$$\int_0^T \left(\sum_{i=0}^{N-1} H(t_i, s)(W_{t_{i+1}} - W_{t_i})\right) ds =$$

$$\sum_{i=0}^{N-1} \left(\int_0^T H(t_i, s) ds\right) (W_{t_{i+1}} - W_{t_i})$$

and justify why we can take the limit to obtain the desired equality.

Exercise 41   In the Heath-Jarrow-Morton model, we assume that the function σ is a positive constant.

1. Show that the solution of (6.22) is given by $f(t,T) = f(0,T) + \sigma^2 t(T - t/2) + \sigma \tilde{W}_t$.

2. Compute the volatilities of the zero-coupon bonds $(\sigma_t^T, 0 \leq t \leq T)$.

3. Find the price at time 0 of a call with maturity θ and strike price K, on a zero-coupon bond with maturity $T > \theta$.

Exercise 42

1. Assume, as in the BGM model, that we have (6.23), with γ deterministic and bounded. Show that the price at time t of a caplet that pays $\delta(L(T,T) - K)_+$ at time $T + \delta$ is given by

$$C_t = \delta P(t, T + \delta) \left[L(t,T) N(d_1(t, L(t,T))) - K N(d_2(t, L(t,T))) \right],$$

where
$$d_1(t, x) = \frac{\log(x/K) + \frac{1}{2} \int_t^T \gamma^2(s,T) ds}{\sqrt{\int_t^T \gamma^2(s,T) ds}}$$

and $d_2(t, x) = d_1(t, x) - \sqrt{\int_t^T \gamma^2(s,T) ds}$.

2. Denote by $(\sigma(t,T))_{0 \leq t \leq T}$ the volatility process of the zero-coupon bond with maturity T. Prove that

$$\sigma_t^T - \sigma_t^{T+\delta} = \frac{\delta \gamma(t,T) L(t,T)}{1 + \delta L(t,T)}.$$

3. Let n be a positive integer, and, for $i = 0, \ldots, n$, let $T_i = i\delta$. Assume that we have (6.23) for $T = T_i$, $i = 1, \ldots, n$, with deterministic bounded non-negative functions $\gamma(t, T_i)$. Prove that, for $i = 1, \ldots, n-1$, and $0 \leq t \leq T_i$,

$$dL(t, T_i) = \gamma(t, T_i) L(t, T_i) \left[-\sum_{j=i+1}^{n} \frac{\delta \gamma(t, T_j) L(t, T_j)}{1 + \delta L(t, T_j)} dt + dW_t^{T_n + \delta} \right].$$

Chapter 7

Asset models with jumps

In the Black-Scholes model, the stock price is a continuous function of time and this property is one of the characteristics of the model. But some rare events (release of an unexpected economic figure, major political changes or even a natural disaster in a major economy) can lead to brusque variations in prices. To model this kind of phenomena, we have to introduce discontinuous stochastic processes.

Most of these models 'with jumps' have a striking feature that distinguishes them from the Black-Scholes model: they are incomplete market models, and there is no perfect hedging of options in this case. It is no longer possible to price options using a replicating portfolio, and the set of probability measures under which the discounted stock price is a martingale is infinite. The usual approach to pricing and hedging in this context consists of choosing one of these probability measures and taking it as a pricing measure.

In this chapter, we will study the simplest models with jumps. The description of these models requires a review of the main properties of the Poisson process; this is the objective of the first section. We will then investigate the dynamics of the risky asset, discuss the computation of European option prices and examine hedging strategies that minimize the quadratic risk under the pricing measure.

7.1 Poisson process

Definition 7.1.1. *Let $(T_i)_{i \geq 1}$ be a sequence of independent, identically, exponentially distributed random variables with parameter λ ($\lambda > 0$), i.e. their density is equal to $\lambda e^{-\lambda x} 1_{\{x > 0\}}$, $\lambda > 0$. Let $\tau_n = \sum_{i=1}^{n} T_i$. We call the* Poisson process with intensity λ *the process N_t defined by*

$$N_t = \sum_{n \geq 1} 1_{\{\tau_n \leq t\}} = \sum_{n \geq 1} n 1_{\{\tau_n \leq t < \tau_{n+1}\}}.$$

Remark 7.1.2. Note that N_t represents the number of points of the sequence $(\tau_n)_{n \geq 1}$ which are smaller than or equal to t. We also have

$$\tau_n = \inf\{t \geq 0, N_t = n\}.$$

The following proposition gives an explicit expression for the law of N_t for a given t.

Proposition 7.1.3. *If $(N_t)_{t \geq 0}$ is a Poisson process with intensity λ then, for any $t > 0$, the random variable N_t follows a Poisson law with parameter λt:*

$$\mathbb{P}(N_t = n) = e^{-\lambda t} \frac{(\lambda t)^n}{n!}, \qquad n \in \mathbb{N}.$$

In particular we have

$$\mathbb{E}(N_t) = \lambda t, \quad \mathrm{Var}(N_t) = \mathbb{E}\left(N_t^2\right) - (\mathbb{E}(N_t))^2 = \lambda t$$

Moreover, for $s > 0$

$$\mathbb{E}(s^{N_t}) = \exp\{\lambda t(s - 1)\}.$$

Proof. We first observe that τ_n has a Gamma distribution with parameters (n, λ), i.e. its density is given by

$$\lambda e^{-\lambda x} \frac{(\lambda x)^{n-1}}{(n - 1)!}, \qquad x > 0.$$

Indeed, the Laplace transform of T_1 is given by

$$\mathbb{E}(e^{-\alpha T_1}) = \frac{\lambda}{\lambda + \alpha}, \qquad \alpha > 0,$$

so that

$$\mathbb{E}(e^{-\alpha \tau_n}) = \left(\mathbb{E}(e^{-\alpha T_1})\right)^n = \left(\frac{\lambda}{\lambda + \alpha}\right)^n.$$

This is the Laplace transform of the Gamma distribution with parameters (n, λ) (cf. Bouleau (1986), Chapter VI, Section 7.12). Now we have, for $n \geq 1$,

$$\mathbb{P}(N_t = n) = \mathbb{P}(\tau_n \leq t) - \mathbb{P}(\tau_{n+1} \leq t)$$
$$= \int_0^t \lambda e^{-\lambda x} \frac{(\lambda x)^{n-1}}{(n - 1)!} dx - \int_0^t \lambda e^{-\lambda x} \frac{(\lambda x)^n}{n!} dx$$
$$= \frac{(\lambda t)^n}{n!} e^{\lambda t}.$$

$\square$

Proposition 7.1.4. *Let $(N_t)_{t \geq 0}$ be a Poisson process with intensity λ and $\mathscr{F}_t = \sigma(N_s; s \leq t)$. The process $(N_t)_{t \geq 0}$ is a process with independent and sta-tionarity increments, i.e.*

- independence: *if $s > 0$, $N_{t+s} - N_t$ is independent of the σ-albebra $\mathscr{F}_t$.*

- stationary: *the law of $N_{t+s} - N_t$ is identical to the law of $N_s - N_0 = N_s$.*

Remark 7.1.5. It is easy to see that the jump times τ_n are stopping times. Indeed, $\{\tau_n \leq t\} = \{N_t \geq n\} \in \mathscr{F}_t$. A random variable T with exponential law satisfies $\mathbb{P}(T \geq t + s | T \geq t) = \mathbb{P}(T \geq s)$. The exponential variables are said to be 'memoryless'. The independence of the increments is a consequence of this property of exponential laws.

Remark 7.1.6. The law of a Poisson process with intensity λ is characterized by either of the following two properties:

- $(N_t)_{t \geq 0}$ is a right-continuous homogeneous Markov process with left-hand limits, such that

$$\mathbb{P}(N_t = n) = e^{\lambda t} \frac{(\lambda t)^n}{n!}, \quad t > 0, n \in \mathbb{N}.$$

- $(N_t)_{t \geq 0}$ is a right-continuous, non-decreasing, process with independent and stationary increments, with $N_0 = 0$ and jumps of amplitude one.

For the first characterization, cf. Bouleau (1988), Chapter III; for the second one, cf. Dacunha-Castelle and Duflo (1986b), Section 6.3.

7.2 Dynamics of the risky asset

The objective of this section is to model a financial market in which there is one riskless assset (with price $S_t^0 = e^{rt}$, at time t) and one risky asset whose price jumps in the proportions $U_1, \ldots, U_j, \ldots$, at times $\tau_1, \ldots, \tau_j, \ldots$ and which, between two jumps, follows the Black-Scholes model. Moreover, we will assume that the τ_j's correspond to the jump times of a Poisson process. To be more rigorous, let us consider a probability space $(\Omega, \mathscr{A}, \mathbb{P})$ on which we define a standard Brownian motion $(W_t)_{t \geq 0}$, a Poisson process $(N_t)_{t \geq 0}$ with intensity λ and a sequence $(U_j)_{j \geq 1}$ of independent, identically distributed random variables taking values in $(-1, +\infty)$. We will assume that the σ-algebras generated respectively by $(W_t)_{t \geq 0}$, $(N_t)_{t \geq 0}$, $(U_j)_{j \geq 1}$ are independent.

For all $t \geq 0$, let us denote by $\mathscr{F}_t$ the σ-algebra generated by the random variables W_s, N_s for $s \leq t$ and $U_j 1_{\{j \leq N_t\}}$ for $j \geq 1$. It can be shown that $(W_t)_{t \geq 0}$ is a standard Brownian motion with respect to the filtration $(\mathscr{F}_t)_{t \geq 0}$, that $(N_t)_{t \geq 0}$ is $(\mathscr{F}_t)$-adapted and that, for all $t > s$, $N_t - N_s$ is independent of the σ-algebra $\mathscr{F}$. The intuitive interpretation of the measurability of the random variables $U_j 1_{\{j \leq N_t\}}$ with respect to $\mathscr{F}_t$ is that, at time t, the relative amplitudes of the jumps taking place before or at time t are known. Observe also that the τ_j's are stopping times of $(\mathscr{F}_t)_{t \geq 0}$, since $\{\tau_j \leq t\} = \{N_t \geq j\} \in \mathscr{F}_t$.

The dynamics of X_t, the price of the risky asset at time t, can now be described in the following manner. The process $(X_t)_{t \geq 0}$ is an adapted, right-continuous process satisfying:

- On the time intervals $[\tau_j, \tau_{j+1})$,

$$dX_t = X_t(\mu dt + \sigma dW_t),$$

where σ and μ are constant, with $\sigma > 0$.

- At time τ_j, the jump of X_t is given by

$$\Delta X_{\tau_j} = X_{\tau_j} - X_{\tau_j^-} = X_{\tau_j^-} U_j,$$

so that $X_{\tau_j} = X_{\tau_j^-}(1 + U_j)$.

We have, for $t \in [0, \tau_1)$,

$$X_t = X_0 e^{(\mu - \sigma^2/2)t + \sigma W_t}.$$

Consequently, the left-hand limit at τ_1 is given by

$$X_{\tau_1^-} = X_0 e^{(\mu - (\sigma^2/2))\tau_1 + \sigma W_{\tau_1}},$$

and

$$X_{\tau_1} = X_0(1 + U_1)e^{(\mu - (\sigma^2/2))\tau_1 + \sigma W_{\tau_1}}.$$

Then, for $t \in [\tau_1, \tau_2)$,

$$\begin{aligned}
X_t &= X_{\tau_1} e^{(\mu - (\sigma^2/2))(t - \tau_1) + \sigma(W_t - W_{\tau_1})}, \\
&= X_{\tau_1^-}(1 + U_1)e^{(\mu - (\sigma^2/2))(t - \tau_1) + \sigma(W_t - W_{\tau_1})} \\
&= X_0(1 + U_1)e^{(\mu - (\sigma^2/2))t + \sigma W_t}.
\end{aligned}$$

Repeating this scheme, we obtain

$$X_t = X_0 \left(\prod_{j=1}^{N_t}(1 + U_j) \right) e^{(\mu - (\sigma^2/2))t + \sigma W_t},$$

with the convention $\prod_{j=1}^{0} = 1$.

The process $(X_t)_{t \geq 0}$ is obviously right-continuous, adapted and has only finitely many discontinuities on each interval $[0, t]$. It can also be proved that it satisfies, for all $t \geq 0$,

$$\mathbb{P} \text{ a.s.,} \quad X_t = X_0 + \int_0^t X_s(\mu ds + \sigma dW_s) + \sum_{j=1}^{N_t} X_{\tau_j^-} U_j. \qquad (7.1)$$

This model is called a *jump-diffusion* model. In differential form, (7.1) can be written

$$\frac{dX_t}{X_{t-}} = \mu dt + \sigma dW_t + dZ_t,$$

where $Z_t = \sum_{j=1}^{N_t} U_j$. The process (Z_t) is called a *compound Poisson process*.

7.3 Martingales in a jump-diffusion model

In this section, we discuss some technical results concerning martingales within the jump-diffusion model presented in the previous section. We first give a necessary and sufficient condition on the parameters for the discounted price process to be a martingale.

Proposition 7.3.1. *Suppose* $\mathbb{E}|U_1| < \infty$. *The process* $(\tilde{X}_t = e^{-rt}X_t)_{0 \leq t \leq T}$ *is a martingale if and only if*

$$\mu = r - \lambda \mathbb{E}(U_1). \tag{7.1}$$

For the derivation of $\mathbb{E}(X_t | \mathscr{F}_s)$ we will need the following lemma, which means intuitively that the relative amplitudes of the jumps that take place after time s together with the increments of W and N after s are independent of the σ-algebra $\mathscr{F}_s$.

Lemma 7.3.2. *For* $s \geq 0$, *denote by* $\mathscr{G}_s$ *the* σ-*algebra generated by the random variables* U_{N_s+j}, $j \geq 1$, $W_{t+s} - W_s$, $N_{t+s} - N_s$, $t \geq 0$. *The* σ-*algebras* $\mathscr{G}_s$ *and* $\mathscr{F}_s$ *are independent.*

Proof. Let $\mathscr{W}_s$ (resp. $\mathscr{N}_s$) be the σ-algebra generated by the random variables W_u (resp. N_u), $0 \leq u \leq s$. We will also need the σ-algebra $\hat{\mathscr{W}}_s$ (resp. $\hat{\mathscr{N}}_s$), generated by the random variables $W_{t+s} - W_s$ (resp. $N_{t+s} - N_s$), $t \geq 0$.

In order to establish the independence of $\mathscr{G}_s$ and $\mathscr{F}_s$, we need to prove $\mathbb{P}(A \cap B) = \mathbb{P}(A)\mathbb{P}(B)$, where A (resp. B) is any element of a class of events that is stable under finite intersections and generates $\mathscr{G}_s$ (resp. $\mathscr{F}_s$). Take

$$A = A_W \cap A_N \cap \{(U_{N_s+1}, \ldots, U_{N_s+k}) \in F\},$$

where $A_W \in \hat{\mathscr{W}}_s$, $A_N \in \hat{\mathscr{N}}_s$, k is a positive integer and F is a Borel subset of $\mathbb{R}^k$, and

$$B = B_W \cap B_N \cap \{(U_1, \ldots, U_d) \in G\} \cap \{d \leq N_s\},$$

where $B_W \in \mathscr{W}_s$, $B_N \in \mathscr{N}_s$, d is a positive integer and G a Borel subset of $\mathbb{R}^d$.

Then we have, using the independence of the processes $(W_t)_{t \geq 0}$, $(N_t)_{t \geq 0}$ and the sequence $(U_j)_{j \geq 1}$,

$$\mathbb{P}(A \cap B) = \mathbb{P}(A_W \cap B_W)\mathbb{P}(A_N \cap B_N \cap C), \tag{7.2}$$

where

$$C = \{(U_{N_s+1}, \ldots, U_{N_s+k}) \in F\} \cap \{(U_1, \ldots, U_d) \in G\} \cap \{d \leq N_s\}.$$

We have

$$\mathbb{P}(A_N \cap B_N \cap C) = \sum_{p=d}^{+\infty} \mathbb{P}(A_N \cap B_N \cap C_p \cap \{N_s = p\}),$$

where $C_p = \{(U_{p+1}, \ldots, U_{p+k}) \in F\} \cap \{(U_1, \ldots, U_d) \in G\}$. Now, using the independence of $(N_t)_{t \geq 0}$ and the sequence $(U_j)_{j \geq 1}$, and $p \geq d$, we get, with the notation

$B_N^p = B_N \cap \{N_s = p\}$,

$$\mathbb{P}(A_N \cap B_N \cap C_p \cap \{N_s = p\})$$
$$= \mathbb{P}(A_N \cap B_N^p)\mathbb{P}((U_{p+1}, \ldots, U_{p+k}) \in F, (U_1, \ldots, U_d) \in G)$$
$$= \mathbb{P}(A_N \cap B_N^p)\,\mathbb{P}((U_{p+1}, \ldots, U_{p+k}) \in F)\,\mathbb{P}((U_1, \ldots, U_d) \in G)$$
$$= \mathbb{P}(A_N)\,\mathbb{P}(B_N^p)\,\mathbb{P}((U_1, \ldots, U_k) \in F)\,\mathbb{P}((U_1, \ldots, U_d) \in G),$$

where we have used the independence of $\mathcal{N}_s$ and $\hat{\mathcal{N}}_s$. Hence

$$\mathbb{P}(A_N \cap B_N \cap C)$$

$$= \mathbb{P}(A_N)\mathbb{P}((U_1, \ldots, U_k) \in F) \sum_{p=d}^{+\infty} \mathbb{P}((U_1, \ldots, U_d) \in G)\,\mathbb{P}(B_N \cap \{N_s = p\})$$

$$= \mathbb{P}(A_N)\mathbb{P}((U_1, \ldots, U_k) \in F)\,\mathbb{P}((U_1, \ldots, U_d) \in G)\,\mathbb{P}(B_N \cap \{N_s \geq d\}).$$

Going back to (7.2) and observing that A_W and B_W are independent, we have

$$\mathbb{P}(A \cap B) = \mathbb{P}(A_W)\mathbb{P}(B_W)\mathbb{P}(A_N) \times$$
$$\times \mathbb{P}((U_1, \ldots, U_k) \in F)\,\mathbb{P}((U_1, \ldots, U_d) \in G)\,\mathbb{P}(B_N \cap \{N_s \geq d\})$$
$$= \mathbb{P}(A_W)\mathbb{P}(A_N)\mathbb{P}((U_1, \ldots, U_k) \in F) \times$$
$$\times \mathbb{P}((U_1, \ldots, U_d) \in G)\,\mathbb{P}(B_N \cap B_W \cap \{N_s \geq d\}).$$

It is now clear that $\mathbb{P}(A \cap B) = \mathbb{P}(A)\mathbb{P}(B)$. A by-product of this proof is the fact that the sequence $(U_{N_s+j})_{j \geq 1}$ has the same distribution as $(U_j)_{j \geq 1}$. □

Proof of Proposition 7.3.1. We have

$$\mathbb{E}(\tilde{X}_t | \mathscr{F}_s) = \tilde{X}_s \mathbb{E}\left(e^{(\mu - r - \sigma^2/2)(t-s) + \sigma(W_t - W_s)} \prod_{j=N_s+1}^{N_t} (1 + U_j) \Big| \mathscr{F}_s\right)$$

$$= \tilde{X}_s \mathbb{E}\left(e^{(\mu - r - \sigma^2/2)(t-s) + \sigma(W_t - W_s)} \prod_{j=1}^{N_t - N_s} (1 + U_{N_s+j}) \Big| \mathscr{F}_s\right)$$

$$= \tilde{X}_s \mathbb{E}\left(e^{(\mu - r - \sigma^2/2)(t-s) + \sigma(W_t - W_s)} \prod_{j=1}^{N_t - N_s} (1 + U_{N_s+j})\right),$$

where we have used Lemma 7.3.2. Hence, using the fact that the sequence $(U_{N_s+j})_{j \geq 1}$ has the same distribution as $(U_j)_{j \geq 1}$,

$$\mathbb{E}(\tilde{X}_t | \mathscr{F}_s) = \tilde{X}_s e^{(\mu - r)(t-s)} \mathbb{E}\left(\prod_{j=N_s+1}^{N_t} (1 + U_j)\right)$$

$$= \tilde{X}_s e^{(\mu - r)(t-s)} e^{\lambda(t-s)\mathbb{E}(U_1)},$$

where the last equality follows from Exercise 43. It is now clear that $(\tilde{X}_t)$ is a martingale if and only if $\mu = r - \lambda\mathbb{E}(U_1)$. □

In order to deal with the terms due to the jumps in the dynamics of hedging strategies, we will need two additional lemmas, whose proofs can be omitted at first reading. We will denote by ν the common law of the random variables U_j.

Lemma 7.3.3. *Let $\Phi(y, z)$ be a measurable function from $\mathbb{R}^d \times \mathbb{R}$ to $\mathbb{R}$, such that for any real number z the function $y \mapsto \Phi(y, z)$ is continuous on $\mathbb{R}^d$, and let $(Y_t)_{t \leq 0}$ be a left-continuous process, taking values in $\mathbb{R}^d$, adapted to the filtration $(\mathscr{F}_t)_{t \geq 0}$. Assume that, for all $t > 0$,*

$$\mathbb{E}\left(\int_0^t ds \int \nu(dz) \Phi^2(Y_s, z) \right) < +\infty.$$

Then the process M_t defined by

$$M_t = \sum_{j=1}^{N_t} \Phi(Y_{\tau_j}, U_j) - \lambda \int_0^t ds \int \nu(dz) \Phi(Y_s, z),$$

is a square-integrable martingale and

$$M_t^2 - \lambda \int_0^t ds \int \nu(dz) \Phi^2(Y_s, z)$$

is a martingale

Note that by convention $\sum_{j=1}^0 = 1$.

Proof. First assume that Φ is bounded and let

$$C = \sup_{(y,z) \in \mathbb{R}^d \times \mathbb{R}} |\Phi(y, z)|.$$

Then we have $\left| \sum_{j=1}^{N_t} \Phi(Y_{\tau_j}, U_j) \right| \leq CN_t$ and $\left| \int_0^t \int \nu(dz) \Phi(Y_s, z) \right| \leq Ct$. Therefore, M_t is square-integrable. Now, fix s and t, with $s < t$, and let

$$Z = \sum_{j=N_s+1}^{N_t} \Phi(Y_{\tau_j}, U_j).$$

Given a partition $\rho = (s_0 = s < s_1 < \cdots < s_m = t)$ of the interval $[s, t]$, define

$$Z^\rho = \sum_{i=0}^{m-1} \sum_{j=N_{s_i}+1}^{N_{s_i+1}} \Phi(Y_{s_i}, U_j).$$

Let $|\rho| = \max_{1 \leq i \leq m}(s_i - s_{i-1})$. The left-continuity of $(Y_t)_{t \geq 0}$ and the continuity of Φ with respect to y imply that Z^ρ converges almost surely to Z as $|\rho| \to 0$. Moreover, $|Z^\rho| \leq C(N_t - N_s)$, so that, as $|\rho| \to 0$, Z^ρ converges to Z in L^2.

We have

$$\mathbb{E}(Z^\rho | \mathscr{F}_s) = \mathbb{E}\left(\sum_{i=0}^{m-1} \mathbb{E}(Z_{i+1} | \mathscr{F}_{s_i}) \bigg| \mathscr{F}_s \right), \tag{7.3}$$

with

$$Z_{i+1} = \sum_{j=N_{s_i}+1}^{N_{s_{i+1}}} \Phi(Y_{s_i}, U_j) = \sum_{j=1}^{N_{s_{i+1}}-N_{s_i}} \Phi(Y_{s_i}, U_{N_{s_i}+j}).$$

Using Lemma 7.3.2 and the fact that Y_{s_i} is $\mathscr{F}_{s_i}$-measurable, we apply Proposition A.2.5 of the Appendix to see that

$$\mathbb{E}(Z_{i+1}|\mathscr{F}_{s_i}) = \bar{\Phi}_i(Y_{s_i}),$$

where $\bar{\Phi}_i$ is defined by

$$\bar{\Phi}_i(y) = \mathbb{E}\left(\sum_{j=1}^{N_{s_{i+1}}-N_{s_i}} \Phi(y, U_{N_{s_i}+j})\right).$$

$\bar{\Phi}_i(y)$ is thus the expectation of a random sum and, from Exercise 44,

$$\bar{\Phi}_i(y) = \lambda(s_{i+1} - s_i) \int d\nu(z)\Phi(y, z).$$

Going back to equation (7.3), we deduce

$$\mathbb{E}(Z^\rho|\mathscr{F}_s) = \mathbb{E}\left(\sum_{i=0}^{m-1} \bar{\Phi}_i(Y_{s_i})\bigg|\mathscr{F}_s\right) = \mathbb{E}\left(\sum_{i=0}^{m-1} \lambda(s_{i+1} - s_i) \int d\nu(z)\Phi(Y_{s_i}, z)\bigg|\mathscr{F}_s\right).$$

Taking limits as $|\rho| \to 0$ yields

$$\mathbb{E}\left(\sum_{j=N_s+1}^{N_t} \Phi(Y_{\tau_j}, U_j)\bigg|\mathscr{F}_s\right) = \mathbb{E}\left(\lambda \int_s^t du \int d\nu(z)\Phi(Y_u, z)\bigg|\mathscr{F}_s\right),$$

which proves that M_t is a martingale. Now set $\bar{Z}^\rho = \sum_{i=0}^{m-1} \mathbb{E}(Z_{i+1}|\mathscr{F}_{s_i})$. We can write

$$\bar{Z}^\rho = \sum_{i=0}^{m-1} \bar{\Phi}_i(Y_{s_i}) = \sum_{i=0}^{m-1} \lambda(s_{i+1} - s_i) \int d\nu(z)\Phi(Y_{s_i}, z).$$

Moreover,

$$\mathbb{E}\left((Z^\rho - \bar{Z}^\rho)^2\bigg|\mathscr{F}_s\right) = \mathbb{E}\left[\left(\sum_{i=o}^{m-1}[Z_{i+1} - \mathbb{E}(Z_{i+1}|\mathscr{F}_{s_i})]\right)^2\bigg|\mathscr{F}_s\right]$$

$$= \mathbb{E}\left(\sum_{i=o}^{m-1}[Z_{i+1} - \mathbb{E}(Z_{i+1}|\mathscr{F}_{s_i})]^2\bigg|\mathscr{F}_s\right)$$

$$+ 2\sum_{i<j} \mathbb{E}\left((Z_{i+1} - \mathbb{E}(Z_{i+1}|\mathscr{F}_{s_i}))(Z_{j+1} - \mathbb{E}(Z_{j+1}|\mathscr{F}_{s_j}))\bigg|\mathscr{F}_s\right).$$

Taking the conditional expectation with respect to $\mathscr{F}_{s_j}$ and using the fact that Z_{i+1} is $\mathscr{F}_{s_{i+1}}$ and hence $\mathscr{F}_{s_j}$-measurable, we see that the second sum is zero. Whence

$$\mathbb{E}\left((Z^\rho - \bar{Z}^\rho)^2\bigg|\mathscr{F}_s\right) = \mathbb{E}\left(\sum_{i=0}^{m-1}(Z_{i+1} - \mathbb{E}(Z_{i+1}|\mathscr{F}_{s_i}))^2\bigg|\mathscr{F}_s\right)$$

$$= \mathbb{E}\left(\sum_{i=o}^{m-1} \mathbb{E}\left([Z_{i+1} - \mathbb{E}(Z_{i+1}|\mathscr{F}_{s_i})]^2 \mid \mathscr{F}_{s_i}\right)\bigg|\mathscr{F}_s\right).$$

Using Lemma 7.3.2 once again,

$$\mathbb{E}\left([Z_{i+1} - \mathbb{E}(Z_{i+1}|\mathscr{F}_{s_i})]^2 \,|\mathscr{F}_{s_i}\right) = V(Y_{s_i}),$$

where the function V is defined by

$$V(y) = \text{Var}\left(\sum_{j=1}^{N_{s_{i+1}}-N_{s_i}} \Phi(y, U_{N_{s_i}+j})\right)$$

and, from Exercise 44,

$$V(y) = \lambda(s_{i+1} - s_i) \int d\nu(z)\Phi^2(y, z).$$

Therefore

$$\mathbb{E}\left((Z^\rho - \bar{Z}^\rho)^2 \,\bigg|\mathscr{F}_s\right) = \mathbb{E}\left(\sum_{i=0}^{m-1} \lambda(s_{i+1} - s_i) \int d\nu(z)\Phi^2(Y_{s_i}, z) \,\bigg|\mathscr{F}_s\right),$$

and so, as $|\rho| \to 0$,

$$\mathbb{E}\left[(M_t - M_s)^2|\mathscr{F}_s\right] = \mathbb{E}\left[\lambda \int_s^t du \int d\nu(z)\Phi^2(Y_u, z) \,\bigg|\mathscr{F}_s\right]. \tag{7.4}$$

Since $(M_t)_{t\geq 0}$ is a square-integrable martingale, we have

$$\mathbb{E}\left[(M_t - M_s)^2|\mathscr{F}_s\right] = \mathbb{E}\left(M_t^2 + M_s^2 - 2M_t M_s|\mathscr{F}_s\right) = \mathbb{E}\left(M_t^2 - M_s^2|\mathscr{F}_s\right),$$

and (7.4) implies that $M_t^2 - \lambda \int_0^t du \int d\nu(z)\Phi^2(Y_u, z)$ is a martinagle.

If we do not assume that Φ is bounded, but instead

$$\mathbb{E}\left(\int_0^t ds \int d\nu(z)\Phi^2(Y_s, z)\right) < +\infty,$$

for any t, we can introduce the (bounded) functions Φ^n defined by $\Phi^n(y, z) = \inf(n, \sup(-n, \Phi(y, z)))$, and the martingales $(M_t^n)_{t\geq 0}$ defined by

$$M_t^n = \sum_{j=1}^{N_t} \Phi^n(Y_{\tau_j}, U_j) - \lambda \int_0^t ds \int d\nu(z)\Phi^n(Y_s, z).$$

It is easily seen that $\mathbb{E}\left(\int_0^t ds \int d\nu(z)(\Phi^n(Y_s, z) - \Phi(Y_s, z))^2\right)$ tends to 0 as $n \to \infty$. It follows that the sequence $(M_t^n)_{n\geq 1}$ is Cauchy in L^2 and as M_t^n converges to M_t a.s., M_t is square-integrable and taking the limit, the lemma is satisfied for Φ. □

Lemma 7.3.4. *We keep the hypotheses and notations of Lemma 7.3.3. Let* $(A_t)_{t\geq 0}$ *be an adapted process such that* $\mathbb{E}\left(\int_0^t A_s^2 ds\right) < +\infty$ *for all* $t > 0$. *Let* $L_t = \int_0^t A_s dW_s$ *and, as in Lemma 7.3.3,*

$$M_t = \sum_{j=1}^{N_t} \Phi(Y_{\tau_j}, U_j) - \lambda \int_0^t ds \int d\nu(z)\Phi(Y_s, z).$$

Then the product $(L_t M_t)_{t \geq 0}$ is a martingale.

Proof.   It is sufficient to prove the lemma for Φ bounded (the general case is proved by approximating Φ by $\Phi^n = \inf(n, \sup(-n, \Phi))$, as in the proof of Lemma 7.3.3). Fix $s < t$ and denote by $\rho = (s_0 = s < s_1 < \cdots < s_m = t)$ a partition of the interval $[s, t]$. We have

$$\mathbb{E}\left(L_t M_t - L_s M_s | \mathscr{F}_s\right) = \mathbb{E}\left[\sum_{i=0}^{m-1} \mathbb{E}\left(L_{s_{i+1}} M_{s_{i+1}} - L_{s_i} M_{s_i} \mid \mathscr{F}_{s_i}\right) \middle| \mathscr{F}_s\right].$$

On the other hand, since $(L_t)_{t \geq 0}$ and $(M_t)_{t \geq 0}$ are martingales,

$$\mathbb{E}\left(L_{s_{i+1}} M_{s_{i+1}} - L_{s_i} M_{s_i} \mid \mathscr{F}_{s_i}\right) = \mathbb{E}\left((L_{s_{i+1}} - L_{s_i})(M_{s_{i+1}} - M_{s_i}) | \mathscr{F}_{s_i}\right).$$

Whence

$$\mathbb{E}(L_t M_t - L_s M_s \mid \mathscr{F}_s) = \mathbb{E}(\Lambda^\rho \mid \mathscr{F}_s),$$

with

$$\Lambda^\rho = \sum_{i=0}^{m-1} (L_{s_{i+1}} - L_{s_i})(M_{s_{i+1}} - M_{s_i}).$$

We have

$$|\Lambda^\rho| \leq \sup_{0 \leq i \leq m-1} \left|L_{s_{i+1}} - L_{s_i}\right| \sum_{i=0}^{m-1} \left|M_{s_{i+1}} - M_{s_i}\right|.$$

Observe that

$$\sum_{i=0}^{m-1} \left|M_{s_{i+1}} - M_{s_i}\right| \leq \sum_{j=N_s+1}^{N_t} \left|\Phi(Y_{\tau_j}, U_j)\right| + \lambda \int_s^t du \int d\nu(z) \left|\Phi(Y_s, z)\right|$$

$$\leq C(N_t - N_s) + \lambda C(t - s),$$

with $C = \sup_{y,z} |\Phi(y, z)|$. From the continuity of $t \mapsto L_t$, we see that Λ^ρ converges to 0 almost surely as $|\rho| \to 0$. Moreover,

$$|\Lambda^\rho| \leq 2C \sup_{s \leq u \leq t} |L_u| (N_t - N_s + \lambda(t - s)).$$

The random variable $\sup_{s \leq u \leq t} |L_u|$ is in L^2 (cf. Doob's inequality, Chapter 3, Theorem 3.3.7). We also have $N_t - N_s \in L^2$. We deduce that Λ^ρ converges to 0 in L^1, and consequently

$$\mathbb{E}\left(L_t M_t - L_s M_s \mid \mathscr{F}_s\right) = 0.$$

$\square$

7.4   Pricing options in a jump-diffusion model

We go back to the jump-diffusion model, as introduced in Section 7.2. We want to price and hedge European options with maturity T in this model. We first observe that under the probability $\mathbb{P}^*$ with density

$$\frac{d\mathbb{P}^*}{d\mathbb{P}} = e^{\theta W_T - \frac{\theta^2}{2} T}, \quad \text{where} \quad \theta = \frac{r - \mu - \lambda \mathbb{E}(U_1)}{\sigma},$$

the discounted stock price process $(\tilde{X}_t)_{t\geq 0} = (e^{-rt}X_t)_{t\geq 0}$ is a martingale. Indeed, under $\mathbb{P}^*$, the process $(W_t^*)_{0\leq t\leq T}$ with $W_t^* = W_t - \theta t$ is a standard Brownian motion. Moreover, under $\mathbb{P}^*$, the processes $(W_t^*)_{0\leq t\leq T}$, $(N_t)_{0\leq t\leq T}$ and the sequence $(U_j, j\geq 1)$ are independent, and the laws of $(N_t)_{0\leq t\leq T}$ and the U_j's remain unchanged. On the other hand, we have

$$X_t = X_0 \left(\prod_{j=1}^{N_t}(1 + U_j) \right) e^{(\mu^* - (\sigma^2/2))t + \sigma W_t^*},$$

with $\mu^* = \mu + \sigma\theta$, so that $\mu^* = r - \lambda\mathbb{E}(U_1) = r - \lambda\mathbb{E}^*(U_1)$, and the martingale property follows from Proposition 7.3.1.

It turns out that, in a general jump-diffusion model, the set of probability measures under which the discounted stock price process is a martingale is infinite (see Exercise 46). Throughout this section, we will assume that

$$\mu = r - \lambda\mathbb{E}(U_1) = r - \lambda \int z d\nu(z). \tag{7.1}$$

In other words, we are choosing a risk neutral probability measure, and we assume that under this risk neutral measure, the stock price follows a jump-diffusion model. The purpose of this section is to define option prices and hedging strategies in reference to this particular risk neutral measure. For technical reasons, we will also assume that the random variables U_j are square-integrable. Note that

$$\mathbb{E}\left(X_t^2\right) = X_0^2 \mathbb{E}\left(\exp\left(\left(\mu - \frac{\sigma^2}{2} \right) t + \sigma W_t \right) \prod_{j=1}^{N_t}(1 + U_j) \right)^2$$

and consequently, using Exercise 43,

$$\mathbb{E}\left(X_t^2\right) = X_0^2 \exp((\sigma^2 + 2r)t) \exp\left(\lambda t \mathbb{E}\left(U_1^2\right) \right).$$

Therefore the process $(\tilde{X}_t)_{0\leq t\leq T}$ is a square-integrable martingale.

7.4.1 Admissible strategies

In the following, we fix a finite horizon T. A trading strategy will be defined, as in the Black-Scholes model, by an adapted process $\phi = ((H_t^0, H_t))_{0\leq t\leq T}$, taking values in $\mathbb{R}^2$, representing the amounts of assets held over time; but, to take the jumps into account, we will constrain the processes (H_t^0) and (H_t) to be *left-continuous*. Since the process (X_t) itself is right-continuous, this means, intuitively, that one can react to the jumps only after their occurrence. This condition is the counterpart of the condition of *predictability* that is found in the discrete models (cf. Chapter 1) and which is slightly more prickly to define in continuous-time.

The value at time t of the strategy ϕ is given by $V_t = H_t^0 e^{rt} + H_t X_t$, and the strategy is said to be *self-financing* if

$$dV_t = H_t^0 r e^{rt} dt + H_t dX_t,$$

i.e. taking into account equation (7.1), $dV_t = H_t^0 r e^{rt} dt + H_t X_t(\mu dt + \sigma dW_t)$ between the jump times and, at a jump time τ_j, V_t jumps by an amount $\Delta V_{\tau_j} = H_{\tau_j} \Delta X_{\tau_j} = H_{\tau_j} U_j X_{\tau_j^-}$. Precisely, the self-financing condition can be written as

$$V_t = H_t^0 e^{rt} + H_t X_t = V_0 + \int_0^t H_s^0 r e^{rs} ds$$

$$+ \int_0^t H_s X_s (\mu ds + \sigma dW_s)$$

$$+ \sum_{j=1}^{N_t} H_{\tau_j} U_j X_{\tau_j^-}. \qquad (7.2)$$

For this equation to make sense, it suffices to impose the condition

$$\int_0^T \left| H_s^0 \right| ds + \int_0^T H_s^2 ds < \infty, \text{a.s.}$$

(it is easily seen that $s \mapsto X_s$ is almost surely bounded). Actually, in order to make sure that the value of a hedging strategy is square-integrable, we will impose a stronger condition of integrability on the process $(H_t)_{0 \leq t \leq T}$, thus restricting the class of admissible strategies as follows.

Definition 7.4.1. An *admissible strategy* is defined by a process

$$\phi = ((H_t^0, H_t))_{0 \leq t \leq T}$$

adapted, left-continuous, with values in $\mathbb{R}^2$, satisfying equality (7.2) a.s. for all $t \in [0, T]$ and such that $\int_0^T \left| H_s^0 \right| ds < +\infty$ $\mathbb{P}$ a.s. and $\mathbb{E} \left(\int_0^T H_s^2 X_s^2 ds \right) < +\infty$.

Note that we do not impose any non-negativity condition on the value of admissible strategies. The following proposition is the counterpart of Proposition 4.1.2 of Chapter 4.

Proposition 7.4.2. *Let $(H_t)_{0 \leq t \leq T}$ be an adapted, left-continuous process such that*

$$\mathbb{E} \left(\int_0^T H_s^2 X_s^2 ds \right) < \infty,$$

and let $V_0 \in \mathbb{R}$. There exists a unique process $(H_t^0)_{0 \leq t \leq T}$ such that the pair $((H_t^0, H_t))_{0 \leq t \leq T}$ defines an admissible strategy with initial value V_0. The discounted value at time t of this strategy is given by

$$\tilde{V}_t = V_0 + \int_0^t H_s \tilde{X}_s \sigma dW_s + \sum_{j=1}^{N_t} H_{\tau_j} U_j \tilde{X}_{\tau_j^-} - \lambda \mathbb{E}(U_1) \int_0^t ds H_s \tilde{X}_s.$$

Proof. If the pair $\left(H_t^0, H_t\right)_{0 \leq t \leq T}$ defines an admissible strategy, its value at time t is given by $V_t = Y_t + Z_t$, with $Y_t = V_0 + \int_0^t H_s^0 r e^{rs} ds + \int_0^t H_s X_s(\mu ds + \sigma dW_s)$ and $Z_t = \sum_{j=1}^{N_t} H_{\tau_j} U_j X_{\tau_j^-}$. Differentiating the product $e^{-rt} Y_t$, we have

$$e^{-rt} V_t = V_0 + \int_0^t (-re^{-rs}) Y_s ds + \int_0^t e^{-rs} dY_s + e^{-rt} Z_t. \qquad (7.3)$$

Moreover, the product $e^{-rt} Z_t$ can be written as follows:

$$e^{-rt} Z_t = \sum_{j=1}^{N_t} e^{-rt} H_{\tau_j} U_j X_{\tau_j^-}$$

$$= \sum_{j=1}^{N_t} \left(e^{-r\tau_j} + \int_{\tau_j}^t (-re^{-rs}) ds \right) H_{\tau_j} U_j X_{\tau_j^-}$$

$$= \sum_{j=1}^{N_t} e^{-r\tau_j} H_{\tau_j} U_j X_{\tau_j^-}$$

$$+ \sum_{j=1}^{N_t} \int_0^t ds 1_{\{\tau_j \leq s\}} (-re^{-rs}) H_{\tau_j} U_j X_{\tau_j^-}$$

$$= \sum_{j=1}^{N_t} e^{-r\tau_j} H_{\tau_j} U_j X_{\tau_j^-} + \int_0^t ds(-re^{-rs}) \sum_{j=1}^{N_s} H_{\tau_j} U_j X_{\tau_j^-}$$

$$= \sum_{j=1}^{N_t} e^{-r\tau_j} H_{\tau_j} U_j X_{\tau_j^-} + \int_0^t (-re^{-rs}) Z_s ds.$$

Writing this in (7.3) and expressing dY_s, we obtain

$$\tilde{V}_t = V_0 + \int_0^t (-re^{-rs}) V_s ds + \int_0^t H_s^0 r ds + \int_0^t H_s \tilde{X}_s(\mu ds + \sigma dW_s)$$

$$+ \sum_{j=1}^{N_t} H_{\tau_j} U_j \tilde{X}_{\tau_j^-}$$

$$= V_0 - \int_0^t r \left(H_s^0 + H_s \tilde{X}_s \right) ds + \int_0^t H_s^0 r ds + \int_0^t H_s \tilde{X}_s(\mu ds + \sigma dW_s)$$

$$+ \sum_{j=1}^{N_t} H_{\tau_j} U_j \tilde{X}_{\tau_j^-}$$

$$= V_0 + \int_0^t H_s \tilde{X}_s((\mu - r) ds + \sigma dW_s) + \sum_{j=1}^{N_t} H_{\tau_j} U_j \tilde{X}_{\tau_j^-},$$

which, taking into account equality (7.1), yields

$$\tilde{V}_t = V_0 + \int_0^t H_s \tilde{X}_s \sigma dW_s + \sum_{j=1}^{N_t} H_{\tau_j} U_j \tilde{X}_{\tau_j^-} - \lambda \mathbb{E}(U_1) \int_0^t ds H_s \tilde{X}_s.$$

It is clear then that if V_0 and (H_t) are given, the unique process (H_t^0) such that $((H_t^0, H_t))_{0 \le t \le T}$ is an admissible strategy with initial value V_0 is given by

$$H_t^0 = \tilde{V}_t - H_t \tilde{X}_t$$

$$= -H_t \tilde{X}_t + V_0 + \int_0^t H_s \tilde{X}_s \sigma dW_s + \sum_{j=1}^{N_t} H_{\tau_j} U_j \tilde{X}_{\tau_j^-} - \lambda \mathbb{E}(U_1) \int_0^t ds H_s \tilde{X}_s.$$

From this formula, we see that the process (H_t^0) is adapted, has a left-hand limit at any point and is such that $H_t^0 = H_{t-}^0$. This last property is straightforward if t is not a jump time τ_j, and if t is some τ_j, we have

$$H_{\tau_j}^0 - H_{\tau_j^-}^0 = -H_{\tau_j} \Delta \tilde{X}_{\tau_j} + H_{\tau_j} U_j \tilde{X}_{\tau_j^-} = 0.$$

It is also obvious that $\int_0^T |H_t^0| dt < \infty$ almost surely. Moreover, writing $H_t^0 e^{rt} + H_t X_t = e^{rt} \left(H_t^0 + H_t \tilde{X}_t \right)$ and integrating by parts as above, we see that the pair $(H_t^0, H_t)_{0 \le t \le T}$ defines an admissible strategy with initial value V_0. $\square$

Remark 7.4.3. The condition $\mathbb{E} \left(\int_0^T H_s^2 \tilde{X}_s^2 ds \right) < \infty$ implies that the discounted value $(\tilde{V}_t)$ of an admissible strategy is a square-integrable martingale. This results from the expression in Proposition 7.4.2 and Lemma 7.3.3, applied with the $\mathbb{R}^2$-valued continuous process with left-hand limits defined by $Y_t = (H_t, \tilde{X}_{t-})$ (note that in the integral with respect to ds, one can substitute $\tilde{X}_s$ for $\tilde{X}_{s-}$ because there are only finitely many discontinuities).

7.4.2 Pricing

Let us consider a European option with maturity T, defined by a random variable h, $\mathscr{F}_T$-measurable and square-integrable. To be more specific, let us stand from the writer's point of view. He sells the option at a price V_0 at time 0 and then follows an admissible strategy between times 0 and T. According to Proposition 7.4.2, this strategy is completely determined by the process $(H_t)_{0 \le t \le T}$ representing the amount of the risky asset. If V_t represents the value of this strategy at time t, the hedging mismatch at maturity is given by $h - V_T$. If this quantity is non-negative, the writer of the option loses money, otherwise he earns some. A way of measuring the risk is to introduce the quantity

$$R_0^T = \mathbb{E} \left((e^{-rT}(h - V_T))^2 \right).$$

Since, according to Remark 7.4.3, the discounted value $(\tilde{V}_t)$ is a martingale, we have $\mathbb{E}(e^{-rT} V_T) = V_0$. Applying the identity $\mathbb{E}(Z^2) = (\mathbb{E}(Z))^2 + \mathbb{E}([Z - \mathbb{E}(Z)]^2)$ to the random variable $Z = e^{-rT}(h - V_T)$, we obtain

$$R_0^T = \left(\mathbb{E}(e^{-rT} h) - V_0 \right)^2 + \mathbb{E} \left[e^{-rT} h - \mathbb{E}(e^{-rT} h) - (\tilde{V}_T - V_0) \right]^2. \qquad (7.4)$$

Proposition 7.4.2 shows that the quantity $\tilde{V}_T - V_0$ depends only on the process (H_t) (and not on V_0). If the writer of the option tries to minimize the risk R_0^T, he will ask for a premium $V_0 = \mathbb{E}(e^{-rT}h)$. So it appears that $\mathbb{E}(e^{-rT}h)$ is the initial value of any strategy designed to minimize the risk at maturity, and this is what we will take as the definition of the price of the option associated with h. By a similar argument, we see that an agent selling the option at time $t > 0$, who wants to minimize the quantity $R_t^T = \mathbb{E}\left((e^{-r(T-t)}(h - V_T))^2|\mathscr{F}_t\right)$, will ask for a premium $V_t = \mathbb{E}(e^{-r(T-t)}h|\mathscr{F}_t)$. We will take this quantity to define the price of the option at time t.

7.4.3 Prices of calls and puts

Before tackling the problem of hedging, we try to give an explicit expression for the price of the call or the put with strike price K. We will assume, therefore, that h can be written as $f(X_T)$, with $f(x) = (x - K)_+$ or $f(x) = (K - x)_+$. As we saw earlier, the price of the option at time t is given by

$$\mathbb{E}(e^{-r(T-t)}f(X_T)|\mathscr{F}_t)$$

$$= \mathbb{E}\left(e^{-r(T-t)}f\left(X_t e^{(\mu-(\sigma^2/2))(T-t)+\sigma(W_T-W_t)}\prod_{j=N_t+1}^{N_T}(1+U_j)\right)\Bigg|\mathscr{F}_t\right)$$

$$= \mathbb{E}\left(e^{-r(T-t)}f\left(X_t e^{(\mu-(\sigma^2/2))(T-t)+\sigma(W_T-W_t)}\prod_{j=1}^{N_T-N_t}(1+U_{N_t+j})\right)\Bigg|\mathscr{F}_t\right).$$

From Lemma 7.3.2 and this equality, we deduce that

$$\mathbb{E}\left(e^{-r(T-t)}f(X_T)|\mathscr{F}_t\right) = F(t, X_t),$$

with

$$F(t,x) = \mathbb{E}\left(e^{-r(T-t)}f\left(xe^{(\mu-(\sigma^2/2))(T-t)+\sigma W_{T-t}}\prod_{j=1}^{N_{T-t}}(1+U_j)\right)\right)$$

$$= \mathbb{E}\left(e^{-r(T-t)}f\left(xe^{(r-\lambda\mathbb{E}(U_1)-(\sigma^2/2))(T-t)+\sigma W_{T-t}}\prod_{j=1}^{N_{T-t}}(1+U_j)\right)\right).$$

Note that if we introduce the function

$$F_0(t,x) = \mathbb{E}\left(e^{-r(T-t)}f\left(xe^{(r-(\sigma^2/2))(T-t)+\sigma W_{T-t}}\right)\right),$$

which gives the price of the option for the Black-Scholes model, we have

$$F(t,x) = \mathbb{E}\left(F_0\left(t, xe^{-\lambda(T-t)\mathbb{E}(U_1)}\prod_{j=1}^{N_{T-t}}(1+U_j)\right)\right). \tag{7.5}$$

Since the random variable N_{T-t} is independent of the U_j's, and has a Poisson distribution with parameter $\lambda(T-t)$, we can also write

$$F(t,x) = \sum_{n=0}^{\infty} \mathbb{E}\left(F_0\left(t, xe^{-\lambda(T-t)\mathbb{E}(U_1)} \prod_{j=1}^{n}(1+U_j)\right)\right) \frac{e^{-\lambda(T-t)}\lambda^n(T-t)^n}{n!}.$$

Each term of this series can be computed numerically if we know how to simulate the law of the U_j's. For some laws, the mathematical expectation in the formula can be calculated explicitly (cf. Exercise 47).

7.4.4 Hedging calls and puts

Let us examine the hedging problem for an option $h = f(X_T)$, with $f(x) = (x - K)_+$ or $f(x) = (K - x)_+$. We have seen that the initial value of any admissible strategy aiming at minimizing the risk R_0^T at maturity is given by $V_0 = \mathbb{E}(e^{-rT}h) = F(0, X_0)$. For such a strategy, equality (7.4) becomes

$$R_0^T = \mathbb{E}(e^{-rT}h - \tilde{V}_T)^2.$$

Now we want to determine a process $(H_t)_{0 \leq t \leq T}$ describing the number of shares of the risky asset over the period $[0, T]$, so as to minimize R_0^T. To do so, we need the following proposition.

Proposition 7.4.4. *Consider an admissible strategy $(H_t^0, H_t)_{0 \leq t \leq T}$, with value V_t at time t. Assume that the initial value satisfies $V_0 = \mathbb{E}(e^{-rT}f(X_T)) = F(0, X_0)$. Then, the quadratic risk at maturity $R_0^T = \mathbb{E}(e^{-rT}(f(X_T) - V_T))^2$ is given by the following formula:*

$$R_0^T = \mathbb{E}\left(\int_0^T \left(\frac{\partial F}{\partial x}(s, X_s) - H_s\right)^2 \tilde{X}_s^2 \sigma^2 ds \right.$$

$$\left. + \int_0^T \lambda \int d\nu(z) e^{-2rs} \left(F(s, X_s(1+z)) - F(s, X_s) - H_s z X_s\right)^2 ds\right).$$

Proof. From Proposition 7.4.2, we have, for $t \leq T$,

$$\tilde{V}_t = F(0, X_0) + \int_0^t \sigma H_s \tilde{X}_s dW_s + \sum_{j=1}^{N_t} H_{T_j} U_j \tilde{X}_{T_j^-} - \lambda \int_0^t ds \tilde{X}_s H_s \mathbb{E}(U_1).$$

$$(7.6)$$

On the other hand, we have $\tilde{h} = e^{-rT}f(X_T) = e^{-rT}F(T, X_T)$. Let us introduce the function $\tilde{F}$ defined by

$$\tilde{F}(t, x) = e^{-rt}F(t, xe^{rt}),$$

so that $\tilde{F}(t, \tilde{X}_t) = \mathbb{E}(\tilde{h} \mid \mathscr{F}_t)$. Note that $\tilde{F}(t, \tilde{X}_t)$ is the discounted price of the option at time t. We deduce easily (exercise) from formula (7.5) that $\tilde{F}(t, x)$

is C^2 on $[0, T) \times \mathbb{R}^+$ and, writing down Itô's formula between the jump times, we obtain

$$
\begin{aligned}
\tilde{F}(t, \tilde{X}_t) = F(0, X_0) &+ \int_0^t \frac{\partial \tilde{F}}{\partial s}(s, \tilde{X}_s)ds \\
&+ \int_0^t \frac{\partial \tilde{F}}{\partial x}(s, \tilde{X}_s)\tilde{X}_s(-\lambda\mathbb{E}(U_1)ds + \sigma dW_s) \\
&+ \frac{1}{2}\int_0^t \frac{\partial^2 \tilde{F}}{\partial x^2}(s, \tilde{X}_s)\sigma^2\tilde{X}_s^2 ds + \sum_{j=1}^{N_t} \tilde{F}(\tau_j, \tilde{X}_{\tau_j}) - \tilde{F}(\tau_j, \tilde{X}_{\tau_j^-}).
\end{aligned}
\tag{7.7}
$$

Observe that the function $\tilde{F}(t, x)$ is Lipschitz-continuous with respect to x. Indeed, using the notation $\alpha = r - \lambda\mathbb{E}(U_1) - (\sigma^2/2)$, $\theta = T - t$ and $P_t = \prod_{j=1}^{N_t}(1 + U_j)$, we have

$$
\begin{aligned}
|F(t, x) - F(t, y)| &\leq \mathbb{E}\left(e^{-r\theta}\left|f\left(xe^{\alpha\theta + \sigma W_\theta}P_\theta\right) - f\left(ye^{\alpha\theta + \sigma W_\theta}P_\theta\right)\right|\right) \\
&\leq |x - y|\,\mathbb{E}\left(e^{\sigma W_{T-t} - (\sigma^2/2)(T-t)}e^{-\lambda\mathbb{E}(U_1)(T-t)}P_{T-t}\right) \\
&= |x - y|.
\end{aligned}
$$

It follows that

$$
\begin{aligned}
\mathbb{E}&\left(\int_0^t ds \int d\nu(z)(\tilde{F}(s, \tilde{X}_s(1 + z)) - \tilde{F}(s, \tilde{X}_s))^2\right) \\
&\leq \mathbb{E}\left(\int_0^t ds\tilde{X}_s^2 \int d\nu(z)z^2\right) < +\infty,
\end{aligned}
$$

which, according to Lemma 7.3.3, implies that the process

$$
\begin{aligned}
M_t = \sum_{j=1}^{N_t} \tilde{F}(\tau_j, \tilde{X}_{\tau_j}) - \tilde{F}(\tau_j, \tilde{X}_{\tau_j^-}) \\
- \lambda\int_0^t ds \int (\tilde{F}(s, \tilde{X}_s(1 + z)) - \tilde{F}(s, \tilde{X}_s))d\nu(z)
\end{aligned}
$$

is a square-integrable martingale. We also know that $\tilde{F}(t, \tilde{X}_t)$ is a martingale. Therefore, the process $\tilde{F}(t, \tilde{X}_t) - M_t$ is also a martingale and, from equality (7.7), it is an Itô process. From Exercise 19 of Chapter 3, it can be written as a stochastic integral. Whence

$$
\tilde{F}(t, \tilde{X}_t) - M_t = F(0, X_0) + \int_0^t \frac{\partial \tilde{F}}{\partial x}(s, \tilde{X}_s)\tilde{X}_s\sigma dW_s.
\tag{7.8}
$$

Putting (7.6) and (7.8) together, we get

$$
\tilde{h} - \tilde{V}_T = \bar{M}_T + \hat{M}_T,
$$

with

$$\bar{M}_t = \int_0^t \left(\frac{\partial \tilde{F}}{\partial x}(s, \tilde{X}_s) - H_s \right) \sigma \tilde{X}_s dW_s$$

and

$$\hat{M}_t = \sum_{j=1}^{N_t} \left(\tilde{F}(\tau_j, \tilde{X}_{\tau_j}) - \tilde{F}(\tau_j, \tilde{X}_{\tau_j^-}) - H_{\tau_j} U_j \tilde{X}_{\tau_j^-} \right)$$

$$- \lambda \int_0^t ds \int d\nu(z)(\tilde{F}(s, \tilde{X}_s(1+z)) - \tilde{F}(s, \tilde{X}_s) - H_s z \tilde{X}_s).$$

According to Lemma 7.3.4, $\bar{M}_t \hat{M}_t$ is a martingale and consequently

$$\mathbb{E}\left(\bar{M}_t \hat{M}_t \right) = \bar{M}_0 \hat{M}_0 = 0.$$

Whence

$$\mathbb{E}\left(\tilde{h} - \tilde{V}_T \right)^2 = \mathbb{E}\left(\bar{M}_T^2 \right) + \mathbb{E}\left(\hat{M}_T^2 \right)$$

$$= \mathbb{E}\left(\int_0^T \left(\frac{\partial \tilde{F}}{\partial x}(s, \tilde{X}_s) - H_s \right)^2 \tilde{X}_s^2 \sigma^2 ds \right) + \mathbb{E}\left(\hat{M}_T^2 \right),$$

and, applying Lemma 7.3.3 again,

$$\mathbb{E}\left(\hat{M}_T^2 \right) = \mathbb{E}\left(\lambda \int_0^T ds \int d\nu(z) \left(\tilde{F}(s, \tilde{X}_s(1+z)) - \tilde{F}(s, \tilde{X}_s) - H_s z \tilde{X}_s \right)^2 \right).$$

The risk at maturity is then given by

$$R_0^T = \mathbb{E}\left(\int_0^T \left(\frac{\partial \tilde{F}}{\partial x}(s, \tilde{X}_s) - H_s \right)^2 \tilde{X}_s^2 \sigma^2 ds \right.$$

$$\left. + \int_0^T \lambda \int d\nu(z) \left(\tilde{F}(s, \tilde{X}_s(1+z)) - \tilde{F}(s, \tilde{X}_s) - H_s z \tilde{X}_s \right)^2 ds \right).$$

$\square$

It follows from Proposition 7.4.4 that the minimal risk is obtained when H_s satisfies $\mathbb{P}$ a.s.:

$$\left(\frac{\partial \tilde{F}}{\partial x}(s, \tilde{X}_s) - H_s \right) \tilde{X}_s^2 \sigma^2 +$$

$$\lambda \int d\nu(z) \left(\tilde{F}(s, \tilde{X}_s(1+z)) - \tilde{F}(s, \tilde{X}_s) - H_s z \tilde{X}_s \right) z \tilde{X}_s = 0.$$

It suffices indeed to minimize the integrand with respect to ds. This yields, since $(H_t)_{0 \leq t \leq T}$ must be left-continuous,

$$H_s = \Delta(s, X_{s-}),$$

with

$$\Delta(s,x) = \frac{1}{\hat{\sigma}^2} \left(\sigma^2 \frac{\partial F}{\partial x}(s,x) + \lambda \int d\nu(z) z \frac{(F(s,x(1+z)) - F(s,x))}{x} \right),$$

where $\hat{\sigma}^2 = \sigma^2 + \lambda \int d\nu(z) z^2$. In this way, we obtain a process that satisfies $\mathbb{E}\left(\int_0^T H_s^2 \tilde{X}_s^2 ds \right) < +\infty$ and which determines therefore an admissible strategy minimizing the risk at maturity. Note that if there is no jump ($\lambda = 0$), we recover the hedging formula for the Black-Scholes model and, in this case, we know that the hedging is perfect, i.e. $R_0^T = 0$. But, when there are jumps, the minimal risk is generally positive (cf. Exercise 48 and Chateau (1997)).

Remark 7.4.5. In practice, the parameters of the model (including the law of the U_i's) need to be identified. As for the determination of the volatility in the Black-Scholes model, we can distinguish two approaches: (1) a statistical approach, from historical data, and (2) an implied approach, from market data, in other words, from the prices of options quoted on an organised market. In the second approach, the models with jumps, which involve several parameters, give a better fit to market prices than the Black-Scholes model.

Notes: Financial models with jumps were introduced by Merton (1976). The approach used in this chapter is based on Föllmer and Sondermann (1986), CERMA (1988) and Bouleau and Lamberton (1989). In our presentation, the hedging strategy is defined in order to minimize the variance under a risk neutral measure. Risk minimizing strategies under the historical probability measure have been investigated in Föllmer and Schweizer (1991), Schweizer (1992, 1994, 1995), and more recently in Gourieroux, Laurent, and Pham (1998). The use of the variance as a measure of risk is questionable, and a number of other approaches can be introduced (see for instance El Karoui and Quenez (1995), Föllmer and Leukert (1999, 2000)). In recent years, general processes with stationary and independent increments (the so-called Lévy processes) have been intensively used in financial modelling (see the recent monograph by Cont and Tankov (2004)).

7.5 Exercises

Exercise 43 Let $(V_n)_{n \geq 1}$ be a sequence of non-negative, independent and identically distributed random variables and let N be a random variable with values in $\mathbb{N}$, following a Poisson distribution with parameter λ, independent of the sequence $(V_n)_{n \geq 1}$. Show that

$$\mathbb{E}\left(\prod_{n=1}^{N} V_n \right) = \exp\left(\lambda(\mathbb{E}(V_1) - 1) \right).$$

Exercise 44 Let $(V_n)_{n \geq 1}$ be a sequence of independent, identically distributed, integrable random variables and let N be a random variable taking values in $\mathbb{N}$, integrable and independent of the sequence (V_n). Let

$$S = \sum_{n=1}^{N} V_n$$

(with the convention $\sum_{n=1}^{0} = 0$).

1. Prove that S is integrable and that $\mathbb{E}(S) = \mathbb{E}(N)\mathbb{E}(V_1)$.

2. Assume N and V_1 are square-integrable. Then show that S is square-integrable and that its variance is

$$\mathrm{Var}(S) = \mathbb{E}(N)\mathrm{Var}(V_1) + \mathrm{Var}(N)(\mathbb{E}(V_1))^2.$$

3. Deduce that if N follows a Poisson law with parameter λ, $\mathbb{E}(S) = \lambda\mathbb{E}(V_1)$ and $\mathrm{Var}(S) = \lambda\mathbb{E}\left(V_1^2\right)$.

Exercise 45   The hypotheses and notations are those in Exercise 44. Suppose that the V_j's take values in $\{\alpha, \beta\}$, with $\alpha, \beta, \in \mathbb{R}$, and let $p = \mathbb{P}(V_1 = \alpha) = 1 - \mathbb{P}(V_1 = \beta)$. Prove that S has the same law as $\alpha N_1 + \beta N_2$, where N_1 and N_2 are two independent random variables with Poisson distributions with respective parameters λp and $\lambda(1 - p)$.

Exercise 46   Consider a probability space $(\Omega, \mathscr{A}, \mathbb{P})$, on which we have a standard Brownian motion $(W_t)_{t\geq 0}$, a Poisson process with intensity $\lambda > 0$ $(N_t)_{t\geq 0}$ and a sequence $(U_j)_{j\geq 1}$ of independant, identically distributed, real valued integrable random variables. Assume that the σ-algebras $\mathscr{W}$, $\mathscr{N}$ and $\mathscr{U}$, respectively, generated by $(W_t)_{0\leq t\leq T}$, $(N_t)_{t\geq 0}$ and $(U_j)_{j\geq 1}$ are independent. For $t \geq 0$, define $\mathscr{F}_t$ as in Section 7.2. For $\theta \in \mathbb{R}$ and $u > -1$, let

$$Z_t^{\theta,u} = e^{\theta W_t - \frac{\theta^2}{2}t}(1 + u)^{N_t}e^{-\lambda ut}.$$

1. Prove that $(Z_t^{\theta,u})_{t\geq 0}$ is a martingale.

2. Fix $T > 0$ and define the probability $\hat{\mathbb{P}}$ with density $d\hat{\mathbb{P}}/d\mathbb{P} = Z_T^{\theta,u}$. Prove that, under $\hat{\mathbb{P}}$, the σ-algebras $\mathscr{W}$, $\mathscr{N}$ and $\mathscr{U}$ remain independent, and that the process $(N_t)_{0\leq t\leq T}$ is a Poisson process with intensity $\hat{\lambda} = \lambda(1 + u)$.

3. In the jump-diffusion model, the discounted price at time t is given by $\tilde{X}_t = X_0 e^{(\mu - r - (\sigma^2/2))t + \sigma W_t} \prod_{j=1}^{N_t}(1 + U_j)$. Prove that $(\tilde{X}_t)_{0\leq t\leq T}$ is a $\hat{\mathbb{P}}$-martingale if and only if

$$\mu + \sigma\theta = r - \lambda(1 + u)\mathbb{E}(U_1).$$

If $\sigma > 0$ and $\mathbb{E}(U_1) \neq 0$, this equality is satisfied for infinitely many pairs (θ, u). Therefore, the set of risk neutral probability measures is infinite.

Exercise 47

1. Suppose, with the notations of Section 7.4, that U_1 takes values in $\{a, b\}$, with $p = \mathbb{P}(U_1 = a) = 1 - \mathbb{P}(U_1 = b)$. Write the price formula (7.5) as a double series where each term is calculated from the Black-Scholes formulae. (Hint: use Exercise 45.)

2. Suppose that U_1 has the same law as $e^g - 1$, where g is a normal variable with mean m and variance σ^2. Write the price formula (7.5) as a series of terms calculated from the Black-Scholes formulae (for some interest rates and volatilities to be given).

Exercise 48 The objective of this exercise is to show that there is no perfect hedging of calls and puts for the models with jumps we studied in this chapter. We consider a model in which $\sigma > 0, \lambda > 0$ and $\mathbb{P}(U_1 \neq 0) > 0$.

1. Using Proposition 7.4.4, show that if there is a perfect hedging scheme, then, for ds almost every s and for $d\nu$ almost every z, we have

$$\mathbb{P} \text{ a.s.} \quad zX_s \frac{\partial F}{\partial x}(s, X_s) = F(s, X_s(1 + z)) - F(s, X_s).$$

2. Show that, for $s > 0$, the random variable X_s has a positive density on $[0, \infty)$. (Hint: argue that if Y has a density g and if Z is a random variable independent of Y with values in $(0, \infty)$, the random variable YZ has the density $\int d\mu(z)(1/z)g(y/z)$, where μ is the law of Z.)

3. Under the same assumptions as in the first question, show that there exists $z \neq 0$ such that for $s \in [0, T)$ and $x \in (0, \infty)$,

$$\frac{\partial F}{\partial x}(s, x) = \frac{F(s, x(1 + z)) - F(s, x)}{zx}.$$

Deduce (using the convexity of F with respect to x) that for $s \in [0, T]$, the function $x \mapsto F(s, x)$ is linear.

4. Conclude. (Hint: note that, in the case of the put, the function $x \mapsto F(s, x)$ is non-negative and decreasing on $(0, \infty)$.)

Chapter 8

Credit risk models

In the last few years, the market of credit derivative instruments has developped dramatically. This chapter offers a rather sketchy introduction to the basic concepts of credit risk modelling. Credit risk is associated with the risk of default of a counterparty. In the first section, we present the so-called *structural* models in which the default time is defined as a stopping time with respect to the filtration of the *firm value*. In the second section, we introduce intensity models, which consider the default time as an exogenous random time, characterized by its *hazard rate*. We then describe the valuation of credit default swaps (CDS). The last section is devoted to the concept of *copula*, which is very useful in models involving several default times. For more information on credit risk models, we refer the reader to the recent second edition of Brigo and Mercurio (2006) and, for mathematical developments, to Bielecki and Rutkowski (2002).

8.1 Structural models

Structural models (also called firm value models) propose to model a default event by relating it to the *value* of the firm. We will limit our presentation to Merton's model (see Merton (1974)), which appears as the pioneering model of this approach.

In this model, the value V_t of the firm at time t follows a geometric Brownian motion. The debt of the company is modeled by a zero-coupon bond with maturity T and face value $L > 0$. More precisely, we assume that

$$dV_t = V_t \left((r - k)dt + \sigma dW_t\right), \quad V_0 > 0, \tag{8.1}$$

where $(W_t)_{t \geq 0}$ is a standard Brownian motion under the risk neutral probability measure, which, in this section, we denote by $\mathbb{P}$. The constant r is the instantaneous interest rate and the constant k is an expenditure rate. The volatility $\sigma > 0$ is also assumed constant. In this model, the company defaults if, at time T, the firm value is smaller than its debt. In case of default, the debtors (represented by the bond holder) take control of the company. In other words, the bond holder's payoff at maturity is $V_T \wedge L$. The value at time $t < T$ of this *defaultable* coupon, which can be seen as the *debt* value of the firm, is given by

$$D_t = \mathbb{E}\left(e^{-r(T-t)} V_T \wedge L \mid \mathscr{F}_t\right).$$

Here, the filtration $(\mathscr{F}_t)_{0 \le t \le T}$ is the natural filtration of the process $(V_t)_{0 \le t \le T}$ (and, also, the natural filtration of the Brownian motion). Note that, since $V_T \wedge L = L - (L - V_T)_+$, we have

$$D_t = L e^{-r(T-t)} - \mathbb{E}\left(e^{-r(T-t)}(L - V_T)_+ \mid \mathscr{F}_t\right),$$

so that the computation of D_t is equivalent to the computation of the price of a put option on the firm value (see Exercise 50). In this setting, the *equity* value of the firm at time T (denoted by E_T) is the difference between the firm value and the debt value,

$$E_T = V_T - V_T \wedge L = (V_T - L)_+,$$

so that the shareholders of the firm can be viewed as the holder of a call option on the firm value of the company.

In Merton's model, default may occur at the deterministic time T only. In *first passage* models, default occurs when the firm value hits a possibly time-dependent barrier $H(t)$. The default time is then given by

$$\tau = \inf\{t \ge 0 \mid V_t \le H(t)\}. \tag{8.2}$$

Black and Cox (1976) proposed a model with $H(T) = L$ and $H(t) = \tilde{L} e^{-\gamma(T-t)}$, for $t \in [0, T)$, where $\tilde{L} < L$ and γ is a positive constant. This choice implies that the firm has more flexibility to face default before maturity. Within this model, the computation of the price of a defaultable zero-coupon bond is similar to the pricing of a barrier option (see Exercise 51).

8.2   Intensity-based models

8.2.1   Hazard rate of a random time

Proposition 8.2.1. *Let τ be a real random variable with $\mathbb{P}(\tau > 0) = 1$ and $\mathbb{P}(\tau > t) > 0$ for all $t > 0$. If the cumulative distribution function of τ is of class C^1, there exists a unique, continuous non-negative function λ such that*

$$\forall t > 0, \quad \mathbb{P}(\tau > t) = e^{-\int_0^t \lambda(s)ds}.$$

The function λ is called the *hazard rate* of the random time τ.

Proof. Denote by F the distribution function of τ. We have $\mathbb{P}(\tau > t) = 1 - F(t)$. The function $t \mapsto \ln(1 - F(t))$ is continuously differentiable and non-increasing and the function λ must satisfy

$$\lambda(t) = -\frac{d}{dt} \ln(1 - F(t)) = \frac{F'(t)}{1 - F(t)}, \quad t \geq 0.$$

With this definition of λ, we do have, using $\mathbb{P}(\tau > 0) = 1$,

$$\mathbb{P}(\tau > t) = e^{-\ln(1 - F(t))} = e^{-\int_0^t \lambda(s)ds}.$$

$\square$

Note that in order for the random time τ to be finite, the hazard rate must satisfy $\int_0^{+\infty} \lambda(t)dt = +\infty$. We also have, for $\delta > 0$,

$$\mathbb{P}(\tau \leq t + \delta \mid \tau > t) = \frac{\mathbb{P}(t < \tau \leq t + \delta)}{\mathbb{P}(\tau > t)} = 1 - e^{-\int_t^{t+\delta} \lambda(s)ds},$$

so that

$$\lambda(t) = \lim_{\delta \to 0} \frac{\mathbb{P}(\tau > t + \delta \mid \tau > t)}{\delta}.$$

If τ models a default time, its hazard rate at time t can thus be seen as the rate of occurence of default just after time t, given default has not occurred before time t.

Remark 8.2.2. A random time τ has an exponential distribution if and only if its hazard rate is a constant. On the other hand, if ξ is an exponential random variable with parameter 1, and if λ is a non-negative continuous function on $[0, +\infty)$, the random variable

$$\tau := \inf \left\{ t \geq 0 \mid \int_0^t \lambda(s)ds \geq \xi \right\} \tag{8.1}$$

satisfies

$$\mathbb{P}(\tau \leq t) = \mathbb{P}\left(\int_0^t \lambda(s)ds \geq \xi \right) = 1 - e^{\int_0^t \lambda(s)ds},$$

so that the hazard rate of τ is λ, and (8.1) provides a way of constructing a random time with a given hazard rate.

8.2.2 Intensity and defaultable zero-coupon bonds

The general framework of intensity models can be described as follows. Let $(\Omega, \mathscr{A}, \mathbb{P})$ be a probability space. The information relative to the default free market is denoted by $(\mathscr{F}_t)_{t \geq 0}$. Typically, this filtration incorporates the history of interest rates, so that if we denote by (S_t^0) the price process of the riskless asset and by $(r(t))$ the instantaneous interest rate process (recall that $S_t^0 = \exp(\int_0^t r(s)ds)$), these two processes are $(\mathscr{F}_t)$-adapted.

The default time is modelled by a random variable τ, which may *not* be a stopping time with respect to the default free filtration $(\mathscr{F}_t)$. This is in contrast to the structural approach. In the intensity approach, the default time appears as an exogenous variable and default may occur as a surprise.

On the other hand, at a given date t, investors know if default has occurred or not. Therefore, the total information available at time t can be described by the σ-field $\mathscr{G}_t$, generated by $\mathscr{F}_t$ together with the events $\{\tau \leq s\}$, for $s \in [0, t]$:

$$\mathscr{G}_t = \mathscr{F}_t \vee \sigma(\{\tau \leq s\}, s \leq t).$$

Note that τ is a stopping time with respect to the filtration $(\mathscr{G}_t)$. A *pricing measure* in this setting is a probability measure $\mathbb{P}^*$, equivalent to $\mathbb{P}$, under which discounted prices of risky assets are $(\mathscr{G}_t)$-martingales. Note that the prices of risky assets that may be affected by default are adapted to the filtration $(\mathscr{G}_t)$. Now, consider a *defaultable zero-coupon bond*, with maturity T, which, at time T, pays one unit of currency if default has not occurred and nothing in case of default before or at time T. The payoff of this derivative at time T is given by $\mathbf{1}_{\{\tau > T\}}$. So its value at time t should be given by the conditional expectation, given $\mathscr{G}_t$, with respect to the pricing measure of the discounted payoff, i.e.

$$D(t, T) = \mathbb{E}^* \left(\mathbf{1}_{\{\tau > T\}} e^{-\int_t^T r(s)ds} \,\middle|\, \mathscr{G}_t \right), \quad 0 \leq t \leq T.$$

The following proposition relates the computation of conditional expectations given $\mathscr{G}_t$ to conditional expectations given $\mathscr{F}_t$.

Proposition 8.2.3. *For any non-negative random variable X, we have, with probability one,*

$$\mathbf{1}_{\{\tau > t\}} \mathbb{E}^*(X \mid \mathscr{G}_t) = \mathbf{1}_{\{\tau > t\}} \frac{\mathbb{E}^* \left(X \mathbf{1}_{\{\tau > t\}} \mid \mathscr{F}_t \right)}{\mathbb{E}^* \left(\mathbf{1}_{\{\tau > t\}} \mid \mathscr{F}_t \right)}.$$

Proof. Observe that, on the set $\{\tau > t\}$, we have $\mathbb{E}^* \left(\mathbf{1}_{\{\tau > t\}} \mid \mathscr{F}_t \right) > 0$ almost surely (see Exercise 49), so that the random variable

$$Y := \mathbf{1}_{\{\tau > t\}} \frac{\mathbb{E}^* \left(X \mathbf{1}_{\{\tau > t\}} \mid \mathscr{F}_t \right)}{\mathbb{E}^* \left(\mathbf{1}_{\{\tau > t\}} \mid \mathscr{F}_t \right)}$$

is well defined and is $\mathscr{G}_t$-measurable (as the product of two $\mathscr{G}_t$-measurable random variables). On the other hand, we have

$$\mathbf{1}_{\{\tau > t\}} \mathbb{E}^*(X \mid \mathscr{G}_t) = \mathbb{E}^*(X \mathbf{1}_{\{\tau > t\}} \mid \mathscr{G}_t),$$

because $\{\tau > t\} \in \mathscr{G}_t$. In order to prove that $Y = \mathbb{E}^*(X \mathbf{1}_{\{\tau > t\}} \mid \mathscr{G}_t)$, it is enough to show that

$$\forall A \in \mathscr{C}, \quad \mathbb{E}^*(X \mathbf{1}_{\{\tau > t\}} \mathbf{1}_A) = \mathbb{E}^*(Y \mathbf{1}_A),$$

where $\mathscr{C}$ is a subclass of $\mathscr{G}_t$ that is stable under finite intersections and generates $\mathscr{G}_t$ (see, for instance, Jacod and Protter (2003), Chapter 6). Let $\mathscr{C}$ be the class of events A that can be written in the form

$$A = \{\tau \le s\} \cap B,$$

where $B \in \mathscr{F}_t$ and $s \in [0, t] \cup \{+\infty\}$. The class $\mathscr{C}$ is stable under finite intersections and generates $\mathscr{G}_t$. For $s \in [0, t]$, we have $\mathbb{E}^* \left(X 1_{\{\tau > t\}} 1_{\{\tau \le s\} \cap B} \right) = 0$ and $\mathbb{E}^* \left(Y 1_{\{\tau \le s\} \cap B} \right) = 0$. For $s = \infty$, we have $A = B$ and, since $B \in \mathscr{F}_t$,

$$\mathbb{E}^* \left(X 1_{\{\tau > t\}} 1_B \right) = \mathbb{E}^* \left(\mathbb{E}^* \left(X 1_{\{\tau > t\}} \mid \mathscr{F}_t \right) 1_B \right).$$

We also have

$$\mathbb{E}^* (Y 1_B) = \mathbb{E}^* \left(1_{\{\tau > t\}} \frac{\mathbb{E}^* \left(X 1_{\{\tau > t\}} \mid \mathscr{F}_t \right)}{\mathbb{E}^* \left(1_{\{\tau > t\}} \mid \mathscr{F}_t \right)} 1_B \right)$$

$$= \mathbb{E}^* \left[\mathbb{E}^* \left(1_{\{\tau > t\}} \mid \mathscr{F}_t \right) \frac{\mathbb{E}^* \left(X 1_{\{\tau > t\}} \mid \mathscr{F}_t \right)}{\mathbb{E}^* \left(1_{\{\tau > t\}} \mid \mathscr{F}_t \right)} 1_B \right]$$

$$= \mathbb{E}^* \left(X 1_{\{\tau > t\}} 1_B \right).$$

$\square$

Now, suppose that we know the *conditional hazard rate* of τ given the default free filtration. More precisely, assume that there exists a non-negative $(\mathscr{F}_t)$-adapted process $(\lambda(t))_{t \ge 0}$ such that, for all $t \ge 0$,

$$\mathbb{P}^* (\tau > t \mid \mathscr{F}_t) = \mathbb{E}^* \left(1_{\{\tau > t\}} \mid \mathscr{F}_t \right) = e^{-\int_0^t \lambda(s) ds}.$$

Using Proposition 8.2.3 with $X = 1_{\{\tau > T\}} e^{-\int_t^T r(s) ds}$, we obtain the value for the defaultable zero-coupon bond before default:

$$1_{\{\tau > t\}} D(t, T) = 1_{\{\tau > t\}} \frac{\mathbb{E}^* \left(1_{\{\tau > T\}} e^{-\int_t^T r(s) ds} 1_{\{\tau > t\}} \mid \mathscr{F}_t \right)}{\mathbb{E}^* \left(1_{\{\tau > t\}} \mid \mathscr{F}_t \right)}$$

$$= 1_{\{\tau > t\}} \frac{\mathbb{E}^* \left(1_{\{\tau > T\}} e^{-\int_t^T r(s) ds} \mid \mathscr{F}_t \right)}{e^{-\int_0^t \lambda(s) ds}}$$

$$= 1_{\{\tau > t\}} \frac{\mathbb{E}^* \left(e^{-\int_0^T \lambda(s) ds} e^{-\int_t^T r(s) ds} \mid \mathscr{F}_t \right)}{e^{-\int_0^t \lambda(s) ds}},$$

where the last equality follows from conditioning with respect to $\mathscr{F}_T$ and using the $\mathscr{F}_T$-measurability of $e^{-\int_t^T r(s) ds}$. We have proved the following result.

Proposition 8.2.4. *The value at time t of a defaultable zero-coupon bond with maturity T before default is given by*

$$\tilde{D}(t, T) = \mathbb{E}^* \left(e^{-\int_t^T (r(s) + \lambda(s)) ds} \mid \mathscr{F}_t \right), \quad 0 \le t \le T, \tag{8.2}$$

where $(\lambda(t))$ *is the conditional hazard rate of default given the default free filtration.*

The process $(\lambda(t))$ is also called the *intensity* of the default. Recall that the value of the default free zero-coupon bond is given by

$$P(t,T) = \mathbb{E}^* \left(e^{-\int_t^T r(s)ds} \mid \mathscr{F}_t \right),$$

so that the intensity appears as an additional term to the interest rate in (8.2). For this reason, $\lambda(t)$ is also called the *credit spread* at time t. In case of a *deterministic* intensity, we have

$$\tilde{D}(t,T) = e^{-\int_t^T \lambda(s)ds} P(t,T).$$

Remark 8.2.5. The intensity is deterministic if and only if, for each t, the event $\{\tau > t\}$ is independent of $\mathscr{F}_t$. In particular, this is satisfied if the random variable τ is independent of all the σ-algebras $\mathscr{F}_t$. This assumption is often used in practice. More realistic models assume that the dynamics of the pair $(r(t), \lambda(t))$ is governed by a stochastic differential equation, so that the computation of defaultable bond prices can be done using (8.2). Note that, given a continuous nonnegative $(\mathscr{F}_t)$ adapted process $\lambda = (\lambda(t))$, one can construct a random time with intensity λ, using formula (8.1), with ξ *independent of the default free filtration.*

8.2.3   Credit default swaps

A Credit default swap or CDS is a credit derivative that offers protection against default. An investor A, who wants to be protected against default, agrees with a bank B on the following. At dates $T_1, \ldots, T_n$, A will make fixed payments to B, as long as default has not occurred. On the other hand, in case of default before time T_n, B will make a payment to A. The cashflows paid by A and B respectively are described by the *premium leg* and the *protection leg* as follows.

- Premium leg: cashflow $sN(T_i - T_{i-1})\mathbf{1}_{\{\tau > T_i\}}$ at time T_i, $i = 1, \ldots, T_n$ (with the convention $T_0 = 0$). The number N is called the *nominal* of the swap, and s is the *spread* of the swap.

- Protection leg: cashflow $N(1-R)\mathbf{1}_{\{\tau \leq T_n\}}$ at time τ, where R is the *recovery* rate. The idea is that A holds a bond with nominal N issued by a company C that may default and that, in case of default, the holder of the bond recovers NR, instead of N. So the protection payment must be $N(1-R)$. The actual recovery rate is not known: the usual convention in practice is to take $R = 40\%$.

Remark 8.2.6. In the premium leg, we are omitting the so-called *accrual premium* payment $s(\tau - T_{\beta(\tau)})$, which corresponds to the premium for the period from the last payment date before default and until default ($\beta(\tau) = i - 1$ if $T_{i-1} \leq \tau < T_i$). See Brigo and Mercurio (2006) or Overhaus et al. (2007) for more details.

The premium leg can be evaluated at time 0 by taking the expectation of the discounted cashflows under the pricing measure:

$$P_{A \to B} = \mathbb{E}^* \left(\sum_{i=1}^{n} sN(T_i - T_{i-1}) \mathbf{1}_{\{\tau > T_i\}} e^{-\int_0^{T_i} r(s)ds} \right).$$

Similarly, the protection leg can be evaluated at time 0 by

$$P_{B \to A} = \mathbb{E}^* \left(N(1 - R) \mathbf{1}_{\{\tau \leq T_n\}} e^{-\int_0^{\tau} r(s)ds} \right).$$

The *fair value of the spread* makes these two quantities equal, so that

$$s = \frac{\mathbb{E}^* \left((1 - R) \mathbf{1}_{\{\tau \leq T_n\}} e^{-\int_0^{\tau} r(s)ds} \right)}{\mathbb{E}^* \left(\sum_{i=1}^{n} (T_i - T_{i-1}) \mathbf{1}_{\{\tau > T_i\}} e^{-\int_0^{T_i} r(s)ds} \right)}.$$

By conditioning with respect to $\mathscr{F}_{T_i}$ and using the intensity process, we have

$$\mathbb{E}^* \left(\mathbf{1}_{\{\tau > T_i\}} e^{-\int_0^{T_i} r(s)ds} \right) = \mathbb{E}^* \left(e^{-\int_0^{T_i} \lambda(s)ds} e^{-\int_0^{T_i} r(s)ds} \right).$$

Similarly, it can be proved (see Exercise 52) that

$$\mathbb{E}^* \left(\mathbf{1}_{\{\tau \leq T_n\}} e^{-\int_0^{\tau} r(s)ds} \right) = \mathbb{E}^* \left(\int_0^{T_n} \lambda(u) e^{-\int_0^{u} (\lambda(v) + r(v))dv} du \right).$$

It follows that the fair value of the spread is given by

$$s = \frac{(1 - R) \int_0^{T_n} \mathbb{E}^* \left(\lambda(u) e^{-\int_0^{u} (\lambda(v) + r(v))dv} \right) du}{\sum_{i=1}^{n} (T_i - T_{i-1}) \mathbb{E}^* \left(e^{-\int_0^{T_i} (\lambda(s) + r(s))ds} \right)}. \tag{8.3}$$

In the case of a *deterministic* intensity, we have, in terms of default free zero-coupon bond prices,

$$\mathbb{E}^* \left(e^{-\int_0^{T_i} (\lambda(s) + r(s))ds} \right) = e^{-\int_0^{T_i} \lambda(s)ds} P(0, T_i)$$
$$= \mathbb{P}^*(\tau > T_i) P(0, T_i)$$

and

$$\mathbb{E}^* \left(\lambda(u) e^{-\int_0^{u} (\lambda(v) + r(v))dv} \right) = \lambda(u) e^{-\int_0^{u} \lambda(s)ds} P(0, u).$$

Hence

$$\mathbb{E}^*\left(\int_0^{T_n} \lambda(u)e^{-\int_0^u (\lambda(v)+r(v))dv}du\right) = \int_0^{T_n} \lambda(u)e^{-\int_0^u \lambda(s)ds}P(0,u)du.$$

Going back to (8.3), we have

$$s = \frac{(1-R)\int_0^{T_n} \lambda(u)e^{-\int_0^u \lambda(s)ds}P(0,u)du}{\displaystyle\sum_{i=1}^n (T_i - T_{i-1})\mathbb{P}^*(\tau > T_i)P(0,T_i)}. \tag{8.4}$$

Remark 8.2.7. Since credit default swaps are the most liquid credit derivatives, they are often used for calibration purposes. In practice, the spreads of credit default swaps with various maturities are used to derive implied default probabilities. Denote by s_j the spread of a swap with payment dates $T_1,\ldots,$ T_j and use the following variant of (8.4):

$$s_j = \frac{(1-R)\displaystyle\sum_{i=1}^j \left(\mathbb{P}^*(\tau > T_{i-1}) - \mathbb{P}^*(\tau > T_i)\right)P(0,T_i)}{\displaystyle\sum_{i=1}^j (T_i - T_{i-1})\mathbb{P}^*(\tau > T_i)P(0,T_i)}. \tag{8.5}$$

Note that the numerator in (8.5) can be seen as the discretization of the integral in the numerator of (8.4), since

$$\mathbb{P}^*(\tau > T_{i-1}) - \mathbb{P}^*(\tau > T_i) = \int_{t_{i-1}}^{t_i} \lambda(u)e^{-\int_0^u \lambda(s)ds}du.$$

From the values of the spreads s_j $(j=1,\ldots,n)$, we can compute $\mathbb{P}^*(\tau > T_i)$ for $i=1,\ldots,n$ and define a piecewise constant deterministic hazard rate. In fact, (8.5) can also be seen as the spread, assuming that, in case of default, the protection payment is not made at time τ, but at the next payment date (see Exercise 53).

8.3   Copulas

A copula is the joint distribution function of a vector of uniform random variables on $[0,1]$.

Definition 8.3.1. *A function $C:[0,1]^m \to [0,1]$ is called a copula if there exists a random vector $(U_1,\ldots,U_m)$ such that each U_i is uniformly distributed on $[0,1]$ and*

$$C(u_1,\ldots,u_m)=\mathbb{P}(U_1 \le u_1,\ldots,U_m \le u_m), \quad (u_1,\ldots,u_m)\in[0,1]^m.$$

The following result, known as Sklar's theorem, shows that the law of a vector can be characterised in terms of its marginal distributions *and* a copula.

Theorem 8.3.2. *Let $X = (X_1, \ldots, X_m)$ be a random vector with values in $\mathbb{R}^m$. For $i = 1, \ldots, m$, denote by F_i the distribution function of the random variable X_i $(F_i(x) = \mathbb{P}(X_i \leq x))$. There exists a copula C such that, for all $(x_1, \ldots, x_m) \in \mathbb{R}^m$,*

$$\mathbb{P}(X_1 \leq x_1, \ldots, X_m \leq x_m) = C\left(F_1(x_1), \ldots, F_m(x_m)\right).$$

Moreover, if the functions F_i are continuous, the copula C is unique.

For the proof of Sklar's theorem, we will use the following lemma.

Lemma 8.3.3. *If X is a real-valued random variable with a continuous distribution function F, the random variable $F(X)$ is uniformly distributed on $[0, 1]$.*

Proof. For $u \in (0, 1)$, let

$$G(u) = \inf\{x \in \mathbb{R} \mid F(x) \geq u\}.$$

Note that $G(u)$ is well defined because

$$\lim_{x \to -\infty} F(x) = 0 \text{ and } \lim_{x \to +\infty} F(x) = 1.$$

Due to the right-continuity of F, we have $F(G(u)) \geq u$. In fact, since F is continuous, we have $F(G(u)) = u$ (if we had $F(G(u)) > u$, we would have $F(x) > u$ for x close to $G(u)$ and smaller than $G(u)$, in contradiction with the definition of $G(u)$). On the other hand, it is easy to prove that $F(x) \geq u$ if and only if $x \geq G(u)$. Therefore, we have, using $\mathbb{P}(X = G(u)) = 0$,

$$\begin{aligned}
\mathbb{P}(F(X) \geq u) &= \mathbb{P}(X \geq G(u)) \\
&= \mathbb{P}(X > G(u)) \\
&= 1 - \mathbb{P}(X \leq G(u)) \\
&= 1 - F(G(u)) \\
&= 1 - u,
\end{aligned}$$

which proves that $F(X)$ is uniform on $[0, 1]$. $\square$

Proof of Theorem 8.3.2. We only prove the result in the case of continuous marginals. See Sklar (1996) for the general case. We know from Lemma 8.3.3 that the random variables $F_i(X_i)$ are uniformly distributed on $[0, 1]$. Define the copula C by

$$C(u_1, \ldots, u_m) = \mathbb{P}(F_1(X_1) \leq u_1, \ldots, F_m(X_m) \leq u_m), \quad (u_1, \ldots, u_m) \in [0, 1]^m.$$

Since F_i is non-decreasing, we have $\{X_i \leq x_i\} \subset \{F_i(X_i) \leq F_i(x_i)\}$. Moreover, $\mathbb{P}(X_i \leq x_i) = F_i(x_i) = \mathbb{P}(F_i(X_i) \leq F_i(x_i))$, where the last equality follows from Lemma 8.3.3. Therefore, the events $\{X_i \leq x_i\}$ and $\{F_i(X_i) \leq F_i(x_i)\}$ are almost surely the same and we have

$$\mathbb{P}(X_1 \leq x_1, \ldots, X_m \leq x_m) = \mathbb{P}(F_1(X_1) \leq F_1(x_1), \ldots, F_m(X_m) \leq F_m(x_m))$$
$$= C(F_1(x_1), \ldots, F_m(x_m)).$$

Note that the uniqueness of C follows from the fact that any vector $(u_1, \ldots, u_m)$ with $0 < u_i < 1$ can be written $(F_1(x_1), \ldots, F_m(x_m))$ for some x_i's, due to the continuity of the F_i's.   □

Properties of copulas

- If $(X_1, \ldots, X_n)$ has a copula C and if $f_1, \ldots, f_m$ are strictly increasing and continuous functions, the vector $(f_1(X_1), \ldots, f_m(X_m))$ has the same copula C.

- The random variables $X_1, \ldots, X_m$ are independent if and only if the vector $(X_1, \ldots, X_m)$ admits the copula $C(u_1, \ldots, u_m) = \prod_{i=1}^{m} u_i$.

- If $X_1 = \cdots = X_m$, the vector $(X_1, \ldots, X_m)$ has a copula given by $C(u_1, \ldots, u_m) = \min_{1 \leq i \leq m} u_i$.

- Fréchet-Hoeffding bounds (see Exercise 54). For every copula C, we have

$$C^-(u_1, \ldots, u_m) \leq C(u_1, \ldots, u_m) \leq C^+(u_1, \ldots, u_m),$$

  where $C^-(u_1, \ldots, u_m) = (\sum_{i=1}^{m} u_i - (m-1))_+$ and $C^+(u_1, \ldots, u_m) = \min_{1 \leq i \leq m} u_i$. Note that, for $m \geq 3$, the function C^- is not a copula.

Remark 8.3.4. Copulas are used to price credit derivative instruments involving defaults of several firms. The models require the definition of several default times $\tau_1, \ldots, \tau_m$ corresponding to different *names*. A *first to default* contract will typically involve the distribution of $\min(\tau_1, \ldots, \tau_m)$. The marginal distributions of $\tau_1, \ldots, \tau_m$ may be known (by calibration on credit default swaps on single names), and a copula is used to produce the joint distribution. The so-called *collateralized debt obligations* (CDO) produce i.e. another important example (see Brigo and Mercurio (2006)).

For practical purposes, it is important to have parametric families of copulas that can be adjusted to market data. A popular family in finance is the family of Gaussian copulas (ie copulas of Gaussian vectors). They can be characterized by the covariance matrix of the Gaussian vector. Note that the components of the vector can be assumed to be standard normal variables, since the copula is invariant under increasing transformations of the coordinates. So the diagonal entries of the matrix are equal to 1. By taking all off-diagonal entries of the matrix equal to ρ, we get the one factor Gaussian copula. Note that for the m-dimensional Gaussian copula, we have $\rho \in [-1/(m-1), 1]$ (see Exercise 55).

8.4 Exercises

Exercise 49 Let $(\Omega, \mathscr{A}, \mathbb{P})$ be a probability space and $\mathscr{B}$ a sub-σ-algebra of $\mathscr{A}$. Let $A \in \mathscr{A}$ and $X = \mathbb{E}(1_A \mid \mathscr{B})$. Prove that $\mathbb{P}(A \cap \{X > 0\}) = \mathbb{P}(A)$, so that, on the set A, $X > 0$ almost surely.

Exercise 50 In Merton's model, show that the value at time 0 of the defaultable zero-coupon bond is given by

$$D_0 = Le^{-rT} N(d_-) + V_0 e^{-kT} N(-d_+),$$

where N is the standard normal distribution function and

$$d_{\pm} = \frac{\ln(V_0/L) + (r - k \pm \frac{\sigma^2}{2})T}{\sigma\sqrt{T}}.$$

Exercise 51 Consider a first passage structural model, in which the firm value satisfies (8.1) and the default time is given by (8.2), with $H(t) = H$ for $t < T$ and $H(T) = L > H$. The payoff of the bondholder at maturity is given by $D_T = V_T 1_{\{\tau < T\}} + V_T \wedge L 1_{\{\tau \geq T\}} = V_T \wedge L + (V_T - L)_+ 1_{\{\tau < T\}}$. We assume $V_0 > H$.

1. Prove that for all real numbers ρ and a, and for any non-negative Borel function f on $\mathbb{R}$,,

$$\mathbb{E}\left(e^{\rho W_T - \frac{\rho^2}{2}T} f(W_T) 1_{\{\inf_{0 \leq t \leq T} W_t \leq a\}}\right) = \mathbb{E}\left(f(W_T^\rho) 1_{\{\inf_{0 \leq t \leq T} W_t^\rho \leq a\}}\right),$$

where $W_t^\rho = \rho t + W_t$.

2. Use Exercice 30 to prove that the value at time 0 of the defaultable zero-coupon bond is given by

$$D_0 = D_0^M + \Pi_1 - \Pi_2,$$

where D_0^M is the value of the defaultable zero-coupon bond in Merton's model,

$$\Pi_1 = V_0 e^{-kT} \left(\frac{H}{V_0}\right)^{\frac{2(r-k)}{\sigma^2} + 1} N(d_+)$$

and

$$\Pi_2 = Le^{-rT} \left(\frac{H}{V_0}\right)^{\frac{2(r-k)}{\sigma^2} - 1} N(d_-),$$

where N is the standard normal distribution function and

$$d_{\pm} = \frac{\ln\left(H^2/LV_0\right) + (r - k \pm \frac{\sigma^2}{2})T}{\sigma\sqrt{T}}.$$

3. Explain how the formula should be modified in the Black-Cox specification $(H(t) = \tilde{L}e^{-\gamma(T-t)}, H(T) = L)$.

Exercise 52 In the framework of Section 8.2, let (X_t) be an $(\mathscr{F}_t)$-adapted non-negative continuous process and set $\Lambda_t = \int_0^t \lambda(s)ds$, for $t \geq 0$.

1. Prove that if $t_0 = 0 < t_1 < \ldots < t_n = T$, we have

$$\mathbb{E}^*(X_{t_{i-1}}\mathbf{1}_{\{t_{i-1} < \tau \leq t_i\}}) = \mathbb{E}^*\left(X_{t_{i-1}}(e^{-\Lambda_{t_{i-1}}} - e^{-\Lambda_{t_i}})\right).$$

2. Deduce

$$\mathbb{E}^*(X_\tau \mathbf{1}_{\{\tau \leq T\}}) = \int_0^T \mathbb{E}^*\left(X_u \lambda(u)e^{-\int_0^u \lambda(s)ds}\right) du.$$

Exercise 53 Consider a CDS where, in case of default, the quantity $N(1 - R)$ is paid at the first payment date after default.

1. Prove that the fair value of the spread is given by

$$s = \frac{\mathbb{E}^*\left(\sum_{i=1}^n (1-R)\mathbf{1}_{\{T_{i-1} < \tau \leq T_i\}}e^{-\int_0^{T_i} r(s)ds}\right)}{\mathbb{E}^*\left(\sum_{i=1}^n (T_i - T_{i-1})\mathbf{1}_{\{\tau > T_i\}}e^{-\int_0^{T_i} r(s)ds}\right)}.$$

2. Assume that τ is independent of all the σ-algebras $\mathscr{F}_t$. Show that the intensity process is deterministic and explain formula (8.5).

Exercise 54 Let $U_1, \ldots, U_m$ be m random variables with uniform distribution on $[0, 1]$ and let C be the associated copula:

$$C(u_1, \ldots, u_m) = \mathbb{P}(U_1 \leq u_1, \ldots, U_m \leq u_m).$$

1. Prove that $C(u_1, \ldots, u_m) \leq \min_{1 \leq i \leq m} u_i$.

2. Prove that $C(u_1, \ldots, u_m) \geq C^-(u_1, \ldots, u_m)$, where $C^-(u_1, \ldots, u_m) = \left(\sum_{i=1}^m u_i - (m-1)\right)_+$. (Hint: observe that for any events $A_1, \ldots, A_m$, we have, with $A = \bigcap_{i=1}^m A_i$, $\mathbf{1}_A \geq \sum_{i=1}^m \mathbf{1}_{A_i} - (m-1)$.)

3. Assume $m = 2$. Prove that C^- is the distribution function of the vector $(U, 1 - U)$, where U is uniform on $[0, 1]$.

4. Prove that, for $m \geq 3$, C^- is not a copula. (Hint: argue that if C^- were the distribution function of $(U_1, \ldots, U_m)$, any pair (U_i, U_j), with $i \neq j$, would have the distribution of $(U, 1 - U)$, where U is uniform on $[0, 1]$.)

Exercise 55 Given a real number ρ and an integer $m \geq 2$, consider the matrix, $\Gamma_\rho = (a_{ij})_{1 \leq i,j \leq m}$, with $a_{ii} = 1$ and $a_{ij} = \rho$ for $i \neq j$. Prove that Γ_ρ is a semi-definite positive matrix if and only if $\rho \in [-1/(m-1), 1]$. Show that, if $\rho \geq 0$, Γ_ρ is the covariance of the vector $(X_1, \ldots, X_m)$, with $X_i = \sqrt{\rho}Y + \sqrt{1-\rho}Z_i$, where $Y, Z_1, \ldots, Z_m$ are independent standard normal variables.

Chapter 9

Simulation and algorithms for financial models

9.1 Simulation and financial models

In this chapter, we describe methods that can be used to simulate financial models and compute prices. When we can write the option price as the expectation of a random variable that can be simulated, Monte-Carlo methods can be used. Unfortunately, these methods are quite inefficient and so are only used if there is no closed-form solution available for the price of the option. Simulations are also useful to evaluate complex hedging strategies (for example, to find the impact of hedging frequency on the replicating portfolio).

9.1.1 The Monte-Carlo method

The Monte-Carlo method uses the strong law of large numbers to compute an expectation. Consider a random variable X and assume we are able to generate a sequence of independent trials, $X_1, \ldots, X_n, \ldots$ following the distribution of X. Applying the law of large numbers, we can assert that if f satisfies $\mathbb{E}(|f(X)|) < +\infty$, then

$$\lim_{n \to +\infty} \frac{1}{n} \sum_{1 \leq k \leq n} f(X_k) = \mathbb{E}\left(f(X)\right). \tag{9.1}$$

To implement this method on a computer, we proceed as follows. We suppose that we know how to build a sequence of numbers $(U_n)_{n \geq 1}$ that is the realization of a sequence of independent, uniform random variables on the interval $[0, 1]$, and we look for a function $(u_1, \ldots, u_p) \mapsto F(u_1, \ldots, u_p)$ such that the random variable $F(U_1, \ldots, U_p)$ follows the distribution of X. The sequence of random variables $(X_n)_{n \geq 1}$, where $X_n = F(U_{(n-1)p+1}, \ldots, U_{np})$, is then a sequence of independent random variables following the distribution of X.

For example, we can apply (9.1) to the functions $f(x) = x$ and $f(x) = x^2$ to estimate the first and second-order moments of X (provided $\mathbb{E}(|X|^2)$ is finite).

The sequence $(U_n)_{n \geq 1}$ is obtained in practice from successive calls to a pseudo-random number generator. Most languages available on computers provide such a random function, already coded, which returns either a pseudo-random number between 0 and 1, or a random integer in a fixed interval (see, for instance, the `rand` or `random` functions in the `stdlib` C library).

Remark 9.1.1. The function F can depend in some cases (in particular when it comes to simulate stopping times) on the whole sequence $(U_n)_{n \geq 1}$, and not only on a fixed number of U_i's. The previous method can still be used if we can simulate X from an almost surely finite number of U_i's, this number being possibly random and not bounded. This happens, for instance, in the simulation of a Poisson random variable (see page 211).

Convergence rate of Monte-Carlo methods   Let $(X_1, \ldots, X_n, \ldots)$ be a sample following the distribution of X. The error ε_n in the Monte-Carlo method

$$\varepsilon_n = \mathbb{E}(X) - \frac{1}{n}(X_1 + \cdots + X_n),$$

can be estimated using the central limit theorem, which gives the asymptotic distribution of $\sqrt{n}\varepsilon_n$ for n large.

Theorem 9.1.1 (Central limit theorem)   *Let $(X_i, i \geq 1)$ be a sequence of independent, identically distributed random variables such that $\mathbb{E}(X_1^2) < +\infty$. Let σ^2 denote the variance of X_1, that is*

$$\sigma^2 = \mathbb{E}(X_1^2) - \mathbb{E}(X_1)^2 = \mathbb{E}\left((X_1 - \mathbb{E}(X_1))^2\right).$$

Then

$$\left(\frac{\sqrt{n}}{\sigma}\varepsilon_n\right) \ \text{converges in distribution to } G,$$

where G is a Gaussian random variable with mean 0 and variance 1.

Remark 9.1.3.   From this theorem it follows that for all $c_1 < c_2$,

$$\lim_{n \to +\infty} \mathbb{P}\left(\frac{\sigma}{\sqrt{n}}c_1 \leq \varepsilon_n \leq \frac{\sigma}{\sqrt{n}}c_2\right) = \int_{c_1}^{c_2} e^{-\frac{x^2}{2}} \frac{dx}{\sqrt{2\pi}}.$$

In practice, one applies an approximate rule: for n large enough, the distribution of ε_n is a Gaussian random variable with mean 0 and variance σ^2/n.

Note that it is impossible to bound the error, using this theorem, since the support of any Gaussian random variable is $\mathbb{R}$. Nevertheless, the preceding rule allows one to define a confidence interval. Observing that

$$\mathbb{P}(|G| \leq 1.96) \approx 0.95,$$

with a probability close to 0.95, for n large enough, we have:

$$|\varepsilon_n| \leq 1.96 \frac{\sigma}{\sqrt{n}}.$$

How to estimate the variance The previous result shows that it is crucial to estimate the standard deviation σ of the random variable. It is easy to do this by using the same samples as for the expectation. Let X be a random variable and $(X_1, \ldots, X_n)$ a sample drawn according to the distribution of X. We will denote by $\bar{X}_n$ the Monte-Carlo estimator of $\mathbb{E}(X)$, given by

$$\bar{X}_n = \frac{1}{n} \sum_{i=1}^{n} X_i.$$

A standard estimator for the variance is given by

$$\bar{\sigma}_n^2 = \frac{1}{n-1} \sum_{i=1}^{n} \left(X_i - \bar{X}_n\right)^2.$$

$\bar{\sigma}_n^2$ is often called the empirical variance of the sample. Note that $\bar{\sigma}_n^2$ can be rewritten as

$$\bar{\sigma}_n^2 = \frac{n}{n-1} \left(\frac{1}{n} \sum_{i=1}^{n} X_i^2 - \bar{X}_n^2 \right).$$

On this last formula, it is obvious that $\bar{X}_n$ and $\bar{\sigma}_n^2$ can be computed using only $\sum_{i=1}^{n} X_i$ and $\sum_{i=1}^{n} X_i^2$.

Moreover, one can prove, when $\mathbb{E}(X^2) < +\infty$, that $\lim_{n \to +\infty} \bar{\sigma}_n^2 = \sigma^2$, almost surely, and that $\mathbb{E}\left(\bar{\sigma}_n^2\right) = \sigma^2$ (the estimator is unbiased). This leads to an (approximate) confidence interval by replacing σ by $\bar{\sigma}_n$ in the standard confidence interval. With a probability close to 0.95, $\mathbb{E}(X)$ belongs to the (random) interval given by

$$\left[\bar{X}_n - \frac{1.96\bar{\sigma}_n}{\sqrt{n}}, \bar{X}_n + \frac{1.96\bar{\sigma}_n}{\sqrt{n}} \right].$$

So, with very little additional computations (we only have to compute $\bar{\sigma}_n$ on an already drawn sample) we can give a reasonable estimate of the error done by approximating $\mathbb{E}(X)$ with $\bar{X}_n$. The possibility to give an error estimate with a small numerical cost is an extremely useful feature of Monte-Carlo methods.

9.1.2 Simulation of a uniform distribution on [0, 1]

We now very briefly explain how standard random number generators are built. A simple and very common method is to use a linear congruential generator. A sequence $(x_n)_{n \geq 0}$ of integers between 0 and $m-1$ is generated as follows:

$$\begin{cases} x_0 = \text{initial value} \in \{0, 1, \ldots, m-1\} \\ x_{n+1} = ax_n + b(\text{ modulo } m), \end{cases}$$

a, b, m being integers to be chosen cautiously in order to obtain satisfactory characteristics for the sequence. This method enables us to simulate pseudo-random integers between 0 and $m - 1$; to obtain a random real-valued number between 0 and 1, we can divide this random integer by m.

Sedgewick (1987) advocates the following choice:

$$\begin{cases} a & = & 31415821 \\ b & = & 1 \\ m & = & 10^8. \end{cases}$$

This generator provides reasonable results in common cases. However, it might happen that its period (here equal to $m = 10^8$) is not big enough. Then it is possible to create random number generators with an arbitrary long period by increasing m.

The reader will find much more complete information on random number generators and computer procedures in Knuth (1981) and L'Ecuyer (1990). The reader is also referred to the following Web site entirely devoted to Monte-Carlo simulation: `http://random.mat.sbg.ac.at/links/`.

9.1.3   Simulation of random variables

The probability distributions used for financial models are mainly Gaussian distributions (in the case of continuous models) and exponential and Poisson distributions (in the case of models with jumps). We recall here some elementary methods used to simulate each of these distributions.

Simulation of a Gaussian distribution

A classical method to simulate Gaussian random variables is based on the observation (see Exercise 56) that if (U_1, U_2) are two independent uniform random variables on $[0, 1]$,

$$\sqrt{-2 \log(U_1)} \cos(2\pi U_2)$$

follows a standard Gaussian distribution (i.e. zero-mean and with variance 1).

To simulate a Gaussian random variable with mean m and variance σ, it suffices to set $X = m + \sigma g$, where g is a standard Gaussian random variable.

Simulation of an exponential distribution

We recall that a random variable X follows an exponential distribution with parameter λ if its distribution is

$$1_{\{x \geq 0\}} \lambda e^{-\lambda x} dx.$$

We can simulate X noticing that, if U follows a uniform distribution on $[0, 1]$, $\log(U)/\lambda$ follows an exponential distribution with parameter λ.

Remark 9.1.4. This method of simulation of the exponential distribution is a particular case of the so-called 'inverse distribution function' method (for this matter, see Exercise 57).

Simulation of a Poisson random variable

A Poisson random variable is a variable with values in $\mathbb{N}$ such that

$$\mathbb{P}(X = n) = e^{-\lambda}\frac{\lambda^n}{n!}, \quad n \geq 0.$$

We have seen in Chapter 7 that if $(T_i, i \geq 1)$ is a sequence of exponential random variables with parameter λ, then the distribution of

$$N_t = \sum_{n \geq 1} n 1_{\{T_1 + \cdots + T_n \leq t < T_1 + \cdots + T_{n+1}\}}$$

is a Poisson distribution with parameter λt. Thus N_1 has the same distribution as the variable X we want to simulate. On the other hand, it is always possible to write exponential variables T_i as $-\log(U_i)/\lambda$ where the $(U_i)_{i \geq 1}$ are independent random variables following the uniform distribution on $[0, 1]$. N_1 can be written as

$$N_1 = \sum_{n \geq 1} n 1_{\{U_1 U_2 \ldots U_{n+1} \leq e^{-\lambda} < U_1 U_2 \ldots U_n\}}.$$

This leads to an algorithm to simulate a Poisson random variable.

For the simulation of distributions not mentioned above, or for other methods of simulation of the previous distributions, one may refer to Rubinstein (1981).

Simulation of Gaussian vectors

Multidimensional models will generally involve Gaussian processes with values in $\mathbb{R}^n$. The problem of simulating Gaussian vectors (see Section A.1.2 of the Appendix for the definition of a Gaussian vector) is then essential. We now give a method of simulation for these kinds of random variables.

Let $X = (X_1, \ldots, X_n)$ be a Gaussian vector. Its distribution is characterized by the vector of its means $m = (m_1, \ldots, m_n) = (\mathbb{E}(X_1), \ldots, \mathbb{E}(X_n))$ and its variance-covariance matrix $\Gamma = (\Gamma_{ij})_{1 \leq i \leq n, 1 \leq j \leq n}$, where $\Gamma_{ij} = \mathbb{E}(X_i X_j) - \mathbb{E}(X_i)\mathbb{E}(X_j)$.

The matrix Γ is positive semi-definite and we will assume that it is invertible. We can find the square root of Γ, that is to say a matrix A, such that $A \times {}^t A = \Gamma$ (where ${}^t A$ is the transpose of the matrix A). As Γ is invertible, so is A. Let Z be the vector $A^{-1}(X - m)$. It is easily checked that this vector is a Gaussian vector with zero-mean. Moreover, its variance matrix is given by

$$\mathbb{E}(Z_i Z_j) = \sum_{1 \leq k \leq n, 1 \leq l \leq n} \mathbb{E}(A_{ik}^{-1}(X_k - m_k)A_{jl}^{-1}(X_l - m_l))$$

$$= \sum_{1 \le k \le n, 1 \le l \le n} \mathbb{E}\left(A_{ik}^{-1} A_{jl}^{-1} \Gamma_{kl}\right)$$
$$= (A^{-1}\Gamma({}^tA)^{-1})_{ij} = (A^{-1}A^t A({}^tA)^{-1})_{ij} = Id.$$

This implies that the coordinates of Z are n independent standard normal variables.

The distribution of the vector $X = m + AZ$ can be simulated as follows:

1. Derive the square root of the matrix Γ, say A.

2. Simulate n independent standard normal variables $G = (g_1, \ldots, g_n)$.

3. Return $m + AG$.

Remark 9.1.5. To derive the square root of Γ, we may assume that A is upper-triangular; then there is a unique solution to the equation $A \times {}^tA = \Gamma$. This method of calculation of the square root is called Cholesky's method (for a complete algorithm, see Ciarlet (1988) or Press et al. (1992)).

9.1.4   Simulation of stochastic processes

The methods delineated previously enable us to simulate a random variable, in particular the value of a stochastic process at a given time. Sometimes we need to know how to simulate the whole path of a process (for example, when we are studying the dynamics through time of the value of a portfolio of options; see Exercise 47). This section suggests some simple tricks to simulate paths of processes.

Simulation of a Brownian motion

We distinguish two methods to simulate a Brownian motion $(W_t)_{t \ge 0}$. The first one consists in renormalizing' a random walk. Let $(X_i)_{i \ge 0}$ be a sequence of independent, identically distributed random walks with distribution $\mathbb{P}(X_i = 1) = 1/2$, $\mathbb{P}(X_i = -1) = 1/2$. Then we have $\mathbb{E}(X_i) = 0$ and $\mathbb{E}(X_i^2) = 1$. We set $S_n = X_1 + \cdots + X_n$; then we can approximate the Brownian motion by the process $(X_t^n)_{t \ge 0}$, where

$$X_t^n = \frac{1}{\sqrt{n}} S_{[nt]},$$

where $[x]$ is the largest integer less than or equal to x. This method of simulation of the Brownian motion is partially justified in Exercise 48.

In the second method, we notice that, if $(G_i)_{i \ge 0}$ is a sequence of independent standard normal random variables, if $\Delta t > 0$ and if we set

$$\begin{cases} S_0 = 0 \\ S_{n+1} - S_n = G_n, \end{cases}$$

then the distribution of $(\sqrt{\Delta t}S_0, \ldots, \sqrt{\Delta t}S_n)$ is identical to the distribution of

$$(W_0, W_{\Delta t}, W_{2\Delta t}, \ldots, W_{n\Delta t}).$$

The Brownian motion can be approximated by $X_t^n = \sqrt{\Delta t}\, S_{[t/\Delta t]}$.

Simulation of stochastic differential equations

There are many methods, some of them very sophisticated, to simulate the solution of a stochastic differential equation; the reader is referred to Pardoux and Talay (1985) or Kloeden and Platen (1992) for a review of these methods. Here we present only the simplest method, the so-called 'Euler approximation'. The principle is the following: consider the stochastic differential equation

$$\begin{cases} X_0 &= x \\ dX_t &= b(X_t)dt + \sigma(X_t)dW_t. \end{cases}$$

We discretize time by a fixed mesh Δt. Then we can construct a discrete-time process $(S_n)_{n\geq 0}$ approximating the solution of the stochastic different equation at times $n\Delta t$, setting

$$\begin{cases} S_0 &= x \\ S_{n+1} - S_n &= \{b(S_n)\Delta t + \sigma(S_n)(W_{(n+1)\Delta t} - W_{n\Delta t})\}. \end{cases}$$

If $X_t^n = S_{[t/\Delta t]}$, the process $(X_t^n)_{t\geq 0}$ approximates $(X_t)_{t\geq 0}$ in the following sense:

Theorem 9.1.6. *For any $T > 0$,*

$$\mathbb{E}\left(\sup_{t\leq T} |X_t^n - X_t|^2\right) \leq C_T \Delta t,$$

C_T *being a constant depending only on T.*

A proof of this result (as well as other schemes of discretization of stochastic differential equations) can be found in Chapter 7 of Gard (1988).

The distribution of the sequence $(W_{(n+1)\Delta t} - W_{n\Delta t})_{n\geq 0}$ is the distribution of a sequence of independent normal random variables with zero-mean and variance Δt. In a simulation, we substitute $G_n\sqrt{\Delta t}$ for $(W_{(n+1\Delta t} - W_{n\Delta t})$, where $(G_n)_{n\geq 0}$ is a sequence of independent standard normal variables. The approximating sequence $(S_n')_{n\geq 0}$ is in this case defined by

$$\begin{cases} S_0' &= x \\ S_{n+1}' &= S_n' + \Delta t\, b(S_n') + \sigma(S_n')\, G_n\sqrt{\Delta t}. \end{cases}$$

Remark 9.1.7. We can substitute for the sequence of independent Gaussian random variables $(G_i)_{i\geq 0}$ a sequence of independent random variables $(U_i)_{i\geq 0}$, such that $\mathbb{P}(U_i = 1) = \mathbb{P}(U_i = -1) = 1/2$. Nevertheless, in this case, it must be noticed that the convergence is different from that found in Theorem 9.1.6: in this case the convergence is known as *weak convergence* (or *convergence in distribution* instead of a *strong (or pathwise) convergence* in Theorem 9.1.6. There is still a convergence theorem, but it applies to the distributions of the processes. Kushner (1977) and Pardoux and Talay (1985) can be consulted for explanations on weak convergence and many results on discretization in distribution for stochastic differential equations.

An application to the Black-Scholes model

In the case of the Black-Scholes model, we want to simulate the solution of the equation

$$\begin{cases} X_0 = x \\ dX_t = X_t(rdt + \sigma dW_t). \end{cases}$$

Two approaches are available. The first consists in using the Euler approximation. We set

$$\begin{cases} S_0 = x \\ S_{n+1} = S_n(1 + r\Delta t + \sigma G_n\sqrt{\Delta t}), \end{cases}$$

and simulate X_t by $X_t^n = S_{[t/\Delta t]}$. The other method consists in using the explicit expression of the solution

$$X_t = x\exp\left(rt - \frac{\sigma^2}{2}t + \sigma W_t\right)$$

and simulating the Brownian motion by one of the methods presented previously. In the case where we simulate the Brownian motion by $\sqrt{\Delta t}\sum_{i=1}^n G_i$, we obtain

$$S_n = x\exp\left((r - \sigma^2/2)n\Delta t + \sigma\sqrt{\Delta t}\sum_{i=1}^n G_i\right). \tag{9.2}$$

We always approximate X_t by $X_t^n = S_{[t/\Delta t]}$.

Remark 9.1.8. We can also replace the Gaussian random variables G_i by some Bernoulli variables with values $+1$ or -1 with probability $1/2$ in (9.2); we obtain a binomial-type model close to the Cox-Ross-Rubinstein model used in Section 1.4 of Chapter 1.

Simulation of models with jumps

We have investigated in Chapter 7 an extension of the Black-Scholes model with jumps; we now describe a method for simulating this process. We take the notations and the hypothesis of Chapter 7, Section 7.2. The process $(X_t)_{t\geq 0}$ describing the dynamics of the asset is

$$X_t = x\left(\prod_{j=1}^{N_t}(1 + U_j)\right)e^{(\mu - \sigma^2/2)t + \sigma W_t}, \tag{9.3}$$

where $(W_t)_{t\geq 0}$ is a standard Brownian motion, $(N_t)_{t\geq 0}$ is a Poisson process with intensity λ, and $(U_j)_{j\geq 1}$ is a sequence of independent, identically distributed random variables, with values in $[-1, +\infty)$ and distribution $\mu(dx)$. The σ-algebras generated by $(W_t)_{t\geq 0}, (N_t)_{t\geq 0}, (U_j)_{j\geq 1}$ are supposed to be independent.

In order to simulate this process at times $n\Delta t$, we notice that

$$X_{n\Delta t} = x \times (X_{\Delta t}/x) \times (X_{2\Delta t}/X_{\Delta t}) \times \cdots \times (X_{n\Delta t}/X_{(n-1)\Delta t}).$$

If we note $Y_k = X_{k\Delta t}/X_{(k-1)\Delta t}$, we can prove, from the properties of $(N_t)_{t\geq 0}, (W_t)_{t\geq 0}$ and $(U_j)_{j\geq 1}$, that $(Y_k)_{k\geq 1}$ is a sequence of independent random variables with the same distribution. Since $X_{n\Delta t} = xY_1 \ldots Y_n$, the simulation of X at times $n\Delta t$ comes down to the simulation of the sequence $(Y_k)_{k\geq 1}$. This sequence being independent and identically distributed, it suffices to know how to simulate $Y_1 = X_{\Delta t}/x$. Then we operate as follows:

- We simulate a standard Gaussian random variable G.

- We simulate a Poisson random variable with parameter $\lambda\Delta t$: N.

- If $N = n$, we simulate n random variables following the distribution $\mu(dx): U_1, \ldots, U_n$.

All these variables are assumed to be independent. Then, from equation (8.3), it is clear that the distribution of

$$\left(\prod_{j=1}^{N}(1 + U_j)\right) e^{(\mu - \sigma^2/2)\Delta t + \sigma\sqrt{\Delta t}G}$$

is identical to the distribution of Y_1.

9.2 Introduction to variance reduction methods

All the results of the preceding section show that the ratio $\sigma/\sqrt{n}$ governs the accuracy of a Monte-Carlo method with n simulations. An obvious consequence of this fact is that one always has interest to rewrite the quantity to compute as the expectation of a random variable that has a smaller variance: this is the basic idea of variance reduction techniques. For complements, we refer the reader to Kalos and Whitlock (1986), Hammersley and Handscomb (1979), Rubinstein (1981) and Ripley (1987) for general books and to Glasserman (2004) for a book devoted to financial applications only.

Suppose that we want to evaluate $\mathbb{E}(X)$. We try to find an alternative representation for this expectation as

$$\mathbb{E}(X) = \mathbb{E}(Y) + C,$$

using a random variable Y with lower variance and C a known constant. A lot of techniques are known in order to implement this idea. We give an introduction to some standard methods.

9.2.1   Control variates

The basic idea of a control variate is to write the expectation we want to compute $\mathbb{E}(X)$ as

$$\mathbb{E}(X) = \mathbb{E}(X - Y) + \mathbb{E}(Y),$$

where $\mathbb{E}(Y)$ can be explicitly computed and $\text{Var}(X - Y)$ is smaller than $\text{Var}(X)$. In these circumstances, we can use a Monte-Carlo method to estimate $\mathbb{E}(X - Y)$ and add the exact value of $\mathbb{E}(Y)$. Let us illustrate this principle by several financial examples.

Using call-put parity for variance reduction   Let S_t be the price at time t of a risky asset and denote by C the price of the European call option

$$C = \mathbb{E}\left(e^{-rT}\left(S_T - K\right)_+\right),$$

and by P the price of the European put option

$$P = \mathbb{E}\left(e^{-rT}\left(K - S_T\right)_+\right).$$

The 'call-put parity formula' gives

$$C - P = \mathbb{E}\left(e^{-rT}\left(S_T - K\right)\right) = S_0 - Ke^{-rT}.$$

This formula can be used to replace the computation of a call option by that of a put option since

$$C = \mathbb{E}\left(e^{-rT}\left(K - S_T\right)_+\right) + S_0 - Ke^{-rT}.$$

Remark 9.2.1.   For the Black-Scholes model, explicit formulae for the variance of the put and the call options can be obtained. The variance of the put option is often (but not always) smaller than the variance of the call. In these cases, one should compute put option prices even when one needs call prices.

The Kemna and Vorst method for Asian options   The price of an average (or Asian) put option with fixed strike is given by

$$\mathbb{E}\left(e^{-rT}\left(K - \frac{1}{T}\int_0^T S_s ds\right)_+\right),$$

where $(S_t, t \geq 0)$ denotes the Black Scholes process

$$S_t = x \exp\left(\left(r - \frac{\sigma^2}{2}\right)t + \sigma W_t\right).$$

If σ and r are small enough, we can hope that

$$\frac{1}{T}\int_0^T S_s ds \text{ "is not too far from" } \exp\left(\frac{1}{T}\int_0^T \log(S_s)ds\right).$$

This heuristic argument suggests that we use Y,

$$Y = e^{-rT} \left(K - e^Z \right)_+,$$

with $Z = \frac{1}{T} \int_0^T \log(S_s) ds$ as a control variate. As the random variable Z is Gaussian, we can explicitly compute

$$\mathbb{E} \left(e^{-rT} \left(K - e^Z \right)_+ \right).$$

This is done by using the formula (equivalent to the Black-Scholes formula)

$$\mathbb{E} \left(\left(K - e^Z \right)_+ \right) = K N(-d) - e^{\mathbb{E}(Z) + \frac{1}{2} \mathrm{Var}(Z)} N(-d - \sqrt{\mathrm{Var}(Z)}),$$

where $d = \frac{\mathbb{E}(Z) - \log(K)}{\sqrt{\mathrm{Var}(Z)}}$.

This method is proposed in Kemna and Vorst (1990) and is very efficient for standard values ($\sigma \approx 0.3$ by year, $r \approx 0.1$ by year and $T \approx 1$ year are typical in financial applications). When the values of σ and r are larger, the control variate can be less efficient but still remains useful.

Basket options. A very similar idea can be used for pricing basket (or index) options. The simplest model used to price basket options is the multi-dimensional Black-Scholes model. Let σ be an $n \times d$ matrix and $(W_t, t \geq 0)$ a d-dimensional Brownian motion. Denote by $(S_t, t \geq 0)$ the solution of

$$\begin{cases} dS_t^1 = S_t^1 \left(r dt + [\sigma dW_t]_1 \right) \\ \quad \cdots \\ dS_t^n = S_t^n \left(r dt + [\sigma dW_t]_n \right), \end{cases}$$

where $[\sigma dW_t]_i = \sum_{j=1}^d \sigma_{ij} dW_t^j$. Note that this multidimensional stochastic differential equation can be solved as

$$S_T^i = S_0^i \exp \left(\left(r - \frac{1}{2} \sum_{j=1}^d \sigma_{ij}^2 \right) T + \sum_{j=1}^d \sigma_{ij} W_T^j \right).$$

Moreover, denote by I_t the value of an index

$$I_t = \sum_{i=1}^n a_i S_t^i,$$

where $a_1, \ldots, a_n$ is a given set of positive numbers such that $\sum_{i=1}^n a_i = 1$.

Suppose that we want to compute the price of a European put option with a payoff at time T given by

$$h = (K - I_T)_+.$$

Here, the idea is to approximate

$$\frac{I_T}{m} = \frac{a_1 x_1}{m} e^{r_1 T + \sum_{j=1}^{p} \sigma_{1j} W_T^j} + \cdots + \frac{a_d x_d}{m} e^{r_d T + \sum_{j=1}^{p} \sigma_{dj} W_T^j},$$

where $m = a_1 x_1 + \cdots + a_d x_d$ and $r_i = r - \frac{1}{2} \sum_{j=1}^{p} \sigma_{ij}^2$, by Y, where Y is the lognormal random variable

$$Y = e^{\sum_{i=1}^{d} \frac{a_i x_i}{m} \left(r_i T + \sum_{j=1}^{p} \sigma_{ij} W_T^j \right)}.$$

As we can compute an explicit formula for

$$\mathbb{E}\left[(K - mY)_+ \right],$$

we can use the control variate $Z = (K - mY)_+$ and sample $(K - X)_+ - (K - mY)_+$ (see Section 9.4.5 for more details).

9.2.2   Importance sampling

Importance sampling is another variance reduction procedure. It is obtained by changing the sampling distribution. We introduce this method in a very simple context. Suppose we want to compute

$$\mathbb{E}(f(X)),$$

X being a real random variable following the density $p(x)$ on $\mathbb{R}$; then

$$\mathbb{E}(f(X)) = \int_{\mathbb{R}} f(x) p(x) dx.$$

Let $\tilde{p}$ be another density such that $\tilde{p}(x) > 0$ and $\int_{\mathbb{R}} \tilde{p}(x) dx = 1$. Clearly one can write $\mathbb{E}(f(X))$ as

$$\mathbb{E}(f(X)) = \int_{\mathbb{R}} \frac{f(x) p(x)}{\tilde{p}(x)} \tilde{p}(x) dx = \mathbb{E}\left(\frac{f(Y) p(Y)}{\tilde{p}(Y)} \right),$$

where Y has density $\tilde{p}(x)$ under $\mathbb{P}$. We thus can approximate $\mathbb{E}(f(X))$ by an alternative Monte-Carlo estimator

$$\frac{1}{n} \left(\frac{f(Y_1) p(Y_1)}{\tilde{p}(Y_1)} + \cdots + \frac{f(Y_n) p(Y_n)}{\tilde{p}(Y_n)} \right),$$

where $(Y_1, \ldots, Y_n)$ are independent copies of Y. Set $Z = f(Y) p(Y) / \tilde{p}(Y)$. The variance of the new estimator will be better than the standard one if $\mathrm{Var}(Z) < \mathrm{Var}(f(X))$. Note that it is easy to compute the variance of Z as

$$\mathrm{Var}(Z) = \int_{\mathbb{R}} \frac{f^2(x) p^2(x)}{\tilde{p}(x)} dx - \mathbb{E}(f(X))^2.$$

From this and an easy computation it follows that if $f(x) > 0$ and $\tilde{p}(x) = f(x) p(x) / \mathbb{E}(f(X))$, then $\mathrm{Var}(Z) = 0$! Of course this result can seldom be used in practice as it relies on the exact knowledge of $\mathbb{E}(f(X))$, which is exactly what we want to compute. Nevertheless, it leads to a heuristic approach: take

$$\tilde{p}(x) = \frac{q(x)}{\int_{\mathbb{R}} q(x)dx},$$

where $q(x)$ is an approximation of $|f(x)p(x)|$ such that $\tilde{p}(x)$ can be sampled easily.

An elementary financial example   Suppose that G is a Gaussian random variable with zero-mean and unit variance, and that we want to compute

$$\mathbb{E}\left(f(G)\right),$$

for some function f. Here, we choose the distribution of $\tilde{G} = G + m$ as the new sampling distribution, m being a real constant that has to be determined carefully. We have

$$\mathbb{E}\left(f(G)\right) = \mathbb{E}\left(f(\tilde{G})\frac{p(\tilde{G})}{\tilde{p}(\tilde{G})}\right) = \mathbb{E}\left(f(\tilde{G})e^{-m\tilde{G}+\frac{m^2}{2}}\right).$$

This equality can be rewritten as

$$\mathbb{E}\left(f(G)\right) = \mathbb{E}\left(f(G+m)e^{-mG-\frac{m^2}{2}}\right). \tag{9.1}$$

To be more specific, suppose that we want to compute a European call option in the Black-Scholes model. We have

$$f(G) = \left(\lambda e^{\sigma G} - K\right)_+,$$

and assume that $\lambda \ll K$. In this case, $\mathbb{P}(\lambda e^{\sigma G} > K)$ is very small and it is very unlikely that the option will be exercised. This fact can lead to a very large error in a standard Monte-Carlo method. In order to increase the probability of exercise, we can use the equality (9.1),

$$\mathbb{E}\left(\left(\lambda e^{\sigma G} - K\right)_+\right) = \mathbb{E}\left(\left(\lambda e^{\sigma(G+m)} - K\right)_+ e^{-mG-\frac{m^2}{2}}\right),$$

and choose $m = m_0$ with $\lambda e^{\sigma m_0} = K$, since

$$\mathbb{P}\left(\lambda e^{\sigma(G+m_0)} > K\right) = \frac{1}{2}.$$

This choice of m is certainly not optimal; however, it drastically improves the efficiency of the Monte-Carlo method when $\lambda \ll K$ (see Exercise 58 for a mathematical hint of this fact).

The multidimensional case   Monte-Carlo simulations are really useful for problems involving a large number of assets, and thus we have to extend the previous method to a multidimensional setting.

We consider here the multidimensional Black-Scholes model already introduced on page 217. Let σ be an $n \times d$ matrix and $(W_t, t \geq 0)$ a d-dimensional Brownian motion. Denote by $(S_t, t \geq 0)$

$$S_T^i = S_0^i \exp\left(\left(r - \frac{1}{2}\sum_{j=1}^d \sigma_{ij}^2\right)T + \sum_{j=1}^d \sigma_{ij}W_T^j\right),$$

and by I_t the value of the index $I_t = \sum_{i=1}^n a_i S_t^i$, where $a_1, \ldots, a_n$ is a given set of positive numbers such that $\sum_{i=1}^n a_i = 1$. Suppose that we want to compute the price of any European non-path-dependent option with payoff at time T given by

$$h = f(I_T).$$

Obviously, there exists a function ϕ such that

$$h = \phi(G_1, \ldots, G_d),$$

where $G_j = W_T^j/\sqrt{T}$ are independent standard normal random variables. So, the price of this option can be rewritten as,

$$\mathbb{E}\left(\phi(G)\right)$$

where $G = (G_1, \ldots, G_d)$ is a d-dimensional Gaussian vector with unit covariance matrix.

As in the one-dimensional case, it is easy (by a change of variable) to prove that, if $m = (m_1, \ldots, m_d)$,

$$\mathbb{E}\left(\phi(G)\right) = \mathbb{E}\left(\phi(G+m)e^{-m.G-\frac{|m|^2}{2}}\right), \tag{9.2}$$

where $m.G = \sum_{i=1}^d m_i G_i$ and $|m|^2 = \sum_{i=1}^d m_i^2$. In view of (9.2), the variance $V(m)$ of the random variable

$$X_m = \phi(G+m)e^{-m.G-\frac{|m|^2}{2}}$$

can be computed as

$$V(m) = \mathbb{E}\left(\phi^2(G+m)e^{-2m.G-|m|^2}\right) - \mathbb{E}\left(\phi(G)\right)^2,$$

$$= \mathbb{E}\left(\phi^2(G+m)e^{-m.(G+m)+\frac{|m|^2}{2}}e^{-m.G-\frac{|m|^2}{2}}\right) - \mathbb{E}\left(\phi(G)\right)^2,$$

$$= \mathbb{E}\left(\phi^2(G)e^{-m.G+\frac{|m|^2}{2}}\right) - \mathbb{E}\left(\phi(G)\right)^2.$$

The reader is referred to Glasserman et al. (1999) for an almost optimal way to choose the parameter m based on this representation and to Arouna (2003/4) for a method using stochastic algorithms to compute the optimal parameter.

We now extend these techniques to the case of path-dependent options. We use the Girsanov theorem 4.2.2.

The Girsanov theorem and path-dependent options Let $(S_t, t \geq 0)$ be the solution of

$$dS_t = S_t \left(r dt + \sigma dW_t \right), S_0 = x,$$

where $(W_t, t \geq 0)$ is a Brownian motion under a probability $\mathbb{P}$. We want to compute the price of a path-dependent option with payoff given by

$$\phi(S_t, t \leq T) = \psi(W_t, t \leq T).$$

Common examples of such a situation are

- Asian options, with payoff given by $f(S_T, \int_0^T S_s ds)$,

- Maximum options, with payoff given by $f(S_T, \max_{s \leq T} S_s)$.

We start by considering an elementary importance sampling technique. It is a straightforward extension of the technique used in the preceding example. For every real number λ, define the process $(W_t^\lambda, t \leq T)$ as

$$W_t^\lambda := W_t + \lambda t.$$

According to Girsanov theorem (see Theorem 4.2.2), $(W_t^\lambda, t \leq T)$ is a Brownian motion under the probability distribution $\mathbb{P}^\lambda$ defined by

$$\mathbb{P}^\lambda(A) = \mathbb{E}(L_T^\lambda \mathbf{1}_A), \quad A \in \mathscr{F}_T,$$

where $L_T^\lambda = e^{-\lambda W_T - \frac{\lambda^2 T}{2}}$. Denote by $\mathbb{E}^\lambda$ the expectation under this new probability $\mathbb{P}^\lambda$. For every bounded function ψ we have

$$\mathbb{E}\left(\psi(W_t, t \leq T)\right) = \mathbb{E}^\lambda\left(\psi(W_t^\lambda, t \leq T)\right) = \mathbb{E}\left(L_T^\lambda \psi(W_t^\lambda, t \leq T)\right),$$

and thus

$$\mathbb{E}\left(\psi(W_t, t \leq T)\right) = \mathbb{E}\left(e^{-\lambda W_T - \frac{\lambda^2 T}{2}} \psi(W_t + \lambda t, t \leq T)\right). \tag{9.3}$$

For example, if we want to compute the price of a fixed strike Asian option given by

$$P = \mathbb{E}\left(e^{-rt} \left(\frac{1}{T} \int_0^T x e^{\left(r - \frac{\sigma^2}{2}\right)s + \sigma W_s} ds - K\right)_+\right),$$

we can use the equality (9.3) to obtain

$$P = \mathbb{E}\left(e^{-rt - \lambda W_T - \frac{\lambda^2 T}{2}} \left(\frac{1}{T} \int_0^T x e^{\left(r - \frac{\sigma^2}{2}\right)s + \sigma(W_s + \lambda s)} ds - K\right)_+\right).$$

This representation can be used in case of a deep out of the money option (i.e. when $x \ll K$). Then λ can be chosen such that

$$\frac{x}{T} \int_0^T e^{\left(r - \frac{\sigma^2}{2}\right)s + \sigma \lambda s} ds = K,$$

in order to increase the exercise probability.

9.2.3   Antithetic variables

The use of antithetic variables is widespread in Monte-Carlo simulations. This technique can be efficient, but its gains are less dramatic than other variance reduction techniques. We begin by considering a very simple but instructive example. Let

$$I = \int_0^1 f(x)dx.$$

If U follows a uniform distribution on the interval $[0,1]$, then $1 - U$ has the same distribution as U, and thus

$$I = \frac{1}{2}\int_0^1 (f(x) + f(1-x))dx = \mathbb{E}\left(\frac{1}{2}(f(U) + f(1-U))\right).$$

Therefore one can draw n independent random variables $U_1, \ldots, U_n$ following a uniform distribution on $[0,1]$, and approximate I by

$$I_{2n} = \frac{1}{n}\left(\frac{1}{2}(f(U_1) + f(1-U_1)) + \cdots + \frac{1}{2}(f(U_n) + f(1-U_n))\right)$$
$$= \frac{1}{2n}\left(f(U_1) + f(1-U_1) + \cdots + f(U_n) + f(1-U_n)\right).$$

We need to compare the efficiency of this Monte-Carlo method with the standard one with $2n$ drawings:

$$I_{2n}^0 = \frac{1}{2n}\left(f(U_1) + f(U_2) + \cdots + f(U_{2n-1}) + f(U_{2n})\right)$$
$$= \frac{1}{n}\left(\frac{1}{2}(f(U_1) + f(U_2)) + \cdots + \frac{1}{2}(f(U_{2n-1}) + f(U_{2n}))\right).$$

We will now compare the variances of I_{2n} and I_{2n}^0. Observe that in doing this we assume that most of the numerical work relies on the evaluation of f and the time devoted to the simulation of the random variables is negligible. This is often a realistic assumption.

An easy computation shows that the variance of the standard estimator is

$$\mathrm{Var}(I_{2n}^0) = \frac{1}{2n}\mathrm{Var}\left(f(U_1)\right),$$

whereas

$$\mathrm{Var}(I_{2n}) = \frac{1}{n}\mathrm{Var}\left(\frac{1}{2}(f(U_1) + f(1-U_1))\right) \tag{9.4}$$

$$= \frac{1}{4n}\left(\mathrm{Var}(f(U_1)) + \mathrm{Var}(f(1-U_1)) + 2\mathrm{Cov}(f(U_1), f(1-U_1))\right) \tag{9.5}$$

$$= \frac{1}{2n}\left(\mathrm{Var}(f(U_1) + \mathrm{Cov}(f(U_1), f(1-U_1))\right). \tag{9.6}$$

Obviously, $\mathrm{Var}(I_{2n}) \leq \mathrm{Var}(I_{2n}^0)$ if and only if $\mathrm{Cov}(f(U_1), f(1-U_1)) \leq 0$. One can prove that if f is a monotonic function, this is always true and thus the Monte-Carlo method using antithetic variables is better than the standard one.

These ideas can be generalized in dimensions greater than 1, in which case we use the transformation

$$(U_1, \ldots, U_d) \to (1 - U_1, \ldots, 1 - U_d).$$

More generally, if X is a random variable taking its values in $\mathbb{R}^d$ and T is a transformation of $\mathbb{R}^d$ such that the distribution of $T(X)$ is the same as the distribution of X, we can construct a generalized antithetic method by using the equality

$$\mathbb{E}(f(X)) = \frac{1}{2} \mathbb{E} \left(f(X) + f(T(X)) \right).$$

Namely, if $(X_1, \ldots, X_n)$ are independent and sampled along the distribution of X, we can consider the estimator

$$I_{2n} = \frac{1}{2n} \left(f(X_1) + f(T(X_1)) + \cdots + f(X_n) + f(T(X_n)) \right)$$

and compare it to

$$I_{2n}^0 = \frac{1}{2n} \left(f(X_1) + f(X_2) \right) + \cdots + f(X_{2n-1}) + f(X_{2n}) \right).$$

The same computations as before prove that the estimator I_{2n} is better than the crude one I_{2n}^0 if and only if $\mathrm{Cov}(f(X), f(T(X))) \leq 0$ (which is true when f is monotonous). We now give an elementary example that can be useful in finance.

A toy financial example. Let G be a standard Gaussian random variable and consider the call option

$$\mathbb{E} \left(\left(\lambda e^{\sigma G} - K \right)_+ \right).$$

Clearly the distribution of $-G$ is the same as the distribution of G, and thus we can consider the transformation $T(x) = -x$. As the payoff is increasing as a function of G, the following antithetic estimator certainly reduces the variance:

$$I_{2n} = \frac{1}{2n} \left(f(G_1) + f(-G_1) + \cdots + f(G_n) + f(-G_n) \right),$$

where $f(x) = (\lambda e^{\sigma x} - K)_+$.

Extension to path-dependent options The previous example can be extended to European path-dependent options with payoff given by

$$\psi(W_s, s \leq T).$$

We can use the fact that the distribution of $(W_t, t \leq T)$ and $(-W_t, t \leq T)$ are identical to construct an antithetic method, namely

$$I_{2n} = \frac{1}{2n} \left(\psi(W_s^1, s \leq T) + \psi(-W_s^1, s \leq T) \right.$$
$$\left. + \cdots + \psi(W_s^n, s \leq T) + \psi(-W_s^n, s \leq T) \right),$$

where $W^1, \ldots, W^n$ are independant samples along the distribution of a Brownian motion.

9.2.4 Mean value or conditioning

This method uses the well-known fact that conditioning reduces the variance. Indeed, for any integrable random variable Z, we have

$$\mathbb{E}(Z) = \mathbb{E}(\mathbb{E}(Z|Y)),$$

where Y is any random variable defined on the same probability space as Z. It is well known that $\mathbb{E}(Z|Y)$ can be written as

$$\mathbb{E}(Z|Y) = \phi(Y),$$

for some measurable function ϕ. Suppose in addition that Z is square-integrable. As the conditional expectation is a projection in the Hilbert space L^2, we have

$$\mathbb{E}\left(\phi(Y)^2\right) \leq \mathbb{E}(Z^2),$$

and thus $\mathrm{Var}(\phi(Y)) \leq \mathrm{Var}(Z)$.

Of course the practical efficiency of simulating $\phi(Y)$ instead of Z relies on obtaining an explicit formula for the function ϕ. This can be achieved when $Z = f(X, Y)$, where X and Y are independent random variables. In this case, we have

$$\mathbb{E}(f(X, Y)|Y) = \phi(Y),$$

where $\phi(y) = \mathbb{E}(f(X, y))$ (see Proposition A.2.5).

A basic example. Suppose that we want to compute $\mathbb{P}(X \leq Y)$, where X and Y are independent random variables. This situation occurs in finance when one computes the hedge of an exchange option (or the price of a digital exchange option). We clearly have

$$\mathbb{P}(X \leq Y) = \mathbb{E}(F(Y)),$$

where F is the distribution function of X. The variance reduction can be significant, especially when the probability $\mathbb{P}(X \leq Y)$ is small.

9.3 Exercises

Exercise 56 Let X and Y be two standard Gaussian random variables; derive the joint distribution of (R, Θ), where $R = \sqrt{X^2 + Y^2}$ and Θ is the polar angle.

Deduce that, if U_1 and U_2 are two independent uniform random variables on $[0, 1]$, the random variables

$$\sqrt{-2\log(U_1)}\cos(2\pi U_2) \quad \text{and} \quad \sqrt{-2\log(U_1)}\sin(2\pi U_2)$$

are independent and follow a standard Gaussian distribution.

Exercise 57 Let f be a function from $\mathbb{R}$ to $\mathbb{R}$, such that $f(x) > 0$ for all x, and such that $\int_{-\infty}^{+\infty} f(x)dx = 1$. We want to simulate a random variable X with distribution $f(x)dx$. Let $F(u) = \int_{-\infty}^{u} f(x)dx$. Prove that if U is a uniform random variable on $[0, 1]$, then the distribution of $F^{-1}(U)$ is $f(x)dx$. Deduce a method of simulation of the distribution of X.

Exercise 58 Let λ and K be two real positive numbers such that $\lambda < K$ and X_m be the random variable

$$X_m = \left(\lambda e^{\sigma(G+m)} - K\right)_+ e^{-mG - \frac{m^2}{2}}.$$

We denote its variance by σ_m^2. Give an expression for the derivative of σ_m^2 with respect to m as an expectation and deduce that σ_m^2 is a decreasing function of m when $m \leq m_0 = \log(K/\lambda)/\sigma$.

Exercise 59 Let $(W_t)_{t \geq 0}$ be a standard Brownian motion and $(U_i)_{i \geq 1}$ be a sequence of independent random variables taking values $+1$ or -1 with probability $1/2$. Let $S_n = X_1 + \cdots + X_n$.

1. Prove that, for a fixed $t > 0$, if $X_t^n = S_{[nt]}/\sqrt{n}$, X_t^n converges in distribution to W_t.

2. Let t and s be non-negative; using the fact that the random variable $X_{t+s}^n - X_t^n$ is independent of X_t^n, prove that the pair $\left(X_{t+s}^n, X_t^n\right)$ converges in distribution to (W_{t+s}, W_t).

3. If $0 < t_1 < \cdots < t_p$, show that

$$\left(X_{t_1}^n, \ldots, X_{t_p}^n\right) \text{ converges in distribution to } (W_{t_1}, \ldots, W_{t_p}).$$

9.4 Computer experiments

In this section, we propose computer experiments to illustrate some of the main ideas of this book. Solutions (for these computer experiments) written in Scilab are available on

$$\texttt{http://cermics.enpc.fr/~bl/scilab}$$

Scilab is a free software that can be downloaded from

$$\texttt{http://www.scilab.org}$$

and its syntax is very close to the MatLab one.

9.4.1 Pricing and hedging in the Cox-Ross-Rubinstein model

Here we consider the Cox-Ross-Rubinstein model. We recall (see Section 1.4 of Chapter 1) that S_n denotes the price at time n of the risky asset and, for a given integer $n \geq 0$,

$$S_{n+1} = S_n U_{n+1},$$

where $(U_n, n \geq 1)$ is a sequence of independent random variables such that $\mathbb{P}(U_n = u) = p$ and $\mathbb{P}(U_n = d) = 1 - p$, where $0 < u < d$ and p are real numbers with $0 < p < 1$.

We denote by S_n^x the value at time n of the Cox-Ross-Rubinstein model starting from x at time 0:

$$S_n^x = xU_1 \ldots U_n.$$

S_n^0 denotes the price of the riskless asset, assumed to be given by $S_n^0 = (1 + R)^n$.

Choosing the parameters In computer experiments, the parameters of the model can be chosen as follows:

- riskless annual interest rate: $r = 5\%$,

- current value of the price: $S_0 = 40$,

- strike: $K = 40$ (or more generally any value near S_0),

- number of time steps: $N = 10$,

- maturity date: $T = 4$ months.

The values of u and d will be chosen to look like a Black-Scholes model, more precisely,

- $\Delta t = T/N$ will be the time-step for the discrete model,

- $R = \exp(r * \Delta t) - 1$ will be the riskless discrete interest rate on the period.

The choices of u and d are related to the discretisation of the Black-Scholes model with volatility σ. Assuming that $\sigma = 0.2$ (for a year), d and u can be chosen as (see Chapter 1, Section 1.4 for a justification)

- $d = (1 + R) * \exp\left(-\sigma * \sqrt{\Delta t}\right)$,

- $u = (1 + R) * \exp\left(+\sigma * \sqrt{\Delta t}\right)$.

Hedging in the Cox-Ross-Rubinstein model

1. Describe a method to sample the vector $(S_0, \ldots, S_N)$.

   Plot typical trajectories of the Cox-Ross-Rubinstein process for different values of p (from 0.1 to 0.9). Note that the price of options *does not depend* on the value of p (see Chapter 1).

2. Assuming that $d \le R \le u$, show that the unique p^* such that $\mathbb{E}(S_n/(1 + R)^n) = 1$ is given by
$$p^* = \frac{u - (1 + R)}{u - d}.$$

   Why *must* we compute the price of an option depending on S as an expectation with $p = p^*$ whatever the value of p?

3. Write an iterative algorithm that will compute the price of a call option with strike K at time 0 as a function

   `Price_at_zero(N,K,R,up,down,x).`

   By a time shift argument write a function `Price(n,N,K,R,up,down,x)` that computes the price at time n, when the asset value is x, at this time.

4. Compute the hedge ratio `Hedge(n,N,K,R,up,down,x)` at time n when the asset price is x using the function `Price(n,N,K,R,up,down,x)`.

5. Sample a trajectory $(S_n, 0 \le n \le N)$ and check that the hedging procedure perfectly replicates the payoff whatever the value of p strictly between 0 and 1.

Asian call in the Cox-Ross-Rubinstein model   Consider an Asian call whose payoff at time N is given by
$$f(S_N, S_1 + \cdots + S_{N-1}) = \left(S_N - \frac{1}{N} (S_0 + \cdots + S_{N-1}) \right)_+ .$$

Take $N = 10$ in numerical examples.

1. Using the results of Chapter 1, prove that the price at time n of this option is given by
$$V_n = \frac{1}{(1 + R)^{N-n}} \mathbb{E}^* \left(f(S_N, S_0 + \cdots + S_{N-1}) | \mathscr{F}_n \right).$$

   Deduce from this equality that if
$$u(n, x, m) = \mathbb{E}^* \left(\frac{1}{(1 + R)^{N-n}} f(S_{N-n}^x, m + S_0^x + \cdots + S_{N-n-1}^x) \right),$$

   then $V_n = u(n, S_n, S_0 + S_1 + \cdots + S_{n-1}).$

2. Show that u is the unique solution to

$$
\begin{cases}
u(n, x, m) = \dfrac{1}{1+R} \left[p^* u(n+1, xu, m+x) \right. \\
\qquad\qquad\qquad \left. + (1 - p^*) u(n+1, xd, m+x) \right], n < N, x, m \in \mathbb{R}, \\
u(n, x, m) = \left(x - \dfrac{m}{N} \right)_+, x, m \in \mathbb{R}.
\end{cases}
$$

3. Write a recursive algorithm to compute $u(0, x, 0)$. What is the complexity of this algorithm with respect to N? Can you improve the complexity of this algorithm?

4. Show that, between time n and time $n+1$, the quantity H_{n+1} of risky assets in the perfect hedging portfolio is given by $H_{n+1} = h(n+1, S_n, S_0 + \cdots + S_{n-1})$, where

$$
h(n+1, x, m) = \frac{u(n+1, xu, m+x) - u(n+1, xd, m+x)}{x(u-d)}.
$$

5. Implement this hedging formula and check, using simulation, that it gives a perfect hedging strategy.

9.4.2   Pricing and hedging in the Black-Scholes model

The aim of this computer experiment is to study the influence of the hedging frequency on the quality of the hedging portfolio in the Black-Scholes model.

Sampling the Black-Scholes model   The risky asset is described by the Black-Scholes model,

$$
\begin{cases}
dS_t &= S_t(\mu dt + \sigma dW_t) \\
S_0 &= x,
\end{cases}
$$

where $(W_t)_{t \geq 0}$ is a standard Brownian motion, σ the annual volatility, μ the trend of the risky asset and r the riskless interest rate. Further on we will fix $r = 0,05\%/\text{year}, \sigma = 20\%/\sqrt{\text{year}} = 0.2$ and $x = 100$. The riskless asset is supposed to be equal to $S_t^0 = e^{rt}$.

1. Explain how to sample a sequence of independent normal random variables with mean 0 and variance 1. Using this sequence, sample a trajectory of a Brownian motion and a trajectory of the Black-Scholes model at times kh, $k = 1, \ldots, n$.

2. For the same Brownian motion sample, plot the trajectories of a Black-Scholes model with $r = 0$, $r = 0.10$, $r = 0.20$ and σ remaining unchanged. This parameter change *has no influence* on the option prices (why?).

   Plot the same trajectories with $\sigma = 0.1$, $\sigma = 0.3$, $\sigma = 0.9$. This parameter change *does have an influence* on the option price.

Approximating option hedging   The Black-Scholes model theory suggests rebalancing the portfolio at every time. This is obviously impossible in
practice. Here, we suggest an approximated hedging strategy in which we re-
balance the portfolio at discrete times $k\Delta t$, $k = 0, 1, \dots$ (typical values of Δt
being 1 hour, 1 day or 1 month).

First, we consider the hedge of a call option with strike $K = S_0 = 100$ and
maturity date $T = 1$ year. We fix Δt to 1 month.

1. Implement the Black-Scholes formula for the price of a call option. Plot
   the price of this option as a function of S_0, K, σ, r.

2. Implement the approximated hedging procedure, which consists in fixing
   the quantity of risky asset to the one given by the Black-Scholes formula
   at the beginning of the time interval $[k\Delta t, (k+1)\Delta t])$. Sample the value
   at time T of this portfolio (which is uniquely defined using the self-
   financing condition).

3. For a hedging strategy, we call *residual risk* the difference between the
   final value of the hedging portfolio and the option payoff. The residual
   risk is identically equal to 0 when using a perfect hedge.

   We are now interested in studying the residual risk of the hedging strategy previously defined when Δt tends to 0. Sample the residual risk,
   under the risk neutral probability (with $\mu = r$). Draw its histogram and
   compute its mean and its variance. Compare the empirical mean with
   the theoretical one (which is 0 when $\mu = r$; why?).

4. We study the residual risk when Δt goes to 0. We compare the following
   strategies:

   (a) We do not hedge: we sell the option, get the premium, we wait for
       three months, we take into account the exercise of the option sold
       and we evaluate the portfolio.

   (b) We hedge immediately after selling the option, then we do nothing.

   (c) We hedge immediately after selling the option, then every month.

   (d) We hedge immediately after selling the option, then every 10 days.

   (e) We hedge immediately after selling the option, then every day.

   Compute, using simulation, the mean and variance of the discounted
   final value of the portfolio for these hedging strategies.

5. Redo the previous question assuming successively that $\mu > r$ and $\mu < r$.
   Are there arbitrage opportunities?

   What happens to the mean of the residual risk when Δt tends to 0? to
   the standard deviation?

6. Use the same hedging strategies for a combination of put and call options. We suggest choosing among the following popular combinations
   and studying the residual risk of the hedging strategy as before:

- **Bull spread**: long a call with strike price 90 (written as 90-call) and short a 110-call with same maturity.
- **Strangle**: short a 90-put and short a 110-call.
- **Condor**: short a 90-call, long a 95-call and a 105-call, and finally short a 110-call.
- **Put ratio backspread**: short a 110-put and long three 90-puts.

9.4.3   Pricing and hedging bonds and options in the Vasicek model

Sampling the short rate   The Vasicek model is presented in Chapter 6. The short rate r_t follows the stochastic differential equation

$$dr(t) = a\,(b - r(t))\,dt + \sigma dW_t, \tag{9.1}$$

where a, b, σ are positive constants and W is a standard Brownian motion under $\mathbb{P}$. In computer experiments, one can choose $a = 10/\text{year}$, $r_0 = b = 0.05/\text{year}$, $\sigma = 0.1/\sqrt{\text{year}}$.

1. Show that r_h follows a Gaussian distribution with mean $b + e^{-ah}(r_0 - b)$ and variance $\sigma^2 \frac{1 - e^{-2at}}{2a}$

2. What is the conditional distribution of r_{t+h} given $r_t = r$?

3. Explain how to sample exactly the vector $(r_{kh}, 0 \le k \le N)$.

4. Implement the suggested algorithm and plot the trajectory $(r_{kh}, 0 \le k \le N)$ for $h = 1$ hour, $h = 1$ day, $h = 1$ week and $N = 100$.

Sampling the zero-coupon bond dynamics   We denote by $P(t, T)$ the price at time t of a zero-coupon with maturity date T. We assume that $\mathbb{P}$ is a probability under which all the discounted bond prices

$$\tilde{P}(t, T) = e^{-\int_0^t r_s ds} P(t, T),$$

are martingales.

1. We know (see Chapter 6) that the zero-coupon bonds can be written as

$$P(t, T) = \exp\left[-(T - t)R(T - t, r(t))\right]$$

with

$$R(\theta, r) = R_\infty - \frac{1}{a\theta}\left((R_\infty - r)\left(1 - e^{-a\theta}\right) - \frac{\sigma^2}{4a^2}\left(1 - e^{-a\theta}\right)^2\right)$$

and $R_\infty = \lim_{\theta \to \infty} R(\theta, r) = b - \frac{\sigma^2}{2a^2}$.

Sample the discretized trajectory price of a bond with maturity $T = 1$ $(P(kh, T), 0 \le k \le N)$, where $h = 1$ day and is such that $Nh = 1$ year.

Pricing a zero-coupon bond option   Here, we consider a call option in the Vasicek model, with maturity θ on a zero-coupon bond with maturity T, $T > \theta$. We want to implement a hedging strategy for this option.

1. Show, using the results of Chapter 6, that

$$\frac{d\tilde{P}(t,T)}{\tilde{P}(t,T)} = \sigma_s^T dW_t,$$

where

$$\sigma_s^T = -\sigma \frac{1 - e^{-a(T-s)}}{a}.$$

2. Show, using Proposition 6.1.13, that the price C_t of the call option at time t is given by

$$C_t = P(t,\theta) B\left(t, \frac{P(t,T)}{P(t,\theta)}\right),$$

with

$$B(t,x) = xN(d_1(t,x)) - KN(d_2(t,x)),$$

where N is the cumulative normal distribution function,

$$d_1(t,x) = \frac{\log(x/K) + (\Sigma^2(t,\theta)/2)}{\Sigma(t,\theta)} \quad \text{and} \quad d_2(t,x) = d_1(t,x) - \Sigma(t,\theta)$$

and

$$\Sigma^2(t,\theta) = \int_t^\theta \left(\sigma_s^T - \sigma_s^\theta\right)^2 ds.$$

Implement this formula and plot the option price at time 0 as a function of the strike K.

3. Using Exercise 36 of this chapter, show that

$$C_t = P(t,T)H_t^T + P(t,\theta)H_t^\theta,$$

$$H_t^T = N\left(d_1\left(t, \frac{P(t,T)}{P(t,\theta)}\right)\right) \quad \text{and} \quad H_t^\theta = -KN\left(d_2\left(t, \frac{P(t,T)}{P(t,\theta)}\right)\right).$$

Implement these formulae and plot the values of H_0^T and H_0^θ as a function of the strike K.

Give a perfect hedging portfolio for the call option using only zero-coupon bonds with maturity T and zero-coupon bonds with maturity θ.

4. We are interested in studying the discrete approximation of this perfect hedging portfolio in which the quantity $\bar{H}_s^T$ of zero-coupon bonds with maturity T remains constant on the interval $[kh, (k+1)h]$ and equal to H_{kh}^T. $\bar{H}_s^\theta$, the quantity of zero-coupon bonds with maturity θ, is determined using the discrete self-financing condition at times kh.

For a given h (successively chosen to be $h=1$ day, $h=1$ week, $h=1$ month) sample the residual risk of this approximated hedging portfolio. Plot a histogram of the residual risk and study the values of its mean and its variance when h decreases to 0.

9.4.4   Monte-Carlo methods for option pricing

Gaussian sampling

1. Write a function that samples a vector of independent normal random variables with mean 0 and variance 1.

   Draw the histogram of the vector and compare it with the exact distribution of a normal random variable with mean 0 and variance 1.

2. We want to compute $\mathbb{E}(e^{\beta G})$ using a Monte-Carlo method, where G is Gaussian with mean 0 and variance 1. We recall that $\mathbb{E}(e^{\beta G}) = \exp(\beta^2/2)$.

   Compute $\mathbb{E}(e^{\beta G})$ using simulation for $\beta = 2, 4, 6, 8, 10 \ldots$. Give a confidence interval in each case. For which values of β can you safely use a Monte-Carlo method?

The Black-Scholes model   We consider the Black-Scholes model:

$$S_t = S_0 \exp\left(\left(r - \frac{\sigma^2}{2}\right)t + \sigma W_t\right).$$

In the following, we assume that $S_0 = 100$, $\sigma = 0.3$ (annual volatility) and $r = 0.05$ (exponential riskless interest rate).

1. Plot a histogram of the distribution of W_T and S_T ($T=1$, $\sigma=0.3$, $r=0.05$).

2. We want to compute the price of a call with strike $K=100$. Compute this price using a Monte-Carlo method with a number of trials equal to $N=1000,1000,10000$. Give a confidence interval in each case.

3. We will now use the random variable S_T as a control variate. Check that $\mathbb{E}(S_T) = e^{rT}S_0$ (give a financial argument for this result).

   Write a program using S_T as a control variate. Compare the precision of this method with the previous one using various values for K and S_0.

   How is this method related to the call-put parity formula?

4. We want to compute the price of a call option with strike K where S_0 is small compared to K.

   Show that the relative precision of the computation decreases with S_0/K (take $S_0 = 100$ and $K = 100, 150, 200, 250, 400$).

5. Prove that

$$\mathbb{E}\left(f(W_T)\right) = \mathbb{E}\left(e^{-\lambda W_T - \frac{\lambda^2 T}{2}} f(W_T + \lambda T)\right).$$

When $S_0 = 100$ and $K = 150$, propose a value for λ that may reduce the simulation variance. Empirically check that the variance is reduced by using simulation.

9.4.5 Basket options and control variates

We consider a d-dimensional basket model. Let $(W_t^1, \ldots, W_t^d, t \geq 0)$ be a vector of independent Brownian motions, Σ be a $d \times d$ matrix, and define σ_i by

$$\sigma_i = \sqrt{\sum_{j=1}^{d} \Sigma_{ij}^2},$$

and $\bar{W}^i$ by

$$\bar{W}_t^i = \frac{[\Sigma W_t]_i}{\sigma_i} = \frac{\sum_{j=1}^{d} \Sigma_{ij} W_t^j}{\sigma_i}.$$

$(\bar{W}_t^i, t \geq 0)$ is then still a Brownian motion (check it) and we assume that each of the d assets has a price S_t^i given by a Black-Scholes model driven by the Brownian motion $\bar{W}^i$,

$$\frac{dS_t^i}{S_t^i} = rdt + \sigma_i d\bar{W}_t^i, S_0^i = x_i.$$

In the numerical experiments we will consider that $d = 10$ and $x_i = 100$ for $i = 1, \ldots, d$.

1. Prove that $\mathbb{E}\left(\bar{W}_t^i \bar{W}_t^j\right) = \rho_{ij} t$, where

$$\rho_{ij} = \frac{\sum_{k=1}^{d} \Sigma_{ik} \Sigma_{jk}}{\sigma_i \sigma_j}.$$

Check that ρ is a symmetric positive matrix (positive means $\lambda . \rho \lambda \geq 0$, for every $\lambda \in \mathbb{R}^d$).

2. In the numerical examples, we will assume that the matrix ρ is equal to ρ^0, where $\rho_{ij}^0 = 0.5$ for $i \neq j$ and $\rho_{ii}^0 = 1$.

Check using Scilab that ρ^0 is a positive matrix.

Find (using Scilab) a matrix Σ such that $\sigma_i = \sigma_i^0 = 0.3$ for $i = 1, \ldots, d$ and $\rho = \rho^0$ (this is equivalent to solving the equation $\Sigma\Sigma^* = (\sigma_i^0 \rho_{ij}^0 \sigma_j^0)_{ij}$).

3. Compute the covariance matrix of the vector $(\bar{W}_t^1, \ldots, \bar{W}_t^d)$. Propose a simulation method for the vectors $(\bar{W}_T^1, \ldots, \bar{W}_T^d)$ and then $(S_T^1, \ldots, S_T^d)$.

4. Consider a basket call option on an index I_t given by

$$I_t = a_1 S_t^1 + \cdots + a_d S_t^d,$$

where $a_i > 0$ and $\sum_{i=1}^{d} a_i = 1$ (in numerical applications let $a_1 = \cdots = a_d = 1/d$).

Compute, using a Monte-Carlo method, the price of a call whose payoff is given at time T by

$$(I_T - K)_+,$$

and give an estimate of the error for different values of K ($K = 0.8 I_0$, $K = I_0$, $K = 1.2 I_0$, $K = 1.5 I_0$).

Do the same computation for an index put with payoff $(K - S_T)_+$.

5. Prove that $\mathbb{E}(I_T) = I_0 \exp(rT)$. How can you use I_T as a control variate? Relate this method to the call-put arbitrage relation. Test the efficiency of the method for various values of K.

6. When r and σ are small, justify the approximation of $\log(I_t/I_0)$ by

$$Z_t = \frac{a_1 S_0^1}{I_0} \log(S_t^1/S_0^1) + \cdots + \frac{a_d S_0^d}{I_0} \log(S_t^d/S_0^d).$$

Prove that Z_T is Gaussian with mean

$$T \sum_{i=1}^{d} \frac{a_i S_0^i}{I_0} \left(r - \sigma_i^2/2\right) \text{ and variance } T \frac{1}{I_0^2} \sum_{i=1}^{d} \sum_{j=1}^{d} J_i \rho_{ij} J_j,$$

where $J_i = a_i S_0^i \sigma_i$.

Using a Black-Scholes like formula, give an explicit expression for $\mathbb{E}\left(\left(e^{Z_T} - K\right)_+\right)$ and propose a control variate technique for the computation of the call option.

Compare this method with the standard one for different values of K.

Appendix

A.1 Normal random variables

In this section, we recall the main properties of Gaussian variables. The following results are proved in Bouleau (1986), Chapter VI, Section 9, or Jacod and Protter (2003), Chapter 16.

A.1.1 Scalar normal variables

A real random variable X is a standard normal variable if its probability density function is equal to

$$n(x) = \frac{1}{\sqrt{2\pi}} \exp\left(-\frac{x^2}{2}\right), \quad x \in \mathbb{R}.$$

If X is a standard normal variable and m and σ are two real numbers, then the variable $Y = m + \sigma X$ is normal with mean m and variance σ^2. Its law is denoted by $\mathscr{N}(m, \sigma^2)$ (it does not depend on the sign of σ since X and $-X$ have the same law). If $\sigma \neq 0$, the density of Y is

$$\frac{1}{\sqrt{2\pi}} \exp\left(-\frac{(x-m)^2}{2\sigma^2}\right).$$

If $\sigma = 0$, the law of Y is the Dirac measure at m and therefore it does not have a density. It is sometimes called a 'degenerate normal variable'.

If X is a standard normal variable, we can prove that for any complex number z, we have

$$\mathbb{E}(e^{zX}) = e^{\frac{z^2}{2}}.$$

Thus, the characteristic function of X is given by $\phi_X(u) = e^{-u^2/2}$ and for $Y = m + \sigma X$, $\phi_Y(u) = e^{ium} e^{-u^2\sigma^2/2}$. It is sometimes useful to know that if X

is a standard normal variable, we have $\mathbb{P}(|X| > 1,96\ldots) = 0,05$ and $\mathbb{P}(|X| > 2,6\ldots) = 0,01$. For large values of $t > 0$, the following estimate is handy:

$$\mathbb{P}(X > t) = \frac{1}{\sqrt{2\pi}} \int_t^\infty e^{-x^2/2} dx \leq \frac{1}{t\sqrt{2\pi}} \int_t^\infty x e^{-x^2/2} dx = \frac{e^{-t^2/2}}{t\sqrt{2\pi}}.$$

Finally, one should know that there exist very good approximations of the cumulative normal distribution.

A.1.2 Approximating the distribution function of a standard Gaussian variable

The distribution function of the standard Gaussian variable is often needed to express the price of many classical options. Due to the importance of this function in the pricing of options, we give two approximation formulae from Abramowitz and Stegun (1970):

$$N(x) = \mathbb{P}(G \leq x),$$

where G is a real Gaussian random variable with mean 0 and variance 1. Obviously

$$N(x) = \int_{-\infty}^x e^{-u^2/2} \frac{du}{\sqrt{2\pi}}.$$

The first approximation is accurate to 10^{-7}, but it uses the exponential function. If $x > 0$,

$$
\begin{aligned}
p &= 0.231\,641\,900 \\
b_1 &= 0.319\,381\,530 \\
b_2 &= -0.356\,563\,782 \\
b_3 &= 1.781\,477\,937 \\
b_4 &= -1.821\,255\,978 \\
b_5 &= 1.330\,274\,429 \\
t &= 1/(1 + px)
\end{aligned}
$$

$$N(x) \approx 1 - \frac{1}{\sqrt{2\pi}} e^{-\frac{x^2}{2}} (b_1 t + b_2 t^2 + b_3 t^3 + b_4 t^4 + b_5 t^5).$$

The second approximation is accurate to 10^{-3}, but it involves only a ratio as opposed to an exponential. If $x > 0$,

$$
\begin{aligned}
c_1 &= 0.196\,854 \\
c_2 &= 0.115\,194 \\
c_3 &= 0.000\,344 \\
c_4 &= 0.019\,527
\end{aligned}
$$

$$N(x) \approx 1 - \frac{1}{2}(1 + c_1 x + c_2 x^2 + c_3 x^3 + c_4 x^4)^{-4}.$$

A.1.3 Multivariate normal random variables

Definition A.1.1. *An $\mathbb{R}^d$-valued random variable $X = (X_1, \ldots, X_d)$ is a Gaussian vector (or a multivariate normal random variable) if, for any sequence of real numbers $a_1, \ldots, a_d$, the scalar random variable $\sum_{i=1}^{d} a_i X_i$ is Gaussian.*

The components $X_1, \ldots, X_d$ of a Gaussian vector are obviously normal, but the fact that each component of a vector is a normal random variable does not imply that the vector is normal. However, if $X_1, X_2, \ldots, X_d$ are real-valued, normal, *independent* random variables, then the vector $(X_1, \ldots, X_d)$ is normal.

The covariance matrix of a random vector $X = (X_1, \ldots, X_d)$ is the matrix $\Gamma(X) = (\sigma_{ij})_{1 \leq i,j \leq d}$, with

$$\sigma_{ij} = \text{cov}(X_i, X_j) = \mathbb{E}[(X_i - \mathbb{E}(X_i))(X_j - \mathbb{E}(X_j))].$$

It is well known that if the random variables $X_1, \ldots, X_d$ are independent, the matrix $\Gamma(X)$ is diagonal, but the converse is generally not true, except in the Gaussian case.

Theorem A.1.2. *Let $X = (X_1, \ldots, X_d)$ be a Gaussian vector in $\mathbb{R}^d$. The random variables $X_1, \ldots, X_d$ are independent if and only if the covariance matrix of the vector X is diagonal.*

The reader should consult Bouleau (1986), Chapter VI, p. 155, or Jacod and Protter (2003), Chapter 16, for a proof of this result.

Remark A.1.3. The importance of normal random variables in modelling comes partly from the Central Limit Theorem (cf. Bouleau (1986), Chapter VII, Section 4, or Jacod and Protter (2003), Chapter 21). The simulation of normal and multivariate normal distributions is discussed in Chapter 8. We refer to Dacunha-Castelle and Duflo (1986b), Chapter 5, for their statistical estimation.

A.2 Conditional expectation

A.2.1 Examples of σ-algebras

Consider a measurable space $(\Omega, \mathscr{A})$ and a partition $B_1, B_2, \ldots, B_n$, with n events in $\mathscr{A}$. The set $\mathscr{B}$ containing the elements of $\mathscr{A}$ that are either empty or that can be written as

$$B_{i_1} \cup B_{i_2} \cup \cdots \cup B_{i_k}, \text{ where } i_1, \ldots, i_k \in \{1, \ldots, n\},$$

is a finite sub-σ-algebra of $\mathscr{A}$. It is the σ-algebra generated by the B_i's.

Conversely, any finite sub-σ-algebra $\mathscr{B}$ of $\mathscr{A}$ is generated by a finite partition $(B_1, \ldots, B_n)$ of Ω, with the B_i's in $\mathscr{A}$. The events $B_1, \ldots, B_n$ are the non-empty elements of $\mathscr{B}$ that contain no element of $\mathscr{B}$, except themselves and the empty set. They are called *atoms* of $\mathscr{B}$. There is a one-to-one mapping from the set of finite sub-σ-algebras of $\mathscr{A}$ onto the set of finite partitions of Ω by elements of $\mathscr{A}$. Note that if $\mathscr{B}$ is a finite sub-σ-algebra of $\mathscr{A}$, a map from Ω into $\mathbb{R}$ (equipped with its Borel σ-algebra) is $\mathscr{B}$-measurable if and only if it is constant on each atom of $\mathscr{B}$.

Now, consider a random variable X, defined on $(\Omega, \mathscr{A})$, with values in a measurable space $(E, \mathscr{E})$. The σ-algebra *generated by* X is the smallest σ-algebra on Ω for which X is measurable. This σ-algebra, which we denote by $\sigma(X)$, is obviously included in $\mathscr{A}$, and it is easy to show that an event $A \in \mathscr{A}$ is in $\sigma(X)$ if and only if $A = X^{-1}(B) = \{X \in B\}$ for some $B \in \mathscr{E}$. It can be proved that a real-valued random variable Y on $(\Omega, \mathscr{A})$ is $\sigma(X)$-measurable if and only if it can be written as

$$Y = f \circ X,$$

where f is a real-valued Borel measurable function on $(E, \mathscr{E})$ (cf. Jacod and Protter (2003), Chapter 23). In other words, $\sigma(X)$-measurable random variables are measurable functions of X.

A.2.2   Properties of the conditional expectation

Let $(\Omega, \mathscr{A}, \mathbb{P})$ be a probability space and $\mathscr{B}$ a sub-σ-algebra of $\mathscr{A}$. The definition of the conditional expectation is based on the following theorem (see Jacod and Protter (2003), Chapter 23).

Theorem A.2.1.   *For any real integrable random variable X, there exists a real integrable $\mathscr{B}$-measurable random variable Y such that*

$$\forall B \in \mathscr{B}, \quad \mathbb{E}(X \mathbf{1}_B) = \mathbb{E}(Y \mathbf{1}_B).$$

If $\tilde{Y}$ is another random variable with these properties, then $\tilde{Y} = Y$ $\mathbb{P}$ a.s.

The almost surely uniquely determined random variable Y is called the *conditional expectation of X given $\mathscr{B}$* and is denoted by $\mathbb{E}(X \mid B)$.

If $\mathscr{B}$ is a finite sub-σ-algebra, with atoms $B_1, \ldots, B_n$, then

$$\mathbb{E}(X|B) = \sum_i \mathbb{E}\left(X \frac{\mathbf{1}_{B_i}}{\mathbb{P}(B_i)}\right) \mathbf{1}_{B_i},$$

where the sum runs on the atoms with strictly positive probability. Consequently, on each atom B_i, $\mathbb{E}(X \mid \mathscr{B})$ is the mean value of X on B_i. For the trivial σ-algebra $(\mathscr{B} = \{\emptyset, \Omega\})$, we have $\mathbb{E}(X \mid \mathscr{B}) = \mathbb{E}(X)$.

The computations involving conditional expectations are based on the following properties:

1. If X is $\mathscr{B}$-measurable, $\mathbb{E}(X \mid \mathscr{B}) = X$, a.s.

2. $\mathbb{E}(\mathbb{E}(X \mid \mathscr{B})) = \mathbb{E}(X)$.

3. For any bounded, $\mathscr{B}$-measurable random variable Z,

$$\mathbb{E}(Z\mathbb{E}(X \mid \mathscr{B})) = \mathbb{E}(ZX).$$

4. Linearity: for all real numbers λ and μ,

$$\mathbb{E}(\lambda X + \mu Y \mid \mathscr{B}) = \lambda \mathbb{E}(X \mid \mathscr{B}) + \mu \mathbb{E}(Y \mid \mathscr{B}) \text{ a.s.}$$

5. Positivity: if $X \geq 0$, then $\mathbb{E}(X|\mathscr{B}) \geq 0$ a.s. and, more generally, $X \geq Y \Rightarrow \mathbb{E}(X|\mathscr{B}) \geq \mathbb{E}(Y|\mathscr{B})$ a.s. It follows from this property that

$$|\mathbb{E}(X|B)| \leq \mathbb{E}(|X||\mathscr{B}) \text{ a.s.,}$$

and therefore $\|\mathbb{E}(X|B)\|_{L^1(\Omega)} \leq \|X\|_{L^1(\Omega)}$.

6. If $\mathscr{C}$ is a sub-σ-algebra of $\mathscr{B}$, then

$$\mathbb{E}\left(\mathbb{E}(X \mid \mathscr{B}) \mid \mathscr{C}\right) = \mathbb{E}(X \mid \mathscr{C}) \text{ a.s.}$$

7. If Z is $\mathscr{B}$-measurable and bounded, $\mathbb{E}(ZX \mid \mathscr{B}) = Z\mathbb{E}(X \mid \mathscr{B})$ a.s.

8. If X is independent of $\mathscr{B}$, then $\mathbb{E}(X \mid \mathscr{B}) = \mathbb{E}(X)$ a.s.

The converse property is not true, but we have the following result.

Proposition A.2.2. *A real random variable X is independent of the σ-algebra $\mathscr{B}$ if and only if*

$$\forall u \in \mathbb{R}, \quad \mathbb{E}\left(e^{iuX} \mid \mathscr{B}\right) = \mathbb{E}(e^{iuX}) \text{ a.s.} \tag{1}$$

Proof. Given Property 8 above, we just need to prove that (1) implies that X is independent of $\mathscr{B}$. If $\mathbb{E}(e^{iuX}|\mathscr{B}) = \mathbb{E}(e^{iuX})$, then, by definition of the conditional expectation, $\mathbb{E}\left(e^{iuX}\mathbf{1}_B\right) = \mathbb{E}(e^{iuX})\mathbb{P}(B)$, for all $B \in \mathscr{B}$. Assuming $\mathbb{P}(B) \neq 0$, we can write

$$\mathbb{E}\left(e^{iuX}\frac{\mathbf{1}_B}{\mathbb{P}(B)}\right) = \mathbb{E}(e^{iuX}).$$

This equality means that the characteristic function of X is identical under measure $\mathbb{P}$ and measure $\mathbb{P}_B$, where the density of $\mathbb{P}_B$ with respect to $\mathbb{P}$ is equal to $\mathbf{1}_B/\mathbb{P}(B)$. The equality of characteristic functions implies the equality of probability laws and consequently

$$\mathbb{E}\left(f(X)\frac{\mathbf{1}_B}{\mathbb{P}(B)}\right) = \mathbb{E}(f(X)),$$

for any bounded Borel function f, hence the independence. □

Remark A.2.3. If X is square-integrable, so is $\mathbb{E}(X \mid \mathscr{B})$, and $\mathbb{E}(X \mid \mathscr{B})$ coincides with the orthogonal projection of X on $L^2(\Omega, \mathscr{B}, \mathbb{P})$, which is a closed subspace of the Hilbert space $L^2(\Omega, \mathscr{A}, \mathbb{P})$, equipped with the scalar product $(X, Y) \mapsto \mathbb{E}(XY)$ (cf. Bouleau (1986), Chapter VIII, Section 2 or Jacod and Protter (2003), Chapter 23). Thus, the conditional expectation of X given $\mathscr{B}$ appears as the least-square best $\mathscr{B}$-measurable predictor of X. In particular, if $\mathscr{B}$ is the σ-algebra generated by a random variable ξ, the conditional expectation $\mathbb{E}(X \mid \mathscr{B})$ (denoted by $\mathbb{E}(X \mid \xi)$, the conditional expectation in this context) is the best approximation of X by a function of ξ, since $\sigma(\xi)$-measurable random variables are measurable functions of ξ. Note that we have $\|\mathbb{E}(X \mid \mathscr{B})\|_{L^2(\Omega)} \le \|X\|_{L^2(\Omega)}$.

Remark A.2.4. The conditional expectation $\mathbb{E}(X \mid \mathscr{B})$ can be defined for any non-negative random variable X (without integrability condition). Then we have $\mathbb{E}(XZ) = \mathbb{E}(\mathbb{E}(X \mid \mathscr{B})Z)$, for any $\mathscr{B}$-measurable non-negative random variable Z. The rules are basically the same as in the integrable case (see Dacunha-Castelle and Duflo (1986b), Chapter 6, or Jacod and Protter (2003)).

A.2.3 Computations of conditional expectations

The following proposition is crucial and is used quite often in this book.

Proposition A.2.5. *Let X and Y be two random variables with values in $(E, \mathscr{E})$ and $(F, \mathscr{F})$, respectively. Assume that X is $\mathscr{B}$-measurable and that Y is independent of $\mathscr{B}$. Then, for any nonnegative (or bounded) Borel function Φ on $(E \times F, \mathscr{E} \otimes \mathscr{F})$, the function φ defined by*

$$\varphi(x) = \mathbb{E}(\Phi(x, Y)), \quad x \in E,$$

is a Borel function on $(E, \mathscr{E})$ and we have, with probability one,

$$\mathbb{E}(\Phi(X, Y) \mid \mathscr{B}) = \varphi(X).$$

In other words, under the previous assumptions, we can compute

$$\mathbb{E}(\Phi(X, Y) \mid \mathscr{B})$$

as if X were a constant.

Proof. Denote by $\mathbb{P}_Y$ the law of Y. We have

$$\varphi(x) = \int_T \Phi(x, y) d\mathbb{P}_Y(y)$$

and the measurability of φ is a consequence of the Fubini theorem. Let Z be a non-negative $\mathscr{B}$-measurable random variable (for example, $Z = \mathbf{1}_B$, with $B \in$

$\mathscr{B}$). If we denote by $\mathbb{P}_{X,Z}$ the law of (X, Z), it follows from the independence between Y and (X, Z) that

$$\mathbb{E}(\Phi(X, Y)Z) = \int \int \Phi(x, y)z d\mathbb{P}_{X,Z}(x, z) d\mathbb{P}_Y(y)$$

$$= \int \left(\int \Phi(x, y)d\mathbb{P}_Y(y) \right) z d\mathbb{P}_{X,Z}(x, z)$$

$$= \int \varphi(x)z d\mathbb{P}_{X,Z}(x, z)$$

$$= \mathbb{E}(\varphi(X)Z),$$

which completes the proof.   □

Remark A.2.6. In the Gaussian case, the computation of a conditional expectation is particularly simple. Indeed, if $(Y, X_1, X_2, \ldots, X_n)$ is a Gaussian vector (in $\mathbb{R}^{n+1}$), the conditional expectation $Z = \mathbb{E}(Y|X_1, \ldots, X_n)$ has the following form:

$$Z = c_0 + \sum_{i=1}^{n} c_i X_i,$$

where $c_0, \ldots, c_n$ are real numbers. This means that the function of the X_i's that approximates Y in the least-square sense is linear. One can compute Z by projecting the random variable Y in L^2 on the linear subspace generated by the constant 1 and the X_i's (cf. Bouleau (1986), Chapter 8, Section 5).

A.3 Separation of convex sets

In this section, we state the convex separation theorem that we use in the first chapter. For more details, the diligent reader can refer to Dudley (2002).

Theorem A.3.1. *Let C be a closed convex set that does not contain the origin. Then there exists a real linear function ξ defined on $\mathbb{R}^n$ and $\alpha > 0$ such that*

$$\forall x \in C, \quad \xi(x) \geq \alpha.$$

In particular, the hyperplane $\xi(x) = 0$ does not intersect C.

Proof. Let λ be a non-negative real number such that the closed ball $B(\lambda)$ with centre at the origin and radius λ intersects C. The set $C \cap B(\lambda)$ is closed and bounded, hence compact. Let x_0 be the point where the map $x \mapsto \|x\|$ achieves its minimum on $C \cap B(\lambda)$ (where $\| \cdot \|$ is the Euclidean norm). For x in the complement of $B(\lambda)$, we have $\|x\| \geq \lambda \geq \|x_0\|$, so that

$$\forall x \in C, \quad \|x\| \geq \|x_0\|.$$

The vector x_0 is nothing but the projection of the origin on the closed convex set C. If we consider $x \in C$, then for all $t \in [0, 1]$, $x_0 + t(x - x_0) \in C$, since C is convex. By expanding the inequality

$$\|x_0 + t(x - x_0)\|^2 \geq \|x_0\|^2,$$

we get $x_0.x \geq \|x_0\|^2 > 0$, for any $x \in C$, where $x_0.x$ denotes the scalar product of x_0 and x. This completes the proof. □

Theorem A.3.2. *Consider a compact convex set K and a vector subspace V of $\mathbb{R}^n$. If V and K are disjoint, there exists a linear functional ξ defined on $\mathbb{R}^n$, satisfying the following conditions:*

1. $\forall x \in K, \quad \xi(x) > 0.$

2. $\forall x \in V, \quad \xi(x) = 0.$

Therefore, the subspace V is included in a hypherplane that does not intersect K.

Proof. The set

$$C = K - V = \{x \in \mathbb{R}^n \mid \exists(y, x) \in K \times V, x = y - z\}$$

is convex, closed (because V is closed and K is compact) and does not contain the origin. By Theorem A.3.1, we can find a linear functional ξ defined on $\mathbb{R}^n$ and some $\alpha > 0$ such that

$$\forall x \in C, \quad \xi(x) \geq \alpha.$$

Hence

$$\forall y \in K, \quad \forall z \in V, \quad \xi(y) - \xi(z) \geq \alpha. \tag{2}$$

By taking $z = 0$, we get $\xi(y) \geq \alpha$ for $y \in K$. Now, fix $y \in K$ and apply (2) to λz instead of z, with $\lambda \in \mathbb{R}$. This yields $\xi(z) = 0$. □

Bibliography

Milton Abramowitz and Irene A. Stegun, editors. *Handbook of mathematical functions with formulas, graphs, and mathematical tables*. Dover Publications Inc., New York, 1992. Reprint of the 1972 edition.

K.I. Amin and A. Khanna. Convergence of american option values from discrete to continuous-time financial models. *Mathematical Finance*, 4:289–304, 1994.

Bouhari Arouna. Adaptative Monte Carlo method, a variance reduction technique. *Monte Carlo Methods and Applications*, 10(1):1–24, 2004.

Bouhari Arouna. Robbins-Monro algorithms and variance reduction in finance. *The Journal of Computational Finance*, 7(2), 2003/4.

P. Artzner and F. Delbaen. Term structure of interest rates: The martingale approach. *Advances in Applied Mathematics*, 10:95–129., 1989.

L. Bachelier. Théorie de la spéculation. *Annales Scientifiques de l'Ecole Normale Supérieure*, 17:21–86, 1900.

G. Barone-Adesi and R. Whaley. Efficient analytic approximation of american option values. *Journal of Finance*, 42:301–320, 1987.

A. Bensoussan. On the theory of option pricing. *Acta Applicandae Mathematicae*, 2:139–158, 1984.

A. Bensoussan and J.-L. Lions. *Applications des inéquations variationnelles en contrôle stochastique*. Dunod, Paris, 1978. Méthodes Mathématiques de l'Informatique, No. 6.

Alain Bensoussan and Jacques-Louis Lions. *Applications of variational inequalities in stochastic control*, volume 12 of *Studies in Mathematics and its Applications*. North-Holland Publishing Co., Amsterdam, 1982. Translated from the French.

Tomasz R. Bielecki and Marek Rutkowski. *Credit risk: modelling, valuation and hedging*. Springer Finance. Springer-Verlag, Berlin, 2002.

F. Black and J. C. Cox. Valuing corporate securities: some effects of bond indenture provisions. *Journal of Finance*, 31:351–367, 1976.

F. Black and M. Scholes. The pricing of options and corporate liabilities. *Journal of Political Economy*, 81:635–654, 1973.

N. Bouleau. *Probabilités de l'Ingénieur*. Hermann, Paris, 1986.

N. Bouleau. *Processus Stochastiques et Applications*. Hermann, Paris, 1988.

N. Bouleau and D. Lamberton. Residual risks and hedging strategies in markovian markets. *Stochastic Processes and their Applications*, 33:131–150, 1989.

A. Brace, D. Gatarek, and M. Musiela. The market model of interest rate dynamics. *Mathematical Finance*, 7:127–155, 1977.

M.J. Brennan and E.S. Schwartz. The valuation of the american put option. *Journal of Finance*, 32:449–462, 1977.

M.J. Brennan and E.S. Schwartz. A continuous time approach to the pricing of bonds. *Journal of Banking and Finance*, 3:133–155, 1979.

Damiano Brigo and Fabio Mercurio. *Interest rate models—theory and practice*. Springer Finance. Springer-Verlag, Berlin, second edition, 2006. With smile, inflation and credit.

CERMA. Sur les risques résiduels des stratégies de couverture d'actifs conditionnels. *Comptes Rendus de l'Académie des Sciences*, 307:625–630, 1988.

O. Chateau. *Quelques remarques sur les processus à accroissements indépendants et stationnaires, et la subordination au sens de Bochner*. PhD thesis, Université de Paris VI, 1997.

D. Chevance. *Résolution numérique des équations différentielles stochastiques rétrogrades*. PhD thesis, Université d'Aix-Marseille I, 1990.

P. G. Ciarlet and J.-L. Lions, editors. *Handbook of numerical analysis. Vol. I*. Handbook of Numerical Analysis, I. North-Holland, Amsterdam, 1990.

P.G. Ciarlet. *Une Introduction à l'analyse numérique matricielle et à l'optimisation*. Masson, Paris, 1988.

R. Cont and P. Tankov. *Financial modelling with jump processes*. Chapman & Hall/CRC Financial Mathematics Series. Chapman & Hall/CRC, Boca Raton, FL, 2004.

G. Courtadon. The pricing of options on default-free bonds. *Journal of Financial and Quantitative Analysis*, 17:301–329, 1982.

J.C. Cox and M. Rubinstein. *Options markets*. Prenctice–Hall, London, 1985.

J.C. Cox, J.E. Ingersoll, and S.A. Ross. A theory of the term structure of interest rates. *Econometrica*, 53:385–407, 1985.

D. Dacunha-Castelle and M. Duflo. *Probability and statistics, volume 2.* Springer-Verlag, New-York, 1986a.

D. Dacunha-Castelle and M. Duflo. *Probability and statistics, volume 1.* Springer-Verlag, New-York, 1986b.

R.C. Dalang, A. Morton, and W. Willinger. Equivalent martingale measures and no-arbitrage in stochastic securities market models,. *Stochastics and Stochastics Reports*, 29(2):185–202, 1990.

F. Delbaen and W. Schachermayer. A general version of the fundamental theorem of asset pricing. *Mathematische Annalen*, 300:463–520, 1994.

Freddy Delbaen and Walter Schachermayer. *The mathematics of arbitrage.* Springer Finance. Springer-Verlag, Berlin, 2006.

E. Derman and I. Kani. Riding on a smile. *RISK*, 7:32–39, 1994.

R. M. Dudley. *Real analysis and probability*, volume 74 of Cambridge Studies in Advanced Mathematics. Cambridge University Press, Cambridge, 2002. Revised reprint of the 1989 original.

D. Duffie. *Security markets, stochastic models.* Academic Press, New York, 1988.

D. Duffie, D. Filipović, and W. Schachermayer. Affine processes and applications in finance. *The Annals of Applied Probability*, 13(3):984–1053, 2003.

B. Dupire. Pricing with a smile. *RISK*, 7:18–20, 1994.

N. El Karoui. *Les aspects probabilistes du contrôle stochastique*, volume 876 of Lecture Notes in Mathematics., pages 72–238,. Springer-Verlag, New York, 1981.

N. El Karoui and M.C. Quenez. Dynamic programming and pricing of contingent claims in an incomplete market. *S.I.A.M. Journal Control and Optimization*, 33:29–66, 1995.

N. El Karoui and J.C. Rochet. A pricing formula for options on coupon-bonds. *Cahier de recherche du GREMAQ-CRES*, 8925, 1989.

Wendell H. Fleming and H. Mete Soner. *Controlled Markov processes and viscosity solutions*, volume 25 of *Stochastic Modelling and Applied Probability*. Springer-Verlag, New York, second edition, 2006.

H. Föllmer and M. Schweizer. Hedging of contingent claims under incomplete information. In M.H.A. Davis and R.J. Elliott, editors, *Applied Stochastic Analysis*, volume 5 of *Stochastics Monographs*, pages 389–414. Gordon and Breach, New York, 1991.

H. Föllmer and D. Sondermann. Hedging of non redundant contingent claims. In W.Hildebrand and A. Mas-Colell, editors, *Contributions to Mathematical Economics in Honor of Gerard Debreu.* North-Holland, Amsterdam, 1986.

Hans Föllmer and Peter Leukert. Quantile hedging. *Finance and Stochastics*, 3(3):251–273, 1999. ISSN 0949-2984.

Hans Föllmer and Peter Leukert. Efficient hedging: cost versus shortfall risk. *Finance and Stochastics*, 4(2):117–146, 2000.

Hans Föllmer and Alexander Schied. *Stochastic finance*, volume 27 of de Gruyter Studies in Mathematics. Walter de Gruyter & Co., Berlin, extended edition, 2004. An introduction in discrete time.

A..Friedman. *Stochastic differential equations and epplications.* Academic Press, New York, 1975.

T. Gard. *Introduction to stochastic differential equation.* Marcel Dekker, 1988.

Hélyette Geman, Nicole El Karoui, and Jean-Charles Rochet. Changes of numéraire, changes of probability measure and option pricing. *Journal of Applied Probability*, 32(2):443–458, 1995.

I.I. Gihman and A.V. Skorohod. *Introduction à la Théorie des Processus Aléatoires.* Mir, 1980.

I. I. Gikhman and A. V. Skorokhod. *Introduction to the theory of random processes.* Translated from the Russian by Scripta Technica, Inc. W. B. Saunders Co., Philadelphia, Pa., 1969.

P. Glasserman, P. Heidelberger, and P. Shahabuddin. Asymptotically optimal importance sampling and stratification for pricing path dependent opions. *Mathematical Finance*, 9(2):117–152, April 1999.

Paul Glasserman. *Monte Carlo methods in financial engineering*, volume 53 of Applications of Mathematics. Springer-Verlag, New York, 2004.

R. Glowinsky, J.L. Lions, and R. Trémolières. *Analyse numérique des inéquations variationnelles.* Dunod, 1976.

Christian Gourieroux, Jean Paul Laurent, and Huyên Pham. Mean-variance hedging and numéraire. *Mathematical Finance*, 8(3):179–200, 1998.

J. Hammersley and D. Handscomb. *Monte Carlo methods.* Chapman and Hall, London, 1979.

M.J. Harrison and D.M. Kreps. Martingales and arbitrage in multiperiod securities markets. *Journal of Economic Theory*, 29:381–408, 1979.

M.J. Harrison and S.R. Pliska. Martingales and stochastic integrals in the theory of continuous trading. *Stochastic Processes and Their Applications*, 11:215–260, 1981.

M.J. Harrison and S.R. Pliska. A stochastic calculus model of continuous trading: complete markets. *Stochastic Processes and Their Applications*, 15:313–316, 1983.

D. Heath, A. Jarrow, and A. Morton. Bond pricing and the term structure of interest rates: a new methodology. *Econometrica*, 60:77–105, 1992.

T.S. Ho and S.B. Lee. Term structure movements and pricing interest rate contingent claims. *Journal of Finance*, 41:1011–1029, 1986.

C.F. Huang and R.H. Litzenberger. *Foundations for financial economics*. North-Holland, New-York, 1988.

N. Ikeda and S. Watanabe. *Stochastic differential equations and diffusion processes*. North-Holland, Tokyo, 1981.

Jean Jacod and Philip Protter. *Probability essentials*. Universitext. Springer-Verlag, Berlin, second edition, 2003.

P. Jaillet, D. Lamberton, and B. Lapeyre. Variationnal inequalities and the pricing of american options. *Acta Applicandae Mathematicae*, 21:263–289, 1990.

F. Jamshidian. An exact bond pricing formula. *Journal of Finance*, 44:205–209, 1989.

Farshid Jamshidian. Libor and swap market models and measures (*). *Finance and Stochastics*, 1(4):293–330, 1997.

M.H. Kalos and P.A. Whitlock. *Monte Carlo methods*. John Wiley & Sons, 1986.

I. Karatzas. On the pricing of american options. *Applied Mathematics and Optimization*, 17:37–60, 1988.

I. Karatzas. Optimization problems in the theory of continuous trading. *SIAM Journal on Control an Optimization*, 27:1221–1259, 1989.

I. Karatzas and S.E. Shreve. *Brownian motion and stochastic calculus*. Springer-Verlag, New York, 1988.

A.G.Z Kemna and A.C.F. Vorst. A pricing method for options based on average asset values. *Journal of Banking and Finance*, pages 113–129, March 1990.

P.E. Kloeden and E. Platen. *Numerical solution of stochastic differential equations*. Springer-Verlag, New York, 1992.

D.E. Knuth. *The Art of Computer programming, Vol. 2, Seminumerical Algorithms*. Addison-Wesley, Reading, Mass., 1981.

H.J. Kushner. *Probability Methods for Approximations in Stochastic Control and for Elliptic Equations*. Academic Press, New York, 1977.

D. Lamberton and G. Pagès. Sur l'approximation des réduites. *Annales de l'IHP*, 26:331–355, 1990.

P. L'Ecuyer. Random numbers for simulation. *Communications of the ACM*, 33, 10 1990.

L. MacMillan. Analytic approximation for the american put price. *Advances in Futures and Options Research*, 1:119–139, 1986.

R.C. Merton. Theory of rational option pricing. *The Bell Journal of Economics and Management Science*, 4:141–183, 1973.

R.C. Merton. On the pricing of corporate debt: The risk structure of interest rates. *Journal of Finance*, 29:449–470, 1974.

R.C. Merton. Option pricing when underlying stock returns are discontinuous. *Journal of Financial Economics*, 3:125–144, 1976.

M. Minoux. *Programmation mathématique, 2 tomes*. Dunod, 1983.

A.J. Morton. *Arbitrage and Martingales*. PhD thesis, Cornell University, 1989.

Marek Musiela and Marek Rutkowski. *Martingale methods in financial modelling*, volume 36 of Stochastic Modelling and Applied Probability. Springer-Verlag, Berlin, second edition, 2005.

J. Neveu. *Martingales à temps discret*. Masson, Paris, 1972.

Marcus Overhaus, Ana Bermúdez, Hans Buehler, Andrew Ferraris, Christopher Jordinson, and Aziz Lamnouar. *Equity hybrid derivatives*. Wiley Finance Series. Springer-Verlag, Hoboken, New-Jersey, John Wiley & Sons edition, 2007.

E. Pardoux and D. Talay. Discretization and simulation of stochastic differential equations. *Acta Applicandae Mathematicae*, 3:23–47, 1985.

Goran Peskir and Albert Shiryaev. *Optimal stopping and free-boundary problems*. Lectures in Mathematics, ETH Zurich. Birkhauser, Berlin, 2006.

S.R Pliska. *Introduction to mathematical finance: discrete time models*. Blackwell, Malden, 1997.

W. H. Press, Saul A. Teukolsky, W. T. Vetterling, and Brian P. Flannery. *Numerical receipes in C*. Cambridge University Press, Cambridge, second edition, 1992. The art of scientific computing.

P.A. Raviart and J.M. Thomas. *Introduction à l'analyse numérique des équations aux dérivées partielles*. Masson, Paris, 1983.

D. Revuz and M. Yor. *Continuous Martingale Calculus*. Springer-Verlag, Berlin, 1990.

B.D. Ripley. *Stochastic Simulation*. Wiley, New York, 1987.

L. C. G. Rogers. Equivalent martingale measures and no-arbitrage. *Stochastics Stochastics Rep.*, 51(1-2):41–49, 1994. ISSN 1045-1129.

L.C.G. Rogers and D. Williams. *Diffusions, Markov processes and martingales, Tome 2, Itô calculus*. John Wiley and Sons, New York, 1987.

R.Y. Rubinstein. *Simulation and the Monte Carlo method*. Wiley Series in Probabilities and Mathematical Statistics. John Wiley & Sons, New York, 1981.

S. Schaefer and E.S. Schwartz. A two-factor model of the term structure: an approximate analytical solution. *Journal of Financial and Quantitative Analysis*, 19:413–424, 1984.

M. Schweizer. Option hedging for semi-martingales. *Stochastic Processes and their Applications*, 1989.

M. Schweizer. Mean-variance hedging for general claims. *Annals of Applied Probability*, 2:171–179, 1992.

M. Schweizer. Approximating random variables by stochastic integrals. *Annals of Probability*, 22(3):1536–1575, 1994.

M. Schweizer. On the minimal martingale measure and the foellmer-schweizer decomposition. *Stochastic Analysis and Applications*, 13(5):573–599, 1995.

R. Sedgewick. *Algorithms*. Addison–Wesley, Reading, MA, 1987.

A. N. Shiryayev. *Optimal stopping rules*. Springer-Verlag, New York, 1978. Translated from the Russian by A. B. Aries, Applications of Mathematics, Vol. 8.

Steven E. Shreve. *Stochastic calculus for' finance. II.* Springer Finance. Springer-Verlag, New York, 2004. Continuous-time models.

A. Sklar. Random variables, distribution functions, and copulas—a personal look backward and forward. In *Distributions with fixed marginals and related topics (Seattle, WA, 1993)*, volume 28 of IMS Lecture Notes Monogr. Ser., pages 1–14. Institute of Mathematical Statistics, Hayward, CA, 1996.

C. Stricker. Arbitrage et lois de martingales. *Annales de l'Institut Henri Poincaré*, 26:451–460, 1990.

D. Talay. Simulation of stochastic differential systems. In Paul Krée and Walter Wedig, editors, *Probablistic methods in applied physics*, volume 451 of Lecture Notes in Physics, pages 54–96, Berlin, Heidelberg, 1995. Springer-Verlag.

David Williams. *Probability with martingales*. Cambridge Mathematical Text-books. Cambridge University Press, Cambridge, 1991. ISBN 0-521-40455-X; 0-521-40605-6.

Xiao Lan Zhang. Numerical analysis of American option pricing in a jump-diffusion model. *Mathematics of Operations Research*, 22(3):668–690, 1997.

Index